T0321350

Waste Technology for Emerging Economies

This unique volume covers many aspects of waste management in developing countries. There is a focus on various sources of waste including the pressing issues of agricultural, medicinal, industrial, and urban waste, and emerging problems with e-waste, nanowaste, and microplastics in marine environments. This volume addresses the critical environmental issues resulting from rapid urbanization and industrialization, particularly in the developing world. High-end technologies that can utilize waste as a resource to generate products, processes, and revenue are also discussed.

Features

- Presents technical perspectives on emerging wastes in developing economies
- Discusses the issues of e-waste, which is growing three times faster than general municipal waste globally
- Covers the spectrum of nanowaste to upcycling in the market
- Discusses management of marine plastic debris and microplastics
- Diverse audience including those in solid waste management, electrical and electronic technology, and the medical industry

Urbanization, Industrialization, and the Environment
Series Editor: Viduranga Waisundara

Crisis Management of Chronic Pollution: Contaminated Soil and Human Health
Edited by Magalie Lesueur Jannoyer, Philippe Cattan, Thierry Woignier, and Florence Clostre

Biochar: Production, Characterization, and Applications
Edited by Yong Sik Ok, Sophie M. Uchimiya, Scott X. Chang, Nanthi Bolan

Phytotechnology: A Sustainable Platform for the Development of Herbal Products
Edited by Wanderley Pereira Oliveira

Waste Technology for Emerging Economies
Edited by T.C. Bamunuarachchige and H.K.S. de Zoysa

Waste Technology for Emerging Economies

Edited By
T. C. Bamunuarachchige
H. K. S. de Zoysa

CRC Press
Taylor & Francis Group
Boca Raton London New York

CRC Press is an imprint of the
Taylor & Francis Group, an **informa** business

Front cover image: SWKStock/Shutterstock

First edition published 2023
by CRC Press
6000 Broken Sound Parkway NW, Suite 300, Boca Raton, FL 33487-2742

and by CRC Press
4 Park Square, Milton Park, Abingdon, Oxon, OX14 4RN

Library of Congress Cataloging-in-Publication Data
Names: Bamunuarachchige, T.C., editor. | Zoysa, H. K .S. de, editor.
Title: Waste technology for emerging economies / edited by T.C.
Bamunuarachchige, Department of Bioprocess Technology, Faculty of Technology,
Rajarata University, Sri Lanka, H.K.S. de Zoysa, Department of Bioprocess Technology,
Faculty of Technology, Rajarata University, Sri Lanka.
Description: First edition. | Boca Raton : CRC Press, 2023. |
Series: Urbanization, industrialization, and the environment |
Includes bibliographical references and index. |
Identifiers: LCCN 2022032009 (print) | LCCN 2022032010 (ebook) |
ISBN 9780367655112 (hb) | ISBN 9780367676889 (pb) | ISBN 9781003132349 (ebk)
Subjects: LCSH: Refuse and refuse disposal–Developing countries.
Classification: LCC TD790 .W37 2023 (print) | LCC TD790 (ebook) |
DDC 628.4/4091724–dc23/eng/20221011
LC record available at https://lccn.loc.gov/2022032009
LC ebook record available at https://lccn.loc.gov/2022032010

ISBN: 978-0-367-65511-2 (HB)
ISBN: 978-0-367-67688-9 (PB)
ISBN: 978-1-003-13234-9 (EB)

DOI: 10.1201/9781003132349

Typeset in Times
by Newgen Publishing UK

Contents

Acknowledgment

The editors wish to thank Ms Apsara Vidanapathirana, Coordinator, English Language teaching Unit, Faculty of Science, University of Peradeniya, Sri Lanka for language editing of chapters.

Editors

T.C. Bamunuarachchige is the present Head of the Department of Bioprocess Technology, Faculty of Technology, Rajarata University of Sri Lanka (RUSL). He has obtained his PhD in microbial biotechnology from the University of Peradeniya (UoP), Sri Lanka. He has been a member of the Working Committee on Biotechnology and Bioethics of the National Science Foundation (NSF) in Sri Lanka and the Program Coordinator for the Technology Degree programs conducted by the RUSL.

Bamunuarachchige is a certified tutor mentor trainer in distance and continuing education. He has served as the Program Coordinator of the distance education program from 2009 to 2012 at RUSL. He has supervised several postgraduate students in the fields of microbial biotechnology and molecular biology. He has a wealth of experience in writing proposals for competitive grants including World Bank and Asian Development Bank-funded projects. Bamunuarachchige is a certified peer reviewer in the Elsevier Researcher Academy and has served as a reviewer for ten WoS journals.

He is also an Editorial Board member of the *Journal of Food Research International*. His research mainly involves microbial biotechnology with a special focus on how to use waste material in the production of commercially important agricultural applications such as biopesticides and biofertilizers.

H.K.S. de Zoysa is a lecturer at the Department of Bioprocess Technology, Faculty of Technology, Rajarata University of Sri Lanka (RUSL). He earned his B.Sc. in applied biology and MPhil in biological chemistry from the RUSL. He has obtained his second master's degree in marine biology from the University of Naples Federico II, Naples, Italy and is currently pursuing a PhD in biology at the same university.

He has considerable experience in applied biology, marine science, environment management, molecular biology, natural products, ecological and biodiversity conservation studies. He is a reviewer for many journals and publishers of repute. Most of Zoysa's current research is concerned with the marine environment and understanding the impact of waste and management to ensure ecosystem stability.

Contributors

Prasanna Abeyrathna
Department of Anatomy, A.T. Still
University of Health Sciences, USA

Gayan Abeysinghe
Graduate School of Science and
Technology, University of
Tsukuba, Japan
Department of Biological Sciences,
Faculty of Applied Sciences, Rajarata
University of Sri Lanka, Mihintale,
Sri Lanka

Rajesh Ahirwar
Department of Environmental
Biochemistry, ICMR-National
Institute for Research in
Environmental Health, Bhopal, India

P. H. L. Arachchige
Monash University, Australia

Tameka Dean
Philadelphia College of Osteopathic
Medicine, USA

D. M. S. B. Dissanayaka
Department of Crop Science, Faculty of
Agriculture, University of Peradeniya,
Sri Lanka

P. A. K. N. Dissanayake
Department of Oceanography and
Marine Geology, Faculty of Fisheries
and Marine Sciences and Technology,
University of Ruhuna, Sri Lanka

M. S. Ekanayake
Department of Aquaculture and
Fisheries, Faculty of Livestock
Fisheries and Nutrition, Wayamba

University of Sri Lanka, Gonawila,
Sri Lanka

Eustace Fernando
Faculty of Applied Sciences, Rajarata
University of Sri Lanka, Mihintale,
Sri Lanka

Dilani K. Hettiarachchi
Department of Biological Sciences,
Faculty of Applied Sciences, Rajarata
University of Sri Lanka, Mihintale,
Sri Lanka

Jagath Illanagsinghe
Lyceum International School, Colombo,
Sri Lanka

Randika Jayasinghe
University of Sri Jayewardenepura,
Pitipana, Sri Lanka

Dinuka Lakmali Jayasuriya
Department of Biological Sciences,
Faculty of Applied Sciences, Rajarata
University of Sri Lanka, Mihintale,
Sri Lanka

Shayan Memar
A.T. Still University of Health
Sciences, USA

E. G. Perera
Department of Bioprocess Technology,
Faculty of Technology, Rajarata
University of Sri Lanka, Mihintale,
Sri Lanka

Shalini Lalanthika Rajakaruna
National Institute of Fundamental
Studies, Hantana, Sri Lanka

Rupa Rani
Department of Environmental
  Biochemistry, ICMR—National
  Institute for Research in
  Environmental Health, Bhopal, India

Lalith M. Rankoth
Department of Crop Science, Faculty of
  Agriculture, University of Peradeniya,
  Sri Lanka

Oshani Ratnayake
Department of Biological Sciences,
  Faculty of Applied Sciences, Rajarata
  University of Sri Lanka, Mihintale,
  Sri Lanka

J. D. M. Senevirathna
Department of Animal Science, Faculty
  of Animal Science and Export
  Agriculture, Uva Wellassa University,
  Badulla, Sri Lanka
Department of Aquatic Bioscience,
  Faculty of Agricultural and Life
  Sciences, The University of Tokyo,
  Tokyo, Japan

Neha Sharma
Department of Biotechnology,
  Maharishi Markandeshwar (Deemed
  to be University), Mullana, Ambala
  (Haryana), India

Sanjay K. Sharma
Green Chemistry and Sustainability
  Research Group, Department of

Chemistry, JECRC University, Jaipur
  (Rajasthan), India

G. G. N. Thushari
Department of Animal Science, Faculty
  of Animal Science and Export
  Agriculture, Uva Wellassa University,
  Badulla, Sri Lanka
Department of Aquatic Bioscience,
  Faculty of Agricultural and Life
  Sciences, The University of Tokyo,
  Tokyo, Japan

N. D. A. D. Wijegunawardana
Department of Bioprocess Technology,
  Faculty of Technology, Rajarata
  University of Sri Lanka, Mihintale,
  Sri Lanka

Naveen M. Wijesena
Department of Biology, University of
  Bergen, Bergen, Norway

Neelamanie Yapa
Department of Biological Sciences,
  Faculty of Applied Sciences, Rajarata
  University of Sri Lanka, Mihintale,
  Sri Lanka

H. K. S. De Zoysa
Department of Bioprocess Technology,
  Faculty of Technology, Rajarata
  University of Sri Lanka, Mihintale,
  Sri Lanka
Department of Biology, University of
  Naples Federico II, Naples, Italy

Introduction

Chapter 1 Waste Management Challenges in Developing Countries

Inadequate systems of waste collection, improper disposal forms such as open dumping or illegal dumping and scavenging of dumping sites pose a threat to the environment and human health in the developing countries. Given the economic growth and increase in consumption levels, it has been assessed that the generation of waste in developing countries is at least two times more than in developed world, and if left unaddressed could very well imperil the future generations. Hence, there is a dire need for waste management practices that aim at sustainable acquiring, disposing, and recycling of waste.

This chapter contemplates on the challenges in the following aspects of waste management in the developing countries:

- Waste generation in developing countries; the structure and issues
- Civic awareness and role in waste management
- Organizational capacity in waste management
- Waste management flow; from generation of waste and collection to recycling and reduction
- Health and environmental factors
- The integrated sustainable waste management approaches
- Future directions in enhancing the capacity development of waste management

Chapter 2 Management of Medical Waste: New Strategies and Techniques

Medical waste management involves processing a complex mix of waste materials, including chemicals, human tissue, pathogens, radioactive materials, and sharps. Due to the high risk posed by these materials to the environment and public health, various complex regulations govern the management of medical wastes. Medical practices can vary in size and scope, making it challenging to use one-size-fits-all strategies in waste management. The chapter discusses the challenges of medical waste management, regulatory framework, techniques, and best practices.

Chapter 3 Management of E-waste

Electronic waste (e-waste) is defined as electronic products that are unserviceable or not in use. This includes a wide range of products from personal items like phones, computers, household items like televisions, video cassette recorders (VCRs), and more specialized equipment from medical and logistics industries.

E-waste contains hazardous materials, which can harm the environment and human health. E-waste is growing three times faster than general municipal waste all over the world, due to increased technology trends, reduced product life span, and consumer demand for new products.

DOI: 10.1201/9781003132349-1

E-waste management broadly covers the following research disciplines.

- Classification techniques of e-waste and legislative framework
- Use of recycled material in manufacturing
- Recyclable product and packaging design (designed to recycle)
- Encouraging recycling within communities and "right to repair" concept
- Collection of e-waste with minimum environmental impact
- Material recovery techniques: ferrous, nonferrous (rare earth material: gold, platinum, molybdenum), glass, and plastics
- Responsible recovery processes and use of automation

Chapter 4 Management of Marine Plastic Debris and Microplastics
The ocean is suffering from a constant and unprecedented accumulation of plastic debris and microplastics globally, emerging as problem for its disposal. This is mainly due to the adoption of plastics as a substitute for traditional materials since the 1950s. The most common feature of plastic materials is their durability, which has created an emerging issue. The chapter will discuss the main threats from plastic debris and microplastics in the marine environment and their management by using new strategies and technological methods with potential reduction measures to mitigate this emerging problem.

Chapter 5 Nanowaste Management
In recent years, nanotechnology has developed numerous new applications such as medical imaging, modification of textiles, nanocomposite scaffolds, construction materials, bioremediation, and biomedicine. Due to the large-sale production of nanomaterials, its environmental impact will be a serious problem in the future. Nanoparticles and nanowaste are more reactive and toxic than regular waste material. Importantly, they do not behave like regular bulk materials. Evidently, they can float in the air and penetrate animal and plant cells easily, giving rise to unknown effects. Moreover, nanoparticles have the capability to transport heavy metals like cadmium and others environmental contaminants through air, soil, and water.

Nanowaste management is difficult due to its small size. The conventional waste treatment processes such as incineration and scrubbing/filtering cannot be used to decompose nanowaste. It is a clear fact that the treatment processes should be designed with a zero direct release into the environment. In order to achieve this, numerous methods have recently been proposed to remove or recycle nanoparticles from waste.

In this chapter, the following areas will be discussed.

- What is nanowaste and how will it affect us?
- Nanowaste classification
- Hazardous nanomaterials
- Nanowaste management
  - Nanowaste treatment process
  - Nanowaste disposal and recycling standards

Chapter 6 Microbiology of Wastewater Management: Challenges, Opportunities, and Innovations

The chapter discusses different wastewater sources such as industrial, agricultural, and municipal waste and ballast including the salient features such as presence and detection of pathogenic microbes and viruses, the factors that are ideal for their survival, and the methods by which their presence and impact could be minimized. The microbial diagnostic techniques focuses mainly on the novel and immerging technologies capable of handling culture independent analysis such as the use of stable isotope probing and several platforms of NGS like high-throughput amplicon sequencing and nanopore sequencing. A section on the impact of constituents in various sources of wastewater on the beneficial microbes focuses on how to mitigate the challenges by identifying, detecting the levels, and integration of technologies to remove them. The chapter also includes a section on safe utilization of various forms of wastewater sources to generate products, processes, and revenues.

Chapter 7 Phytoremediation: A Green Tool to Manage Waste

Intensive farming, industrialization, and urbanization, especially in many developing countries, potentially increases the usage of harmful chemical compounds and heavy metals, leading to the soil and ground water contamination. This creates potential health hazard for humans as well as to the ecosystem. Phytoremediation is increasingly recognized as a promising eco-friendly technique to remediate the chemical contaminants. The chapter includes challenges for phytoremediation such as phytotoxicity of leachate and its pollution index, in situ phytoremediation for industrial waste, and disposal of phytoremediation plant waste. Moreover, the chapter includes relatively new concepts such as use of invasive plants in phytoremediation, combining other strategies such as electro remediation, bioaugmentation, and synergistic biosorption with phytoremediation and enhancing plants for metal transport, transformation, volatilization within the plant, and tolerance to toxicity.

Chapter 8 Green Chemistry and its Applications in Waste Management

Green chemistry includes a set of principles aimed to address challenges to reduce and minimize generation of waste products and hazardous substances in design, manufacturing, and application of chemical products. This comprehensive chapter discusses green chemistry practices for effective waste management as an integrated approach from industrial (commercial), health care settings (biomedical waste), and municipal (domestic) waste view point.

Chapter 9 Bioenergy and Biofuels from Wastes

This chapter offers a comprehensive review of the conversion of current waste to bioenergy and biofuel technologies. Technologies reviewed include waste biomass to biofuels (bio-butanol, biodiesel, bioethanol, and biogas), waste biomass to biogenic electricity (microbial fuel cells), and other novel waste to biofuel conversion technologies (syngas, biomass liquefaction, and waste biomass pyrolysis).

Chapter 10 Upcycling: A New Perspective on Waste Management in a Circular Economy

"Upcycling" has emerged as a new intervention to reduce material and energy use in business processes through reusing, repairing, repurposing, and upgrading waste material in a creative way. However, upcycling is still considered as niche practice and remains unclear for many on how exactly to use upcycling for better management of waste. The objective of this chapter is therefore to explore the opportunities and challenges for upcycling. The chapter highlights technical, social, and economic aspects of upcycling practices and uncovers opportunities for better, streamlined practices to produce value-added products from waste material.

1 Waste Management Challenges in Developing Countries

P. A. K. N. Dissanayake[1], D. M. S. B. Dissanayaka[2], Lalith M. Rankoth[2] and Gayan Abeysinghe [3,4,]*
[1]Department of Oceanography and Marine Geology, Faculty of Fisheries and Marine Sciences and Technology, University of Ruhuna, Sri Lanka
[2]Department of Crop Science, Faculty of Agriculture, University of Peradeniya, Sri Lanka
[3]Graduate School of Science and Technology, University of Tsukuba, Japan
[4]Department of Biological Sciences, Faculty of Applied Sciences, Rajarata University of Sri Lanka, Mihintale, Sri Lanka
*Corresponding author: gayandak1@gmail.com

CONTENTS

DOI: 10.1201/9781003132349-2

1.1 INTRODUCTION: CURRENT STATUS OF WASTE MANAGEMENT, A LOOK BACK AT THE DEVELOPING COUNTRIES

Waste management could be considered as one of the most strenuous environmental issues, especially in urban places in developing countries. It is unquestionably a global issue. If waste is not adequately disposed of, it can threaten public health and the environment. Moreover, it is a growing issue that is closely linked to how society generates and consumes (Nuhu et al., 2019; Abtin et al., 2020). The term "waste" immediately directs our thoughts toward the management of waste or at best the concepts of recycling and reuse, as these terms are promoted even in the smallest sectors of the social cohort. But they are not the only aspects that would have ramifications on the environment and even public health. Furthermore, in the 21st century, waste management is one of the necessary utility services, particularly in residential regions. Providing safe drinking water, shelter, food, energy, transportation, and communications is a fundamental human need that can provide sufficient sanitation and is critical to the society and the economy (Justice et al., 2021; Pitchayanin et al., 2017). Nonetheless, waste management infrastructure in developing nations may not receive the same attention as critical health care facilities or upgrading education infrastructure, and there are numerous reasons why waste management should be a priority in developing countries.

The mismanagement of the waste starting from the point of collection and to its disposal, and the processing in between, has caused global environmental pollution and even caused economic instability (Vitorino et al., 2017; Gupta et al., 2015). Current informal waste management methods have several negative repercussions, ranging from incorrect trash disposal, causing marine pollution, to informal garbage collectors experiencing health and financial hazards while seeking to make a livelihood outside of the regular waste management system (Phosrikham et al., 2020). However, just implementing remedies practiced in developed nations will not suffice to address these challenges. Due to cultural, economic, and resource disparities in developing countries, various problems must be addressed to improve waste management systems. Waste management is critical for creating sustainable and habitable communities, yet it is difficult for many developing countries. Effective waste management is costly, frequently accounting for 20%–50% of municipal expenditure (Rosaria et al., 2017; World Bank., 2011). Operating this critical municipal function necessitates implementing integrated systems that are efficient, sustainable, and socially supportive. Inadequate planning, insufficient funds, obsolete collection fleet and equipment, and use of inappropriate methods for completely different waste characteristics have contributed to the loss of expenditure and energy in the direction of waste management in the developing countries (Zhiyong et al., 2018; Hassan et al., 2016). Furthermore, the most critical needs are lack of political will to deal with the problem, lack of governmental waste management policies, and rules and regulations. No policies related to preserving or creating a" circular economy" are at the top of the list of most pressing needs (Diaz, 2017).

In underdeveloped nations, trash disposal has always been a difficult challenge to solve. Landfills are typically open dumping grounds with no leachate or gas collection systems. Some could be in environmentally or hydrologically vulnerable locations. In most cases, they are run below the necessary hygienic standards. The fiscal provisions for operation and maintenance in the municipal budget are insufficient. As a result, residents are subjected to inadequate and dangerous facilities that pose public health concerns and aesthetic burdens. Besides, most open trash dumps are found on the outskirts of big cities, in open spaces, wetland areas, or near-surface water sources. The location of open dumps is usually determined by the availability of collection trucks rather than hydrological or public health factors. Because there are typically no trucks for collection in rural regions and small towns, unregulated dumping occurs within built-up areas, posing health risks and detrimental environmental impact. Further clarifying, unsustainable methods that increase environmental pollution and disease transmission exacerbate waste management in developing nations. The primary concerns identified are open dumping in unmanaged sites, open burning of waste fractions, and mishandling of leachate produced in final disposal sites. The condition is exacerbated in slum regions characterized by high human density, traffic, and air and water pollution. Uncontrolled dumping in open places near water bodies is a common issue in these settings, leading to public health concerns (Modak et al., 2015; Manaf et al., 2009). These concerns have a noticeable detrimental influence on residents and visitors visiting a nation. The most serious concern is smoke from burning debris, which may blanket portions of residential areas and damage the population's quality of life. Citizens are indeed impacted by the smoke from burning garbage and the odor of rotting waste. The nuisances are most significant during the rainy season when the region gets overrun with flies and insects. Runoff from the dumpsite containing dissolved pollutants enters aquatic bodies, while leachate contaminates the soil and groundwater (Modak et al., 2015; Sanneh et al., 2011). Another environmental crisis that arises from surface water contamination is caused due to poor leachate management and uncontrolled material flows from open dump sites. Marine littering, caused mainly by plastic trash, is a visible influence impacting the world's seas and oceans and the extent and magnitude of the consequences of marine litter vary with environmental, social, economic, and public safety (Andrady et al., 2015; Ivar Do Sul et al., 2014).

This chapter elaborates on the factors that potentially contribute to improper waste management, the particular difficulties faced in developing countries, and discusses the ideal solutions that could be recommended for good waste management practices that apply in a typical setting in developing countries.

1.2 CIVIC AWARENESS AND ROLE IN WASTE MANAGEMENT

Rapid urbanization coupled with changes in consumption patterns of people have made a significant change in the quantity and complexity of the waste generated in many countries (Minn et al., 2010; Moustakas et al., 2020; Ko et al., 2020). Unlimited waste generation in those countries have created a greater demand for waste management services. Therefore, waste management has emerged as a global challenge, particularly in economically developing countries (Aparcana, 2016; Han et al., 2018; Ko et al., 2020). The challenge of waste management has further exacerbated due to the limited human, technical, and financial capacities of the related authorities. Moreover, insufficient capacities of municipalities in delivering waste management services have resulted in dumping of different wastes on streets, public places, and riverbeds, leading to an extensive pollution of land and water bodies while posing a health risk to humans. This has necessitated the seeking of alternative approaches for waste management to keep the cities clean and healthy (Aparcana, 2016; Moustakas et al., 2020; Ko et al., 2020).

Participatory management approach comprising shared roles and responsibilities among related stakeholders has been the often proposed tool as a possible solution to the rapidly intensifying waste problem (Ahmed and Ali, 2004; Zeng et al., 2016; Wang et al., 2018). Most of the gaps in waste management systems in cities can be addressed to a greater extent by proper coordination between stakeholders along the waste management channel where public represents the largest category of stakeholders and has multiple relationships to waste management activities (Figure 1.1, Joseph 2006). Therefore, building civic awareness on their role in waste management is crucial and refers to the information people receive, their concerns about the waste management, and their willingness to take part in favor of the environment through reducing waste generation or proper waste disposal (Lima et al., 2005; Varkuti et al., 2008; Zsoka, 2008; Martínez-Peña et al., 2013).

Sustainable public participation has been the basis for successful waste management in any system (Chung and Poon, 2001; Dhokhikah et al., 2015). The effectiveness and practicality of people's participation in waste management have shown to be successful in several developing countries such as India (Anand, 1999; Rathi, 2006), Thailand (Mongkolchaiarunya, 2005), Bangladesh (Sujauddin et al., 2008), Philippines (Bernardo, 2008), and Nigeria (Ogu, 2000). Given the limited public awareness, policymakers now tend to design appropriate waste management communication policies to ensure the entire cross-section of the population and its diverse demographic profile is educated on waste-handling issues (Kala et al., 2020). People's attitudes and behavior are considered to be the main drivers determining the successful implementation of participatory waste management approaches (Evison and Read, 2001). For example, the long-standing view of the public is that they are only responsible for keeping their home clean, but public places, like streets and drains, are the responsibility of the municipality or other relevant authorities (Minn et al., 2010). This indicates the importance of enhancing awareness by imparting relevant knowledge through environmental education for the residents for selecting the most appropriate solid waste management system (Khoo, 2009; Kala et al., 2020).

Humans are the key component in the waste system through generating a complex stream of waste materials, some of which are highly toxic to the environment. Therefore, everyone needs to have a good understanding of waste management. Even the best conceived waste management plans could become questionable if dwellers in the implementing area are not connected to the plans (Hasan, 2004). Previous case studies from developing world are still valid for today to make the best use of the tool of public awareness and participation in successful waste management programs (Hasan, 2004). Those case studies illustrate how public awareness results in everyone working together toward successful waste management and satisfaction of all the stakeholders involved. For example, educational experts have recognized that one of the most effective ways to create life-long environmental awareness, including waste management among the citizens, is to begin with school children. The United States Environmental Protection Agency (EPA) introduced a funding program to promote a series of short courses for school teachers to raise their awareness on waste management (Hasan, 2004). In China, demonstration projects were shown to impart a significant effect on the perception of environmental pollution, the recyclability of home appliances, and the awareness of characteristic pollution caused by domestic waste (Han et al., 2018). Moreover, campaigns on environmental protection are also useful for improving environmental awareness, pollution caused by waste, and the recyclability of plastic and rubber (Han et al., 2018).

As the public sector in most developing countries fails to cater to the increased demand for waste management services, private–public partnerships could sustain waste management programs through the role of citizens who pay for service delivery charges (Ahmed and Ali, 2006). For example, the Advanced Locality Management Program (ALM) launched in 1997 in India as a citizen–government partnership ensured civic engagement in waste management to a significant level (Basu and Panjabi, 2020). The program comprised a group of citizens (approximately 1,000 citizens in a unit) in buildings and apartments in residential areas in Mumbai city and formed on voluntary agreement between the residents of adjacent housing societies who are committed to organizing door-to-door collection of garbage and waste segregation (Singh and Parthasarathy, 2010). The program was initiated as an informal partnership between Municipal Corporation of Greater Mumbai and residents for minimizing and segregating waste at source. However, it gradually evolved into an institutionalized city-level program, shifting the citizen's role from passive service receivers to active service partners (Ahmed and Ali, 2006; Singh and Parthasarathy, 2010; Basu and Panjabi, 2020). Implementation of successful waste management tools through active involvement of nongovernment organizations (NGOs) and community-based organizations (CBOs) partnering with public institutes has also been implemented in several main cities of Bangladesh (Ahmed and Ali, 2006).

1.3 ORGANIZATIONAL CAPACITY IN WASTE MANAGEMENT

Effective waste management programs require identification of definite roles and legal responsibilities of relevant stakeholders and organizational bodies (Figure 1.1). This is particularly important to avoid ineffectiveness of waste management programs caused by inaction and political instability in developing countries (Schübeler et al., 1996). Although the legal frameworks exist, many of the waste management programs tend to fail as a result of weak institutional structures of governments (Halla and Majani, 1999; Hardoy et al., 2001; Konteh, 2009). The degree of decentralization is a key institutional aspect that needs to be considered for sustainable waste management. Distribution of authority, functions, and responsibilities between governmental institutions are of primary importance (Schübeler et al., 1996; Coffey and Coad, 2010). Moreover, structure of institutional systems and their interactions with other relevant sectors, capacity of institutions, organizational procedures for planning and management, and engagement of private sector and community groups are the determinants of successful delivery of waste management programs (Schübeler et al., 1996).

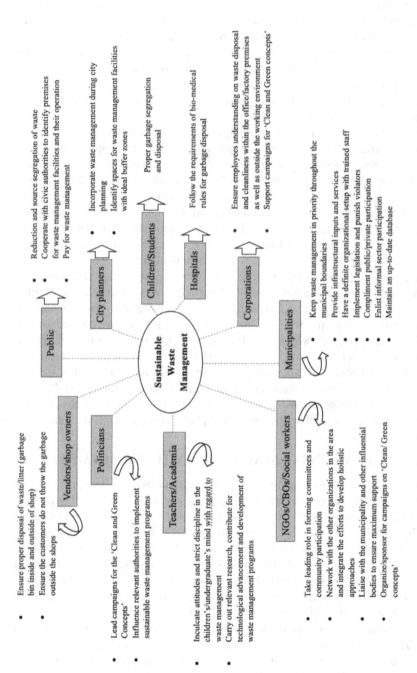

FIGURE 1.1 Stakeholders and institutes involved in waste management and their roles and responsibilities to deliver sustainable waste management systems. (Adapted from Joseph (2006), NGOs, nongovernment organizations; CBOs, community-based organizations.)

Institutional aspects of waste management also comprise different regulations, legislation, and the extent of their enforcement. Regulatory framework should be transparent and need to include inspection and enforcement procedures at local, provincial, and national levels (Schübeler et al., 1996; Coffey and Coad, 2010). In most developing regions, major institutional hindrances to develop and implement sustainable waste management programs are (i) weak institutions, (ii) insufficient and unreliable data on waste generation, and (iii) limited funding (Wilson, 2007). Due to weak institutional support and coordination, enforcement of laws to regulate waste management activities as well as implementation of new projects is often lacking, leading to failures in waste management systems in regions of Asia, Africa, and Latin America (Henry et al., 2006; Coffey and Coad, 2010). The lack of institutional involvement in accurate data management has also caused failures in waste management plans as actual waste generation is significantly higher than the shown data (Shimura et al., 2001; UN-Habitat, 2010; Jha et al., 2011). In many developing countries, waste management is provided limited funding as a result of insufficient tax revenue and mismanagement of funds (Coffey and Coad, 2010). The problem is exacerbated when the central government makes the decision on revenue collection and investment while local authorities are involved in operation and maintenance (Marshall and Farahbakhsh, 2013). Municipalities in developing countries are able to spend only 20–50% of their budget for waste management, which can cater to less than 50% of the municipal population (Henry et al., 2006; Memon, 2010).

It has become extremely difficult for municipalities to find locations for landfill near urban areas due to the rapid increase of land value. Transportation of city-generated wastes to a distant location is even difficult because of the transportation costs and pressure posed by different community groups (Memon, 2010). In addition, technological advancements such as sophisticated vehicles and equipment required for collection, treatment, and disposal of wastes are very expensive and difficult to maintain given their running cost (Coffey and Coad, 2010). In some waste management projects in developing countries, the composition of waste generated is different from those expected and planned to handle before the beginning of the project implementation. This mainly happens in urban areas due to rapidly changing nature (increase population, introduction of new industries etc.) coupled with the complexity of waste generated. This has made the implemented procedures to be of little use (Memon, 2010). These managerial challenges have also limited the institutional capacity, leading to a rapid increase of waste generation, and failures in waste collection, treatment, and disposal in most cities of developing countries (UN-Habitat, 2010). Therefore, strengthening the capacity of organizations involved in waste management and making a proper coordination mechanism along with technical advancements to current waste management systems are of utmost importance. Implementation of participatory approaches with all stakeholders (government, nongovernment, private and community-based organizations, and the public) might provide a viable solution to the limited funding. Having a separate institutional entity with technically capable human resource for data management will ensure accurate and timely statistics along the waste management cycle, thereby planning future waste management programs in marching toward sustainable waste management systems.

1.4 WASTE MANAGEMENT FLOW: FROM THE GENERATION OF WASTE AND COLLECTION TO RECYCLING AND REDUCTION

Waste is generated in many forms such as municipal solid, agricultural and animal, clinical, commercial and industrial, construction and demolition, extraction and mining, oil and gas production, fossil fuel combustion, and sewage sludge (U.S. Environmental Protection Agency (EPA), 2008). The rate of municipal solid waste generation in Asia is 1.2 billion tones in 2016 and the annual waste generation per capita is highest in Hong Kong SAR and China (Kaza et al., 2018). The urban population growth in Asia is estimated to rise from 1.8 billion to 3 billion between 2017 and 2050, which is impossible to cope up with the existing treatment facilities (Hondo et al., 2020).

The process of waste management involves several activities such as collection, transportation, treatment, recycle, reuse, and/or disposal to avert potential health and environmental risks. Perceiving each material's type, composition, and volume is imperative to promote sustainable waste management practices in developed and developing countries. Sustainable development goals (SDGs), targeted in the 2030 agenda, emphasize a necessity of an efficient waste management operation. In addition, SDGs 3, 6, 7, 12, 13, and 14 specifically foreground the consequences of improper waste management practices on the health and safety of public, terrestrial, and marine environments and climate (Hondo et al., 2020). On the other hand, loopholes in waste treatment have impeded its success in many emergent nations based on the case studies cited here.

1.4.1 GENERATION OF WASTE

The population in the developing world is growing at an alarming rate, concomitantly increasing the amount of waste. According to the statistics of the World Bank, the present annual global municipal solid waste generation is about 2.01 billion metric tons. By 2050, it will increase by 70%, reaching 3.40 billion metric tons. Due to the increases in gross national production and population density, the per capita waste production rate in urban areas would rise by 0.2 kg per day and 0.3 kg per day in low-income and middle-income countries. Food and green waste is the principal constituent (44%) in the global waste composition, which is recyclable to produce eco-friendly products (Kaza et al., 2018). In addition, the composition of waste is rapidly changing day by day despite the development of technology, which obstructs sustainable waste management operations (Singh et al., 2014).

1.4.2 WASTE COLLECTION AND TRANSPORTATION

Insufficient collection of waste is a central problem that hampers the efficiency of waste management. On average, the waste collected in low-income countries is about 48% and 26% in urban and rural areas, respectively. The efficacy of waste collection in Sub-Saharan Africa is about 44% and most developed nations such as Europe,

Central Asia, and North America collect at least 90% of waste (Kaza et al., 2018). Approximately 70% is being collected in Kolkata and Malaysia mainly due to the inconsistent collection, less labor power, and insufficient resources (Hazra and Goel, 2009; Manaf et al., 2009). Shortcomings in waste collection mechanisms are evident in Kenya, often resulting from inadequate local infrastructures and logistics. Low-income areas, which have high population densities, are inaccessible and the poor condition of roads makes it hard to collect the waste, especially in rainy seasons. Therefore, local authorities cannot operate at their total capacity in waste collection. Furthermore, poor status and the maintenance of vehicles due to delays in funds worsen the problem (Henry et al., 2006; Muniafu and Otiato, 2010). The amount of daily solid waste generation in Indian cities is about 62 metric tons in 2015 and only 28% has been treated. Inadequate budget and logistic facilities are the major reasons for the low efficacy of waste management (Sharma and Jain, 2018). Furthermore, poor trash segregation at the source point makes it difficult to collect and choose the best treatment option. This is probably due to the unawareness of the impacts of improper waste disposal or ignorance of the public (Hondo et al., 2020). According to the survey carried out in Jakarta (Aprilia et al., 2013), 81% of respondents mentioned that they are not separating their domestic waste. Most respondents (91%) dispose toxic/hazardous and household waste at dumping areas.

1.4.3 Waste Treatment and Disposal

Most developing countries do not follow a preliminary treatment phase before disposal. For example, the Dhapa landfill area in Kolkata receives more than 90% of collected waste without preprocessing (Hazra and Goel, 2009). Proper strategies for collecting hazardous wastes are not treated effectively even with the established incineration systems (Pokhrel and Viraraghavan, 2005). Incinerating waste is not a common practice in developing countries and limited to 1–5%, probably due to the high cost (Khajuria et al., 2010). Open burning and direct disposal of incinerated ashes of hospital waste are regular practices in Ethiopia (Yazie et al., 2019).

Material recovery of e-waste has not received sufficient attention in most developing countries. Developed countries follow the principle of extended producer responsibility (EPR) by setting a take-back system for the used electronic products at their end-of-life (EoL) by the electronic manufacturers and importers. This is a good approach that can be adopted in developing countries as their e-waste management strategies are not up to the standards. The primary reasons behind the inefficient management are the difficulty of recovery of used products, unregulated importation and generation of electronic materials, financial deficits, lack of policies and infrastructure, inconstant data and statistics, and lack of awareness of the consumers regarding the potential impacts of improper disposal of e-waste (Nnorom and Osibanjo, 2008; Adediran and Abdulkarim, 2012; Herat and Agamuthu, 2012). Concerning the nano waste, the possibility of dispersion in an extensive area due to its size makes it challenging to identify the contaminated areas. Moreover, it requires sophisticated technology to detect, treat, and monitor (Musee, 2011), posing a challenge in existing waste management systems in developing countries.

Landfilling is the main choice in most developing nations (65–80%) (Agamuthu, 2012) for waste disposal, as the treatment process is expensive. For instance, the solid waste transported to landfill areas is about 69% in Indonesia, 20–30% in Malaysia, 15% in India, and 9% in Nepal (Khajuria et al., 2010; Dhokhikah and Trihadiningrum, 2012). The collected waste is mostly discarded at dumping sites and burned in the open as a waste management practice, ignoring the environmental and health concerns. Globally, waste disposal at open areas is estimated at 33% (93% in low-income countries and 2% in high-income countries), while 19% of waste is recycled or composted (Kaza et al., 2018). Sanitary landfills (engineered disposal sites) are encouraged over open dumping but are least practiced in underdeveloped countries (Bundhoo, 2018). However, due to the limited land availability, alternative options should be sought to achieve cost-effective and sustainable waste treatment.

Recycling and reuse of waste help to reduce the amount of waste and thereby create a cleaner environment. Bundhoo (2018) reports that only a few developing countries carry out recycling and the majority are collecting the manure for exportation. The fraction of recycling in low-income and middle-income countries remains insignificant and varies between 3.7% and 6% (Kaza et al., 2018).

1.4.4 Management Framework

Local authorities govern waste management under the supervision of a principal government institution in many countries in terms of legal provisions, regulations, and enforcement. For instance, the National Solid Waste Management Department act as the umbrella institution of solid waste management in Malaysia and the relevant local authorities are operated under the apex body (Manaf et al., 2009). However, a centralized organizational structure hampers the effectiveness of the waste management process as it takes a longer time in the decision-making and hinders the participation of relevant stakeholders. Inadequate legal provisions and insufficient enforcement of established laws to address critical issues in waste management are the primary culprits of its weaknesses in Kenya (Henry et al., 2006). Integrated waste management promotes the participation of all stakeholders, including waste generators and the government authorities collaborating with NGOs and the private sector. The decision-making should follow a bottom-up approach to achieve inclusive, sustainable, and transparent solutions. Privatization of solid waste treatment gave promising results in Nairobi (Henry et al., 2006); nevertheless, this should be reinforced by a good legal and financial framework. The government can empower small-scale local groups and CBOs to take part in the treatment process. For instance, City Garbage Recyclers and the Undugu Society in Nairobi actively engage in waste recycling projects while creating livelihood options for the unemployed locals (Henry et al., 2006). Additionally, Zohoori and Ghani (2017) stated that the intervention of political parties jeopardizes the success of waste management as they have their personal interests.

Financial constraints in waste management are a global issue, one of the biggest challenges in low-income developing countries (Mohee et al., 2015; Zohoori and Ghani, 2017). In general, investment of high-income countries in waste management surpasses $100 per ton, while lower income countries only spend $35 per ton. The annual investment of Asia in solid waste management is about $25 billion and is forecast to be approximately $50 billion in 2025 (Hoornweg et al., 1999). Municipalities allocate 20–50% of their recurrent budget in waste treatment in developing countries; nevertheless, only 50% of the population is benefited (Modak et al., 2010). Charging a user fee for the service provided is an option to reduce the budget deficit. According to the survey conducted in Nairobi, Kenya, the residents' willingness to pay for a cleaner environment and a quality service ranges between $20 and $40 per month (Muniafu and Otiato, 2010). However, this is unrealistic in suburban areas where the poverty level is high.

Shortages of skilled staff is another obstacle. This is extremely important in hazardous waste handling and treatment. Proper segregation and storage utilities, standard waste disposal methods, capacity building of sanitary workers, and adequate budgeting are mandatory to improve infectious waste management systems as observed in Pakistan and Mongolia (Ali and Kuroiwa, 2009; Kumar et al., 2015). The medical waste generated in Ethiopia is significant (21–70%). The current waste management system in the country is substandard due to the absence of improper segregation mechanisms, unsanitary disposal, lack of national policy, regulations and enforcement, and meticulous supervision (Yazie et al., 2019).

Waste can be a resource through waste treatment to earn economic benefits (Hondo et al., 2020). For instance, food waste, plastics, and electronic materials have their own added value. Nevertheless, waste reduction and treatment are challenging in developing states since it differs from country to country. The necessity of sound data and information, waste collection mechanism, proper recycling facilities, and technologies is fundamental for making informed decisions and implementing holistic, evidence-based practice for any waste material.

1.5   HEALTH AND ENVIRONMENTAL FACTORS

Improper waste management degenerates the health of the environment endangering the living beings. Lack of environmental awareness of the locals in most low-income and middle-income nations leads to the dumping of litter illicitly at open areas, roadways, and rivers (Bundhoo, 2018). Accumulation of solid waste pollutes the entire environment and deteriorates aesthetic beauty. The people living adjacent to the dumping areas are diagnosed with various health issues such as skin diseases, respiratory diseases, and water-bone diseases. Pungent odor (ammonia and hydrogen sulfide) released by the waste and the noxious gases released by open garbage burning create uninhabitable areas for residents (Henry et al., 2006; Muniafu and Otiato, 2010; Dhokhikah and Trihadiningrum, 2012; Esmaeilizadeh et al., 2020). The greenhouse gases emitted by open burns (carbon dioxide) and anaerobic decomposition of organic trash (methane) contribute to climate change (Agamuthu, 2012). On the other hand, the domestic animals that live adjacent to the dumping site feed on contaminated food. As a result, excessive concentrations of dioxins have been detected in chicken eggs, indicating the risk of cumulation of the toxic elements in biota (Muniafu and Otiato, 2010).

In addition, the litter can block the irrigation canals causing flooding, which was the prime reason for the major flood that happened in Surat, India, in 1994. It resulted in an infestation affecting 1,000 people and death of 56 people. The landslide at the

Payatas dumpsite in Quezon City, Philippines, killed 200 people in 2000, indicating its risk (UN-Habitat, 2010). Poor development of landfill areas irrespective of scientific and environmental standards such as Gokarna and Bishnumati in Nepal are at high risk of groundwater pollution due to the percolation of surface substances through the soil (Pokhrel and Viraraghavan, 2005).

The waste pickers in most developing countries handle the waste manually without any safety gear. Cases of most waste pickers affected by a parasitic disease were recorded in Kolkata (Hazra and Goel, 2009). It is crucial to control the medical and other hazardous wastes that can cause irreversible health impacts. Kumar et al. (2015) have reported that clinical waste management in Pakistan is not performed according to the World Health Organization (WHO) guidelines. The health care staff who handle the waste are not used to wear personal protective equipment. Furthermore, the authors have emphasized the requirement of critical monitoring and supervision when handling infectious waste to ensure staff safety and the environment.

In addition, the toxic heavy metals associated with electronic wastes such as lead (Pb), mercury (Hg), cadmium (Cd), and Chromium (Cr) could cause critical impacts on human health and the environment itself if not properly managed. They can be bioaccumulated and magnified throughout the food chain, resulting in a trophic cascade. Conversely, there is a potential risk of nano waste on the environment and human health. However, the adverse impacts are still unknown due to the limited ecotoxicity research conducted on nanomaterials (Musee, 2011).

1.6   INTEGRATED SUSTAINABLE WASTE MANAGEMENT APPROACHES

As discussed in the previous sections, waste generation in any society is unavoidable and these wastes create numerous environmental, economic, health, ecological, and many other problems to that society. The only way out of such issues is the correct management of waste materials. Henry et al. (2006) reported that waste management includes a series of activities such as collection, transfer, resource recovery, recycling, and treatment in order to prevent health problems to the society and improve environmental quality and support the economy of the society. To achieve those objectives and improve aesthetic and land-use resources, sustainable waste management approaches must be practiced through the integration of different waste management techniques and models (Henry et al., 2006; Nemerow, 2009).

Currently, the integration in waste management more relates to the technological integration taking place, especially in developed countries, for waste management. While the integration waste management (IWM) combines physical and technological components at different hierarchy levels, on the other hand, integrated sustainable waste management (ISWM) addresses both the physical components such as collection, disposal, and recycling together with the governance/management aspects such as finance and economy, institutional capacity, policy changes, and stakeholders' role (Wilson et al., 2013; Wilson et al., 2015).

The availability of high-quality technical packages, adequate organizational capacity, proper planning, and decision-making as well as proper implementation of such set plans have made the developed countries successful in sustainable and integrated waste management (Taelman et al., 2019; Perteghella et al., 2020). Even though the urban solid waste generation in developing countries is lower compared to industrialized countries, waste management in developing nations still remains inadequate (Henry et al., 2006). Rapidly increasing population densities, uplifting living standards, rapid urbanization, and shifting of population more toward urban areas, lack of consistent data, changing waste composition, and generation rates together with lack of organization, poor financial resources, unavailability of proactive policies, and decision-making are some of the reasons for waste management problems faced by developing countries (Agamuthu et al., 2007; Burntley, 2007; Minghua et al., 2009; Guerrero et al., 2013; Wilson et al., 2015). Wilson et al. (2013) further elaborates that the waste collection and controlled disposal of the middle-income and lower income cities are achieved by 95% and 50%, respectively, whereas the recycling rates are between 20% and 30% in many lower income countries. Even with the successful collection of 50–95% of the waste, their sustainable management in developing countries is still questionable.

There are many factors affecting the sustainability in waste management, thus the impact of such systems varies from place to place. Therefore, the use of integrated approaches to achieve sustainability is imperative (Agamuthu et al., 2007).

Agamuthu et al. (2007) have described the drivers of waste management in four broad categories known as *human drivers* (health, education, and awareness), *economic drivers* (profit generation from waste, socioeconomic conditions of the country), *environmental drivers* (effects and impacts on the environment), and *institutional drivers* (legislations, scientific research, and business image and profitability). Considering all such factors and drivers, scientist and researchers have created several integrated sustainable waste management (ISWM) models that are now in use in many developing and developed countries throughout the world.

These models describe waste management through the integration of three major sectors: the flow of materials (waste generation and separation, collection and transport, treatment, recycling, and final disposal), stakeholder involvement (municipality, ministry, NGOs, formal and informal private sector organizations), and process-enabling environment (policy/legal/political, technical, financial, sociocultural, environmental) via which the sustainability in waste management would be feasible (Guerrero et al., 2013). Wilson et al. (2015) also presents a ISWM model as an interconnected "two triangles" concept where one triangle consists of public health (collection of waste), environment (treatment and disposal of waste), and resource value (reduce, reuse, and recycle—3Rs) and the other triangle consists of sound institutions and proactive policies, user and provider inclusivity, and financial sustainability. Anschütz et al. (2004) has reported a similar ISWM model, which identifies three major dimensions in waste management, namely stakeholders, waste system elements (collection, transport, and 3Rs), and aspects (technical, environmental, and financial).

Before identifying these ISWM models, numerous models that addressed specific problems in waste management flow were in use. There were transportation models to address problems in waste material transportation (Truitt et al., 1969), vehicle routine models to minimize the travel distance of wastes (Angelelli and Speranza, 2002), economic evaluation models to check the economy of waste collection and disposal

(Bertazzi and Speranza, 2012), statistical models to estimate the status and quantity of waste generation and utilization (Minghua et al., 2009; Popoviċ et al., 2013), and other models for identifying different composting procedures, energy recovery methods, and recycling techniques (Bidart et al., 2013; Melikoglu, 2013). However, even though each of these models had a specific role in solving a specific problem, none of them lead toward sustainability in waste management. Thus, the integration of a few or as many models as possible to achieve the sustainability of the system is a dire need. When integrating different simple models and conceptualizing new complex ones, three basic categories of analyses can be carried out. They are cost–benefit analysis, life cycle assessment, and multicriteria decision analysis. The cost–benefit analysis allows the decision makers to convert all positive and negative impacts of the models into a monetary value and then assess the productivity based on that whereas the life cycle assessment analyzes the potential impacts and environmental aspects arising due to the flow of wastes from start (generation) to the end (disposal or recycling). Multicriteria decision analysis takes several individual and conflicting aspects into consideration in multidimensional ways and several alternative approaches are used to address the arising problems. The best alternative could be selected depending on the objective, situation, or the place (Morrissey and Browne, 2004). The benefits and drawbacks of these three categories are summarized in Table 1.1.

Based on the resource availability and the objectives, any of the assessment tools mentioned in Table 1.1 can be used to develop ISWM models. The current state of ISWM models is complex, reflect policy changes, closer to reality, target more environmentally useful alternatives, and lead to sustainability (Allaoui et al., 2015). Therefore, these ISWM approaches are complex, but if implemented correctly would provide long-term and promising solutions to waste management problems in developing countries.

As indicated in Figure 1.2, ISWM combine both physical components and governance or management aspects in waste management and thereby convert wastes and energy inputs into useful materials such as compost or biogas and dispose environment friendly products such as treated water.

By practicing ISWM systems, not only environment but also economic and social sustainability can be achieved (Diaz et al., 1996). However, when implementing complex ISWM frameworks such as the model presented in Figure 1.2 in developing countries, some of the components have to be integrated only after proper analysis of the situation. For example, Wilson et al. (2012) reported that the mean organic material content in waste in developing and developed countries is around 67% and 28%, respectively. Thus, when using technological packages, the correct type of machineries must be selected and they should be affordable as well. The costs incurred in maintaining a ISWM cannot be recovered by charging higher service fees due to the lower income levels of the people in developing countries. In such cases, income generation through composting, biogas production, and recycled product development, and subsidizing through government and nongovernment organizations have to be promoted. In such a ISWM system, especially in developing countries where the technology and mechanization are limited compared to developed countries, more emphasis should be placed on the inclusivity of both users and service providers

TABLE 1.1

Benefits and Drawbacks of Cost–Benefit Analysis, Life Cycle Assessment, and Multicriteria Decision Analysis in Waste Management Model Assessment

Benefits	Drawbacks
Cost–benefit analysis	
• Clear results can be presented	• Can raise ethical issues
• All related impacts can be presented in a monetary form	• Uncertainty in converting some environmental and social impacts into monetary form
• Easy to identify the net return of resource use	• Assumptions on price is highly variable
Life cycle assessment	
• Easy to compare between models	• The boundaries are difficult to be assigned
• Trade-offs between models can be identified	• The results for the same waste product may vary often in practice
• Addition of economic assessment of environmental impacts would provide better results	• Use of life cycle assessment alone without considering costs–benefits etc. could provide misleading results
Multicriteria decision analysis	
• Provides a systematic approach in understanding the problem	• Could be very complex and difficult to implement
• Provides a systematic approach in evaluating policy options	• Prioritizing each criterion is very subjective thus could change the result
• Preference of different stakeholders on conflicting matters can be accommodated	• Decision-making highly depends on experience and personal judgment

Source: Finnveden and Ekvall, 1998; Ekvall, 1999; Qureshi et al., 1999; Beynon et al., 2000; Morrissey and Browne, 2004.

where the general public adhere to behavioral policies such as prevention, minimal waste generation, and separate collection for recycling (Wilson et al., 2013). There will not be perfect ISWM models applicable to all the situations in all the developing countries, but ideal solutions have to be provided by addressing local needs of each situation separately. However, the use of integrated and sustainable waste management systems where not only physical but also governance aspects are integrated can easily be practiced and will provide better solutions to waste management problems in developing countries.

1.7 FUTURE DIRECTIONS IN ENHANCING THE CAPACITY DEVELOPMENT OF WASTE MANAGEMENT

The traditional waste management strategies basically focus on improved and efficient waste collection methods, but the real requirement of the present and future world is the identification of value chains in waste management and utilizing them while reducing, removing, reusing, and recycling wastes for achieving economic, environmental, and social sustainability.

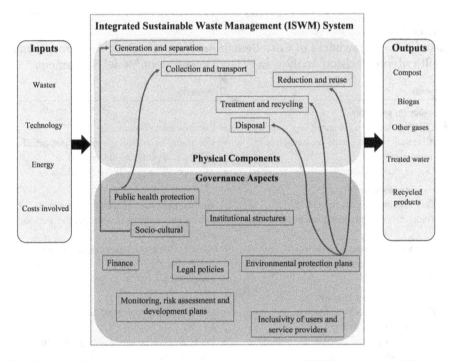

FIGURE 1.2 The integrated sustainable waste management (ISWM) framework. (This is a combined presentation of the findings of Anschütz et al., 2004; Guerrero et al., 2013; Wilson et al., 2013; Allaoui et al., 2015; Wilson et al., 2015; and Basnayake et al., 2020.)

One major problem the developing countries have in terms of waste management is lack of appropriate and reliable baseline data, which leads to difficulties in model creation for sustainable waste management (Zurbrügg et al., 2014). Thus, there is a dire need of data collection by the scientists and academics as well as dissemination of such collected data through a free database. Some of the tools used in waste management strategies in developing countries are not very suitable for the context of a developing country (Zurbrügg et al., 2014). Thus, simple and easy low-cost methods should be developed to address each local situation separately. Most of the studies related to waste management are conducted by researchers in developed countries. The fact that initiation of local research to identify best approaches for waste management by the scientists in developing countries is also required because that would address the real problems in those countries directly. Zurbrügg et al. (2014) further describe that in most of the developing countries, the level of expertise is lower in local stakeholders. Thus, more proper education and training has to be popularized because the designing of ISWM models require expertise in social science, natural science, and engineering. The existing tools should be modified in a way they are less expensive and easy to apply in the future. Additionally, linkages should be created between public and private sector organizations, which will improve the management structure as well as create new job opportunities (Ahmed and Ali, 2004).

The use of Internet of Things (IOT) principles for smart application of waste management strategies would be a timely requirement, which allows the easy integration of technology, data, and human capital (Anagnostopoulos et al., 2015; Deloitte, 2015). Esmaeilian et al. (2018) suggest three smart approaches to future waste management, which describes building infrastructure for collecting life cycle data of the products so that the stakeholders can see the entire life span of a product, building of an intelligent sensor-based mechanism for waste collection and separation, and creation of new business models targeting lower or zero waste generation. The use of tracking databases, IoT technologies, proper knowledge enhancement facilities, establishment of infrastructure, and making the public aware of their responsibility in waste management will enable the production of more holistic ISWM models. Using integrated approaches with modern-day technology and artificial intelligence, with the collective participation of citizens and authorities, would address many present problems in developing countries and help in creating a waste-free environment we all would strive to live in.

1.8 SUMMARY

This chapter presented many aspects of waste management challenges, a multifaceted issue faced especially by the developing countries with an emphasis on the prevailing circumstances of waste management, waste generation in developing countries through management framework, civic responsibilities, and a thorough look at the integrated sustainable waste management processes that are currently being employed. Moreover, the chapter provided insights on the available waste management flow with its impacts on health and environment and tried to bring forth an overview of prevailing regulations in developing countries to come up with potential implementation of effective waste management systems.

REFERENCES

Abtin, M., M. Fatima, and H. Ehsan. 2020. "Environmental and economic assessment of sustainable municipal solid waste management strategies in Iran." *Sustainable Cities and Society* 59: 1–7. doi: 10.1016/j.scs.2020.102161

Adediran, Y. A., and A. Abdulkarim. 2012. "Challenges of electronic waste management in Nigeria." *International Journal of Advances in Engineering & Technology* 4(1): 640–648.

Agamuthu, P. 2012. "Landfilling in developing countries." *Waste Management & Research: The Journal for a Sustainable Circular Economy* 31(1): 1–2. doi: 10.1177/0734242X12469169

Agamuthu, P., S. H. Fauziah, K. M. Khidzir, and A. Noorazamimah Aiza. 2007. "Sustainable waste management-Asian perspectives." *Proceedings of the International Conference on Sustainable Solid Waste Management* 5: 15.

Ahmed, S. A., and M. Ali. 2004. "Partnerships for waste management in developing countries: linking theories to realities." *Habitat International* 28: 467–479. doi:10.1016/S0197-3975(03)00044-4

Ahmed, S. A., and S. M. Ali. 2006. "People as partners: facilitating people's participation in public-private partnerships for solid waste management." *Habitat International* 30: 781–796. doi:10.1016/j.habitatint.2005.09.004

Ali, M., and C. Kuroiwa. 2009. "Status and challenges of hospital solid waste management: case studies from Thailand, Pakistan, and Mongolia." *Journal of Material Cycles and Waste Management* 11(3): 251–257. doi: 10.1007/s10163-009-0238-4

Allaoui, H., A. Choudhary, S. Elsaid, and E.-H. Aghezzaf. 2015. "A framework for sustainable waste management: challenges and opportunities." *Management Research Review* 38(10): 1086–1097.

Anagnostopoulos, T., A. Zaslavsky, and A. Medvedev. 2015. "Robust waste collection exploiting cost efficiency of IoT potentiality in Smart Cities." In *2015 International Conference on Recent Advances in Internet of Things (RIoT)*, Singapore. 1–6.

Anand, P. B. 1999. "Waste management in Madras revisited." *Environment and Urbanization* 11: 161–176. doi:10.1177/095624789901100214

Andrady, A., K. L. Law, C. Wilcox, R. Narayan, J. R. Jambeck, R. Geyer, T. R. Siegler, and M. Perryman. 2015. "Plastic waste inputs from land into the ocean." *Science* 347: 768–771.

Angelelli, E., and M. G. Speranza. 2002. "The application of a vehicle routing model to a waste collection problem: two case studies." *Quantitative Approaches to Distribution Logistics and Supply Chain Management* 53(2): 269–286.

Anschütz, J., J. Ijgosse, and A. Scheinberg. 2004. "Putting integrated sustainable waste management into practice." *Waste Netherland*: 1–102

Aparcana, S. 2016. "Approaches to formalization of the informal waste sector into municipal solid waste management systems in low- and middle-income countries: review of barriers and success factors." *Waste Management* 61: 593–607. doi:10.1016/j.wasman.2016.12.028

Aprilia, A., T. Tezuka, and G. Spaargaren. 2013. "Inorganic and hazardous solid waste management: current status and challenges for Indonesia." *Procedia Environmental Sciences* 17(81): 640–647. doi: 10.1016/j.proenv.2013.02.080

Basnayake, B. F. A., R. T. K. Ariyawansha, A. K. Karunarathna, S. M. Werahera, and N. Mannapperuma. 2020. "Sustainable waste management challenges in Sri Lanka." In *Sustainable Waste Management Challenges in Developing Countries.*. Hershey, PA: IGI Global, 352–381.

Basu, A. K., and S. Punjabi. 2020. "Participation in solid waste management: lessons from the Advanced Locality Management (ALM) programme of Mumbai." *Journal of Urban Management* 9(1): 93–103. doi:10.1016/j.jum.2019.11.002

Bernardo, E. C. 2008. "Solid waste management practices of households in Manila, Philippines." *Annals of the New York Academy of Sciences* 1140: 420–424. doi:10.1196/annals.1454.016

Bertazzi, L., and M. Grazia Speranza. 2012. "Inventory routing problems: an introduction." *Journal on Transportation and Logistics* 1(4): 307–326.

Beynon, M., B. Curry, and P. Morgan. 2000. "The Dempster–Shafer theory of evidence: an alternative approach to multicriteria decision modelling." *Omega* 28(1): 37–50.

Bidart, C., M. Fröhling, and F. Schultmann. 2013. "Municipal solid waste and production of substitute natural gas and electricity as energy alternatives." *Applied Thermal Engineering* 51(1–2): 1107–1115.

Bundhoo, Z. M. A. 2018. "Solid waste management in least developed countries: current status and challenges faced." *Journal of Material Cycles and Waste Management* 20(3): 1867–1877. doi: 10.1007/s10163-018-0728-3

Burnley, S. J. 2007. "A review of municipal solid waste composition in the United Kingdom." *Waste Management* 27(10): 1274–1285.

Chung, S. S., and C. S. Poon. 2001. "A comparison of waste-reduction practices and new environmental paradigm of rural and urban Chinese citizens." *Journal of Environmental Management* 62: 3–19. doi: 10.1006/jema.2000.0408

Coffey, M., and A. Coad. 2010. *Collection of Municipal Solid Waste in Developing Countries.* Malta: UN-HABITAT.

Deloitte. 2015. "Smart cities, how rapid advances in technology are reshaping our economy and society." Available at: www2.deloitte.com/tr/en/pages/public-sector/articles/smart-cities.html

Dhokhikah, Y., and Y. Trihadiningrum. 2012. "Solid waste management in Asian developing countries: Challenges and opportunities." *Journal of Applied Environmental and Biological Science* 2(7): 329–335. Available at: www.researchgate.net/publication/284942823 (Accessed: September 30, 2021).

Dhokhikah, Y., Y. Trihadiningrum, and S. Sunaryo. 2015. "Community participation in household solid waste reduction in Surabaya, Indonesia." *Resources, Conservation and Recycling* 102: 153–162. doi:10.1016/j.resconrec.2015.06.013

Diaz, L. F. 2017. "Waste management in developing countries and the circular economy." *Waste Management and Research: The Journal for a Sustainable Circular Economy* 35(1): 1–2. Available at: https://doi.org/10.1177%2F0734242X16681406

Diaz, L. F., G. M. Savage, and C. G. Golueke. 1996. "Sustainable community systems: the role of integrated solid waste management." In *Proceedings of 19th International Madison Waste Conference*, Madison, WI, 280–291.

Ekvall, T. 1999. "Key methodological issues for life cycle inventory analysis of paper recycling." *Journal of Cleaner Production* 7(4): 281–294.

Esmaeilian, B., B. Wang, K. Lewis, F. Duarte, C. Ratti, and S. Behdad. 2018. "The future of waste management in smart and sustainable cities: a review and concept paper." *Waste Management* 81: 177–195.

Esmaeilizadeh, S., A. Shaghaghi, and H. Taghipour. 2020. "Key informants' perspectives on the challenges of municipal solid waste management in Iran: a mixed method study." *Journal of Material Cycles and Waste Management* 22(4): 1284–1298. doi: 10.1007/s10163-020-01005-6

Evison, T., and A. D. Read. 2001. "Local authority recycling and waste-awareness publicity/promotion." *Resources, Conservation and Recycling* 32: 275–292. doi:10.1016/S0921-3449(01)00066-0

Finnveden, G., and T. Ekvall. 1998. "Life-cycle assessment as a decision-support tool—the case of recycling versus incineration of paper." *Resources, Conservation and Recycling* 24(3–4): 235–256.

Guerrero, L. A., G. Maas, and W. Hogland. 2013. "Solid waste management challenges for cities in developing countries." *Waste Management* 33(1): 220–232.

Gupta, N., K. K. Yadav, and V. Kumar. 2015. "A review on current status of municipal solid waste management in India." *Journal of Environmental Science (China)* 37: 206–217. doi: 10.1016/j.jes2015.01.034

Halla, F., and B. Majani. 1999. "Innovative ways for solid waste management in Dar-Es-Salaam: toward stakeholder partnerships." *Habitat International* 23(3): 351–361. doi:10.1016/S0197-3975(98)00057-5

Han, Z., Q. Duan, Y. Fei, D. Zeng, G. Shi, H. Li, and M. Hu. 2018. "Factors that influence public awareness of domestic waste characteristics and management in rural areas." *Integrated Environmental Assessment and Management* 14: 395–406. doi: 10.1002/ieam.4033

Hardoy, J. E., D. Mitlin, and D. Satterthwaite. 2001. *Environmental Problems in an Urbanizing World: Finding Solutions for Cities in Africa, Asia, and Latin America.* London, Sterling, VA: Earthscan Publications.

Hasan, S. E. 2004. "Public awareness is key to successful waste management." *Journal of Environmental Science and Health* 39: 483–492. doi:10.1081/ese-120027539

Hassan, T., A. Zahra, A. Hassan, F. Armanfar, and R. Dehghanzadeh. 2016. "Characterizing and quantifying solid waste of rural communities." *Journal of Material Cycles and Waste Management* 18: 790–797.

Hazra, T., and S. Goel. 2009. "Solid waste management in Kolkata, India: practices and challenges." *Waste Management* 29(1): 470–478. doi: 10.1016/j.wasman.2008.01.023

Henry, R. K., Z. Yongsheng, and D. Jun. 2006. "Municipal solid waste management challenges in developing countries—Kenyan case study." *Waste Management* 26(1): 92–100. doi: 10.1016/j.wasman.2005.03.007

Herat, S. and Agamuthu, P. 2012. "E-waste: a problem or an opportunity? Review of issues, challenges and solutions in Asian countries." *Waste Management and Research* 30(11): 1113–1129. doi: 10.1177/0734242X12453378

Hondo, D., L. Arthur, and P. J. D. Gamaralalage. 2020. *Solid Waste Management in Developing Asia: Prioritizing Waste Separation.* Tokyo, Japan: Asian Development Bank Institute.

Hoornweg, D., L. Thomas, and K. Varma. 1999. *What a Waste: Solid Waste Management in Asia.* Edited by M. Fossberg et al. Washington, DC: The World Bank. Available at: http://documents1.worldbank.org/curated/en/694561468770664233/pdf/multi-page.pdf

Ivar Do Sul, J. A., and M. F. Costa. 2014. "The present and future of microplastic pollution in the marine environment." *Environmental Pollution* 185: 352–364. doi: 10.1016/j.envpol.2013.10.036

Jha, A., S. K. Singh, G. P. Singh, and P. K. Gupta. 2011. "Sustainable municipal solid waste management in low income group of cities: a review." *Tropical Ecology* 52(1): 123–131.

Joseph, K. 2006. "Stakeholder participation for sustainable waste management." *Habitat International* 30: 863–871. doi:10.1016/j.habitatint.2005.09.009

Jusctice, K. D., G. V. Diogo, and A. P. D. Maria. 2021. "Raising awareness on solid waste management through formal education for sustainability: a developing countries evidence review." *Recycling* 6(1): 6. doi: 10.3390/recycling6010006

Kala, K., B. B. Nomesh, and Sushil. 2020. "Waste management communication policy for effective citizen awareness." *Journal of Policy Modeling* 42: 661–678. doi:10.1016/j.jpolmod.2020.01.012

Kaza, S., et al. 2018. *What a Waste 2.0: A Global Snapshot of Solid Waste Management to 2050.* Washington, DC: World Bank, p. 38. Available at: https://openknowledge.worldbank.org/bitstream/handle/10986/30317/211329ov.pdf

Khajuria, A., Y. Yamamoto, and T. Morioka. 2010. "Estimation of municipal solid waste generation and landfill area in Asian developing countries." *Journal of Environmental Biology* 31(5): 649–654.

Khoo, H. H. 2009. "Life cycle impact assessment of various waste conversion technologies." *Waste Management* 29: 1892–1900. doi: 10.1016/j.wasman.2008.12.020

Ko, S., W. Kim, S. C. Shin, and J. Shin. 2020. "The economic value of sustainable recycling and waste management policies: the case of a waste management crisis in South Korea." *Waste Management* 104: 220–227. doi: 10.1016/j.wasman.2020.01.020

Konteh, F. H. 2009. "Urban sanitation and health in the developing world: reminiscing the nineteenth century industrial nations." *Health and Place* 15 (1): 69–78. doi:10.1016/j.healthplace.2008.02.003

Kumar, R., B. Tasneem Shaikh, R. Somrongthong, and R. S. Chapman. 2015. "Practices and challenges of infectious waste management: a qualitative descriptive study from tertiary care hospitals in Pakistan." *Pakistan Journal of Medical Sciences* 31(4): 795–798. doi: 10.12669/pjms.314.7988

Lima, M. L., J. Barnett, and J. Vala. 2005. "Risk perception and technological development at a societal level." *Risk Analysis* 25: 1229–1239. doi:10.1111/j.1539-6924.2005.00664.x

Manaf, L. A., M. A. A. Samah, and N. I. M. Zukki. 2009. "Municipal solid waste management in Malaysia: practices and challenges." *Waste Management* 29(11): 2902–2906. doi: 10.1016/j.wasman.2008.07.015

Marshall, R. E., and K. Farahbakhsh. 2013. "Systems approaches to integrated solid waste management in developing countries." *Waste Management* 33: 988–1003. doi:10.1016/j.wasman.2012.12.023

Martínez-Peña, R. M., A. L. Hoogesteijn, S. J. Rothenberg, M. D. Cervera-Montejano, and J. G. Pacheco-Ávila. 2013. "Cleaning products, environmental awareness and risk perception in Mérida, Mexico." *PLoS One* 8(8): e74352. doi:10.1371/journal.pone.0074352

Melikoglu, M. 2013. "Vision 2023: assessing the feasibility of electricity and biogas production from municipal solid waste in Turkey." *Renewable and Sustainable Energy Reviews* 19: 52–63.

Memon, M. A. 2010. "Integrated solid waste management based on the 3R approach." *Journal of Material Cycles and Waste Management* 12(1): 30–40. doi:10.1007/s10163-009-0274-0

Minghua, Z., F. Xiumin, A. Rovetta, H. Qichang, F. Vicentini, L. Bingkai, A. Giusti, and L. Yi. 2009. "Municipal solid waste management in Pudong new area, China." *Waste Management* 29(3): 1227–1233.

Minn, Z., S. Srisontisuk, and W. Laohasiriwong. 2010. "Promoting people's participation in solid waste management in Myanmar." *Research Journal of Environmental Sciences* 4: 209–222. doi: 10.3923/rjes.2010.209.222

Modak, P., D. C. Wilson, and C. Velis. 2015. *Global Waste Management Outlook*. Athens, Greece: UNEP, 51–79.

Modak, P., et al. 2010. "Municipal Solid Waste Management: Turning waste in to resources." in *Shanghai Manual: A Guide for Sustainable Urban Development in the 21st Century*, 1–36. Available at: https://issuu.com/zoienvironment/docs/gwmo_report/317 (Accessed: May 5, 2021).

Mohee, R., S. Mauthoor, Z. M. A. Bundhoo, G. Somaroo, N. Soobhany, and S. Gunasee. 2015. "Current status of solid waste management in small island developing states: a review." *Waste Management* 43: 539–549. doi: 10.1016/j.wasman.2015.06.012

Mongkolchaiarunya, J. 2005. "Promoting a community-based solid-waste management initiative in local government: Yala municipality, Thailand." *Habitat International* 29: 27–40. doi:10.1016/S0197-3975(03)00060-2

Morrissey, A. J., and J. Browne. 2004. "Waste management models and their application to sustainable waste management." *Waste Management* 24(3): 297–308.

Moustakas, K., M. Rehan, M. Loizidou, A. S. Nizami, and M. Naqvi. 2020. "Energy and resource recovery through integrated sustainable waste management." *Applied Energy* 261: 114372. doi:10.1016/j.apenergy.2019.114372

Muniafu, M., and E. Otiato. 2010. "Solid waste management in Nairobi, Kenya. A case for emerging economies." *Journal of Language, Technology & Entrepreneurship in Africa. African Journals Online (AJOL)* 2(1): 342–350. doi: 10.4314/jolte.v2i1.52009

Musee, N. 2011. "Nanowastes and the environment: potential new waste management paradigm." *Environment International* 37(1): 112–128. doi: 10.1016/j.envint.2010.08.005

Nemerow, N. L., F. J. Agardy, and J. A. Salvato. 2009. *Environmental Engineering: Environmental Health and Safety for Municipal Infrastructure, Land Use and Planning, and Industry*. Hoboken, NJ: Wiley.

Nnorom, I. C., and O. Osibanjo. 2008. "Overview of electronic waste (e-waste) management practices and legislations, and their poor applications in the developing countries." *Resources, Conservation and Recycling* 52(6): 843–858. doi: 10.1016/j.resconrec.2008.01.004

Nuhu, D. M., I. B. Nawaf, A. N. Ammar, M. A. Isam, and A. Ali. 2019. "Food waste management current practices and sustainable future approaches: a Saudi Arabian perspectives." *Journal of Material Cycles and Waste Management* 21: 678–690.

Ogu, V. I. 2000. "Private sector participation and municipal waste management in Benin City, Nigeria." *Environment and Urbanization* 12: 103–117. doi:10.1177/095624780001200209

Perteghella, A., G. Gilioli, T. Tudor, and M. Vaccari. 2020. "Utilizing an integrated assessment scheme for sustainable waste management in low and middle-income countries: case studies from Bosnia-Herzegovina and Mozambique." *Waste Management* 113: 176–185.

Pitchayanin, S., S. Kunio, and S. Alice. 2017. "Toward effective multi-sector partnership: a case of municipal solid waste management service provision in Bangkok, Thailand." *Kasetsart Journal of Social Sciences* 38(3): 324–330.

Phosrikham, N., W. Laohasiriwong, A. Luenam, and P. S. Keo. 2020. "Health literacy, Occupational health and safety factors and quality of life of municipal waste collectors in the northeast of Thailand." *Indian Journal of Public Health Research and Development* 11(3): 2460–2464.

Pokhrel, D., and T. Viraraghavan. 2005. "Municipal solid waste management in Nepal: practices and challenges." *Waste Management* 25(5): 555–562. doi: 10.1016/j.wasman.2005.01.020

Popović, F. J., J. V. Filipović, and V. N. Božanić. 2013. "Paradigm shift needed: municipal solid waste management in Belgrade, Serbia." *Hemijska industrija* 67(3): 547–557.

Qureshi, M. E., S. R. Harrison, and M. K. Wegener. 1999. "Validation of multicriteria analysis models." *Agricultural Systems* 62(2): 105–116.

Rathi, S. 2006. "Alternative approaches for better municipal solid waste management in Mumbai, India." *Waste Management* 26: 1192–1200. doi:10.1016/j.wasman.2005.09.006

Rosaria, C., L. P. Samuele, M. Shigeru, and T. Tomohiro. 2017. "Does recyclable separation reduce the cost of municipal waste management in Japan?." *Waste Management* 60: 32–41.

Sanneh, E. S., A. H. Hu, Y. M. Chang, and E. Sanyang. 2011. "Introduction of a recycling system for sustainable municipal solid waste management: a case study on the greater Banjul area of the Gambia." *Eviron. Dev. Sustain.* 13: 1065–1080. doi: 10.1007/s10668-011-9305-9

Schübeler, P., K. Wehrle, and Christen J. 1996. *Conceptual Framework for Municipal Solid Waste Management in Low-Income Countries*. Urban management and infrastructure working paper no. 9. Washington, DC: World Bank Group.

Sharma, K. D., and S. Jain. 2018. "Overview of municipal solid waste generation, composition, and management in India." *Journal of Environmental Engineering. American Society of Civil Engineers* 145(3): 04018143. doi: 10.1061/(ASCE)EE.1943-7870.0001490

Shimura, S., I. Yokota, and Y. Nitta. 2001. "Research for MSW flow analysis in developing nations." *Journal of Material Cycles and Waste Management* 3(1): 48–59. doi:10.1007/s10163-000-0038-3

Singh, B., and D. Parthasarathy. 2010. "Civil society organisation partnerships in urban governance: an appraisal of the Mumbai Experience." *Sociological Bulletin* 59(1): 92–110. doi:10.1177/0038022920100105

Singh, J., et al. 2014. "Progress and challenges to the global waste management system." *Waste Management and Research* 32(9): 800–812. doi: 10.1177/0734242X14537868

Sujauddin, M., Huda, S. M. S., and A. T. M. R. Hoque. 2008. "Household solid waste characteristics and management in Chittagong, Bangladesh." *Waste Management* 28(9): 1688–1695. doi: 10.1016/j.wasman.2007.06.013

Taelman, S., D. Sanjuan-Delmás, D. Tonini, and J. Dewulf. 2019. "An operational framework for sustainability assessment including local to global impacts: focus on waste management systems." *Resources, Conservation & Recycling: X* 2: 100005.

Truitt, M. M., J. C. Liebman, and C. W. Kruse. 1969. "Simulation model of urban refuse collection." *Journal of the Sanitary Engineering Division* 95(2): 289–298.

UN-Habitat. 2010. *Solid Waste Management in the World's Cities: Water and Sanitation in the World's Cities 2010*. London, UK: Earthscan.

US Environmental Protection Agency (EPA). 2008. *EPA's 2008 Report on the Environment, EPA's 2008 Report on the Environment*. Washington, DC: National Center for Environmental Assessment. Available at: www.epa.gov/roe

Varkuti, A., K. Kovacs, C. Stenger-Kovacs, and J. Padisak. 2008. "Environmental awareness of the permanent inhabitants of towns and villages on the shores of Lake Balaton with special reference to issues related to global climate change." *Hydrobiologia* 599: 249–257. doi:10.1007/s10750-007-9190-2

Vitorino, de S. M. A., G. S. Montenegro, K. Faceli, and V. Casedi. 2017. "Technologies and decision support systems to aid solid-waste management." *Waste Management* 59: 567–584. doi: 10.1016/j.wasman.2016.10.045

Wang, Z. H., X. Y Dong, and J. H. Yin. 2018. "Antecedents of urban residents' separate collection intentions for household solid waste and their willingness to pay: evidence from China." *Journal of Cleaner Production* 173: 256–264. doi:10.1016/j.jclepro.2016.09.223

Wilson, D. C. 2007. "Development drivers for waste management." *Waste Management and Research* 25(3): 198–207. doi:10.1177/0734242X07079149

Wilson, D. C., L. Rodic, M. J. Cowing, C. A. Velis, A. D. Whiteman, A. Scheinberg, R. Vilches, D. Masterson, J. Stretz, and B. Oelz. 2015. "'Wasteaware' benchmark indicators for integrated sustainable waste management in cities." *Waste Management* 35: 329–342.

Wilson, D. C., L. Rodic, A. Scheinberg, C. A. Velis, and G. Alabaster. 2012. "Comparative analysis of solid waste management in 20 cities." *Waste Management & Research* 30(3): 237–254.

Wilson, D. C., C. A. Velis, and L. Rodic. 2013. "Integrated sustainable waste management in developing countries." *Proceedings of the Institution of Civil Engineers-Waste and Resource Management* 166(2): 52–68.

World Bank. 2011. *Urban Solid Waste Management*.

Yazie, T. D., M. G. Tebeje, and K. A. Chufa. 2019. "Healthcare waste management current status and potential challenges in Ethiopia: a systematic review." *BMC Research Notes* 12(1): 285. doi: 10.1186/s13104-019-4316-y

Zeng, C., D. J. Niu, H. F. Li, T. Zhou, and Y. C. Zhao. 2016. "Public perceptions and economic values of source-separated collection of rural solid waste: a pilot study in China." *Resources, Conservation and Recycling* 107: 166–173. doi:10.1016/j.resconrec.2015.12.010

Zhiyong, H., L. Yong, M. Zhong, G. Shi, Q. Li, D. Zheng, Y. Zhang, Y. Fei, and Y. Xie. 2018. "Influencing factors of domestic waste characteristics in rural areas of developing countries." *Waste Management* 72: 45–54. Available at: https://doi.org/10.1016/j.wasman.2017.11.039

Zohoori, M., and A. Ghani. 2017. "Municipal solid waste management challenges and problems for cities in low-income and developing countries." *International Journal of Science and Engineering Applications* 6(2): 039–048. doi: 10.7753/ijsea0602.1002

Zsoka, A. 2008. "Consistency and 'awareness gaps' in the environmental behavior of Hungarian companies." *Journal of Cleaner Production* 16: 322–329. doi:10.1016/j.jclepro.2006.07.044

Zurbrügg, C., M. Caniato, and M. Vaccari. 2014. "How assessment methods can support solid waste management in developing countries—a critical review." *Sustainability* 6(2): 545–570.

2 Management of Medical Waste

New Strategies and Techniques

Shayan Memar,[1]* M.S.* Tameka Dean,[2]
M.S., D.O., and Prasanna Abeyrathna[3] Ph.D.
[1]A.T. Still University of Health Sciences, USA
[2]Philadelphia College of Osteopathic Medicine, USA
[3]TCU UNTHSC School of Medicine, USA
*Corresponding author: shayan.memar@atsu.edu

CONTENTS

2.1   INTRODUCTION

The disposal and handling of health care waste (HCW) are becoming growing expenditures for nations around the world as access to medical services expand. Although there is no standard globally accepted definition, the term "health care waste" is broadly used to describe waste generated from health care-related activities. HCW is further subdivided into categories based on the contents of the waste or the health risk and pathogen exposure that the waste poses to the general public

DOI: 10.1201/9781003132349-3

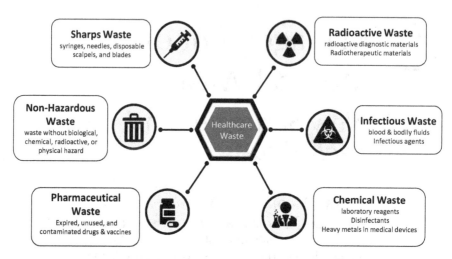

FIGURE 2.1   Categories of healthcare waste.

(Figure 2.1). Each HCW type requires a different method of management for safe handling and disposal. Most waste produced by health care facilities is considered nonhazardous, because it mainly comes from benign sources, such as food preparation, administrative tasks, and housekeeping activities (Yazie, Tebeje and Chufa 2019). Nonhazardous HCW composes approximately 85% of all the HCW generated and can integrate into general domestic waste streams without additional processing (WHO 2014; Windfeld and Brooks 2015).

The remaining 15% of HCW is considered hazardous and consists of infectious, chemical, and radioactive waste (UNEP 2012; WHO 2017). Hazardous HCW is produced from various sources, such as medical laboratories, advanced medical treatments and devices, bodily fluids, and sharps. This chapter focuses on current best practices regarding point-of-generation sorting, waste handling, and waste sterilization or neutralization prior to disposal.

2.2   IMPROVING HEALTH CARE WASTE MANAGEMENT CULTURE AND CAPABILITIES

Training interventions that target medical staff involved in the initial collection and handling of HCW can lead to efficient and cost-effective improvements in HCW management (Hosny et al. 2018; Balushi et al. 2018). However, there remains a great need for standardized guidance, or best practices, for health care workers and waste handlers to sort and identify waste that is hazardous or pathogenic, in order to avoid contamination. Agencies such as the World Health Organization (WHO), Center for Disease Control (CDC), United Nations Environmental Programme (UNEP), or the International Committee of the Red Cross (ICRC) promote awareness of appropriate techniques and guidelines for effective HCW management. The UNEP Compendium of Technologies for the Treatment/Destruction of Health Care Waste has outlined a systemic approach in which one can compare different waste management strategies based on calculated scores for categories, such as environmental and occupational safety, operational costs, and technical comparisons (UNEP 2012).

Educating waste-handling personnel, both clinical and nonclinical workers, on topics such as donning appropriate personal protective equipment (PPE) and the benefits of point-of-generation sorting will, undoubtedly, ease the burden of occupational risk associated with handling hazardous HCW, improve safety, and lower overall costs. The WHO recommends establishing a waste management officer, who leads a waste management committee composed of clinical and nonclinical personnel (2014). The purpose of this committee is to clearly define and delegate all responsibilities and guidelines, as well enforce these policies. The waste management officer and committee are also responsible for ensuring HCW management practices remain operational and up-to-date through regularly timed reviews. This review process may include documentation, interactive workshops, in-person training exercises, regular certification tests, and inspections to promote a culture of awareness and accountability.

Local government agencies may also create guidelines for how waste must be handled and disposed of after it is generated with goals of reducing negative environmental impacts and protecting the health of patients, health care workers, waste handlers, and the general public. As seen in Figure 2.2, a basic schematic for dealing with most forms of HCW is as follows: (1) segregate HCW into adequate containers according to the categories mentioned above; (2) collect HCW using proper protective equipment and methods of transport; (3) store HCW in a secure and temporary storage location, in a utility room away from general access, no longer than 24–48 hours; (4) transport HCW to the treatment or disposal site by using easy-to-clean containers; (5) treatment of HCW should be done in a cost-effective and suitable manner for the specific type of waste; and (6) dispose of HCW in a method specific to the type of waste, as governed by the regulations and laws (Yazie, Tebeje and Chufa 2019).

As newer technologies emerge, the disposal and management of waste can be done more safely and efficiently. However, it is important to note that use of these technologies will be ineffective without adequate training, resources, awareness, commitment, and enforcement to appropriately handle HCW. Therefore, in addition to developing new techniques, it is necessary to properly educate and develop a culture of safe and efficacious HCW management, beginning with point-of-origin segregation of HCW.

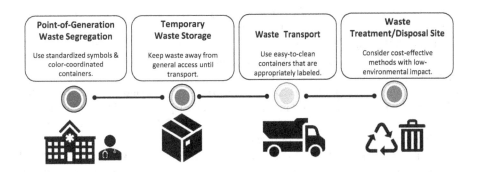

FIGURE 2.2   Health care waste management practices.

2.3 POINT-OF-ORIGIN SEGREGATION OF HEALTH CARE WASTE

Health care facilities that are major sources of potentially hazardous or infectious waste include hospitals and clinics, laboratories and research centers, mortuary and autopsy centers, blood banks and other specimen collection services, and elderly-care homes. The characteristics of waste contents at each source must be taken into consideration when forming a management plan or strategy. A common mistake is to have a single-stream disposal method, in which hazardous wastes were not separated out at the source (Yazie, Tebeje and Chufa 2019; Dehghani et al. 2019; Adu et al. 2020). Early separation is recommended, due to hazardous HCW having the greatest disease-causing potential at the source of its generation (Hoosain et al. 2011; Windfeld and Brooks 2015; Kagonji and Manyele 2016; Olaniyi et al. 2019). Therefore, hazardous HCW must be segregated immediately, as it is generated, in order to avoid cross-contamination with general waste.

Segregation of hazardous HCW at the point of generation, when risks of exposure are the highest, has been shown to significantly reduce disease burden (Windfeld and Brooks 2015; Olaniyi et al. 2019). Consequently, safeguarding hazardous and infectious HCW at the source should be a top priority, due to increased costs of waste management and disposal, in addition to the public health risk associated with the contamination of general waste streams. Point-of-origin segregation of HCW still requires a standardized sorting system utilizing consistent color-coding and appropriate symbols on designated receptacles to be the most effective (Adu et al. 2020).

Sorting should be conducted in a manner that is easily understood and with appropriately coordinated symbols and color schemes. Each category of hazardous waste (infectious, chemical, and radioactive) has designated color-coded containers and bags holding the waste. They are labeled with a standardized symbol, which is easily recognizable and consistent with its disposal requirements and associated handling risks. Figure 2.3 shows common symbols and color systems used in the segregation of hazardous HCW. These symbols and color schemes represent the contents and the associated risk of exposure to waste contents in order to facilitate safe and appropriate segregation and handling.

The primary goals of any HCW management strategy includes minimizing the exposure of hazardous and pathogenic materials capable of threatening the safety and health of the public and reducing the harmful environmental impact of HCW. Although infectious waste makes up a relatively small portion of all HCW, poor segregation practices can contribute to the spread of highly infectious diseases, such as hepatitis, cholera, and dysentery (Coker et al. 2009; Hoosain et al. 2011). Exposure risk can be linked to inappropriate contact with contaminated HCW and runoff from open dump sites containing contaminated HCW (Coker et al. 2009). These exposures can pose considerable harm to health care workers, waste handlers, and the public. The increased cost associated with processing and disposal of mixed, contaminated HCW can be effectively mitigated with appropriate point-of-origin sorting methods by minimizing the total volume of hazardous HCW (Windfeld and Brooks 2015).

To appropriately facilitate downstream management of HCW, medical staff and waste handling personnel should be educated on the best practices of waste-sorting to avoid the accumulation of mixed, contaminated waste that would otherwise not

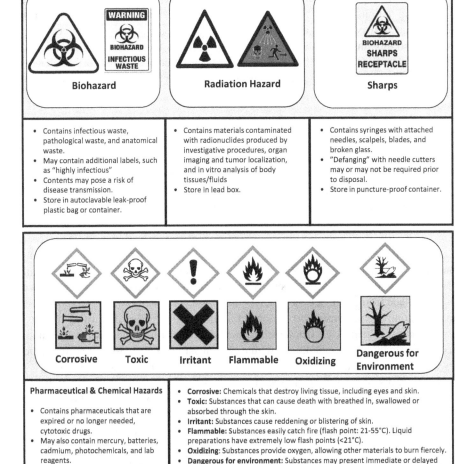

FIGURE 2.3 Common symbols used in hazardous waste segregation.

be considered hazardous or mixing of general waste into segregated infectious waste streams due to unnecessary caution. Even a small amount of contamination in the general waste supply can have negative consequences in regard to health risk, processing and disposal costs, and environmental impact. Therefore, all general HCW contaminated with hazardous must be considered hazardous and undergo additional sterilization treatment and processing (Zimmermann 2017; Adu et al. 2020).

Multiple strategies and technologies have been developed to inactivate segregated HCW that is considered infectious or hazardous to minimize environmental impact and risk to human health. In developing countries, it is common practice for health care facilities to either pretreat waste on-site or use third parties for transport of waste to off-site waste treatment depots before discarding waste into open dumps or landfills

(Hossain et al. 2011; Windfeld and Brooks 2015). Determining the most cost-effective method that minimizes harm to public health and the environment poses challenges in developing nations lacking sufficient regulation and oversight, economic investment for associated disposal fees, and knowledgeable health care staff to implement proper sorting and labeling of segregated hazardous waste. As there is no single method of HCW treatment or disposal capable of eliminating all environmental or health risks, the following sections will review the most up-to-date options available for hazardous HCW management.

2.4 INCINERATORS

Incinerators are commonly used for the combustion of infectious waste and sharps, in which waste is burned between the recommended temperatures of 850°C and 1450°C. Temperature range depends on the type of waste and incinerator technology that is used (UNEP 2012; WHO 2014). Lower temperatures below 450°C should be avoided due to the emission of persistent organic pollutants, such as dioxins (Windfeld and Brooks 2015). Guidelines for the use of incinerators are provided by the Stockholm Convention on Persistent Organic Pollutants, which outlines protective environmental measures and promotes accountability through regular evaluation and reporting (UNEP 2012; WHO 2014). Inexpensive low-technology equipment, such as drums or burners made from brick or concrete, typically burn waste at these lower temperatures in an uncontrolled manner into the open air and are ineffective in burning all waste and pathogenic materials. Combustion technologies with filters burn waste at the recommended higher temperatures with more emission control. They are often fueled by diesel or gas and achieve adequate burning of all materials (Diaz et al. 2005).

Incinerators are becoming phased out due to the risk of the carcinogenic pollutants produced by this method. However, incineration remains a preferred method of HCW disposal management in developing countries (Emmanuel 2007). Whether incineration occurs at on-site hospital facilities or off-site depots, the incinerator used should be sealed off from the environment, without burning into the open air. More modern technologies incorporate a main chamber, a secondary (post-chamber), and emission control systems to filter discharged particles and pollutants.

Air pollution control (APC) systems are designed to reduce harmful emissions of dioxins, furans, and heavy metals that are normally produced at lower operating temperatures. More advanced incinerators have APC systems composed of a two-stage saturator-quencher, a condenser-absorber, a venturi scrubber, a multistage entrainment separator, a wet electrostatic precipitator, and a stack with optional heating for plume suppression (UNEP 2012). All of these components greatly reduce emissions of dioxins, hydrogen chloride, sulfur dioxide, mercury, and cadmium. Furthermore, continuous emission monitoring systems may be installed in order to facilitate efficacy records and maintenance (UNEP 2012). Installing monitoring systems allows for better transparency with the government and the public, which instill public confidence and assurance of their safety.

Newer systems also have a waste feed system, which prevents temperature drops. The waste feed system uses moving grates, reciprocating grates, or rotating drum gates to continually remove ash, as new HCW is added to the incinerator, and to promote combustion (UNEP 2012). However, the higher purchasing and operating costs of these newer technologies may be deterrents, indicating a need for engineers to create new solutions within the existing infrastructure. Additionally, maintenance and service checks are required at regular intervals to maintain optimal functionality (Adu et al. 2020).

Incinerators are used for the disposal of infectious and pathological HCW derived from patient/animal blood or bodily fluids, medical laboratory cultures specimens, and organic human/animal tissue. Culture/specimen waste contains human or animal tissue, urine, or stool. When HCW is properly sorted at the source and only the appropriate HCW is incinerated, unnecessary emissions and environmental risks are avoided. Other examples of pathological waste may include any type of amputation or other body tissue excised from a surgical procedure or autopsy. Due to orthopedic devices and implants becoming more commonplace, extra precautions should be taken to remove metals that would be left behind in incinerators.

Once the waste has been incinerated, it may be transported for disposal and burial in landfills. Prior incineration of HCW reduces the total amount of waste to ashes. This occupies far less space than non-incinerated hazardous HCW and helps combat the ever-increasing size of landfills. Additionally, much of the slag and fly-ash trapped in the incinerator can be recycled into other projects, like road construction or mixed with concrete to help construct landfills (UNEP 2012; Adu et al. 2020). The repurposing of incinerated by-products is highly encouraged as it recycles the waste and further cuts down on landfill usage.

2.5 AUTOCLAVES

Autoclaving involves using high-pressured steam capable of killing harmful bacteria, viruses, fungi, and spores. Autoclaves can be used to sanitize medical equipment and devices that have been contaminated with infectious material. Autoclaving is a more cost-effective alternative to incineration for the treatment of non-reusable items, such as cultures, plates, and vials. The maintenance costs of this system are also lower when compared to incineration (Ferdowsi et al. 2013). Autoclavable infectious waste should be discarded into appropriately labeled biohazard bags to ensure safe transport to disposal sites. PPE recommendations for personnel who are loading and unloading the autoclave include wearing long pants, closed-toed shoes, eye protection, face protection, gloves (including a heat resistant pair), and a laboratory coat.

More advanced, or hybrid autoclaves, combine the technology of other waste management strategies. Hybrid autoclaves may have internal shredders, mixing arms or function as a rotating autoclave. Others remove air before the autoclave process is initiated and include gravity-displacement autoclaves, pre-vacuum/high-vacuum autoclaves, or pressure-pulse autoclaves (WHO 2014). Each of these systems will have different operating times with pre-vacuum autoclaves having the quickest operating time. Autoclaves may contain an outer jacket surrounding the main vessel, in which steam is also loaded into during operation. Injecting steam into the outer jacket in order to heat it is beneficial, because it prevents condensation on the inside chamber and lets operators use steam at lower temperatures. Those without an outer jacket tend to be larger autoclave systems, are cheaper to construct, and are often referred to as a "retort" system (WHO 2014). Since there are many different types of autoclaves, one should take multiple factors into consideration in order to determine the purchase of a suitable autoclave.

Autoclaves must undergo regular inspections and servicing, in order to ensure continued and proper functioning. Along with this, a detailed record log should be kept for information on biological tests, recording thermometers, and servicing (Garibaldi et al. 2017). As each brand will have its own software and controls, the directions to operate the machine, along with any constraints, should be provided in a clear and concise manner for the autoclave operators who, in addition, have been trained to use the autoclave. Other safety measures regarding autoclaves include testing them with a biological indicator, using heat-sensitive autoclave tape, and using chemical indicators in order to assure the machine is functioning properly. Autoclave tape is important in that it verifies the autoclave reached the appropriate temperature; however, it should be noted this does not indicate if the correct sterilization time or pressure were reached in order to kill the targeted organisms. Bowie–Dick test packs are used in autoclaves with vacuum functionality in order to assess whether any air leaks exist and monitor the air removal system (WHO 2014). Lastly, much of this waste will likely cause odors to build up, in which case deodorizers may be useful. Some systems may have mechanisms in place to capture and block the release of odors.

In order to autoclave HCW and medical instruments, it is crucial to properly clean the item so it is freed of dirt, films, or other debris, which might hinder the ability of the steam to penetrate onto the surface of the material. This includes separating dry materials from liquid materials. It is of utmost importance to properly package and contain materials to be autoclaved, in order to ensure proper sterilization. Sterilization can be achieved when the high-pressured steam contacts the microorganisms. This process is facilitated by loosely packing the autoclave chamber and items, so steam can freely reach the materials of interest and the air can be removed (Garibaldi et al. 2017. When transporting waste to the autoclave, it is good practice to transfer the item in a sealed secondary container, with a clean outer surface. The container, or cart, should be metal and lined with a plastic bag in order to prevent any substances from sticking to the container during transportation. This container may then be placed into the autoclave.

According to the autoclave guidelines at the University of Iowa, dry materials must be placed in approved, open autoclave bags, or a bag with holes at the top, filled to no more than 75% capacity. Then 250 mL of water should be added to the bags of solid waste, so more steam is able to contact the material (n.d.). Failure to autoclave with an open bag will block the steam from penetrating, thus preventing adequate sterilization (Garibaldi et al. 2017. As the steam penetrates the bag, it will likely melt or crumble. This issue is dealt with by placing the bags in a shallow, stainless steel pan, because it is able to transfer heat in an efficient manner. A plastic pan will not work, because it will not allow for good transfer of heat. Dry glassware and trash should be autoclaved at 121°C or 134°C for 1 hour with 15 minutes pre-vacuum of 3.6 kPa, while laundry is to be autoclaved at 121°C or 134°C for 30 minutes with 15 minutes of pre-vacuum of 3.6 kPa (Garibaldi et al. 2017; University of Iowa n.d.). Again, these are general guidelines, which may slightly differ between machines as well as vary depending on the targeted microorganisms and degree of desired sterilization.

Liquid materials are to be placed in bottles or flasks. Any cap or stopper should be loosened after the container is placed into the autoclave, in order to prevent pressure build-up and ensure proper sterilization. These containers should not be filled to more than 50% capacity, because the liquids may potentially boil, and spill out of the container, into the autoclave. If it is necessary to sterilize liquids in bulk, certain precautions should be taken to avoid explosions. Usually, the manufacturer will have specific instructions for doing this. This reemphasizes how important it is to keep a protocol near the autoclave for personnel to reference. Generally, for every gallon of infectious liquid, it must be autoclaved for 1 hour at 121°C at 103.4 kPa. After the procedure is over, the liquids must be cooled for at least 20 minutes before being unloaded and transported. Failure to do so could cause an explosion (University of Iowa n.d.). It is good practice to verify the autoclave has cooled and depressurized after reaching the appropriate temperature and operating for the desired length of time before removing any items. Only then can items be safely and carefully removed.

Materials able to undergo autoclaving include biological cultures and stocks; culture dishes and related materials; contaminated solid items, like pipette tips, gloves, and petri dishes; discarded live (including attenuated) viruses and vaccines; polypropylene and polycarbonate plastics; borosilicate glass; and stainless steel (WHO 2014). If any of the above items contains bleach, then the HCW needs to be neutralized using a 1:1 ratio solution of sodium thiosulfate to water.

According to the WHO, materials unable to be autoclaved include those containing solvents, volatile, corrosive, or flammable chemicals; materials contaminated with chemotherapeutic agents or cytotoxic drugs; carcinogens or mutagens; phenol and trizol; plastics composed of polystyrene, polyethylene, and high-density polyethylene; and household glassware (WHO 2014).

2.6 MICROWAVE

Despite a higher entry cost, this technique has the ability to reduce costs and negative environmental impacts when compared to other treatment methods for similar waste products (Zimmermann 2017). Microwaving HCW consists of using electromagnetic radiation at a frequency of 2,450 MHz and a wavelength of 12.24 cm. Microwaves heat the water content in the HCW, generate steam and disinfect materials. This is accomplished as long as there is a sufficient amount of water present, therefore a humidifier may be built into the system (WHO 2014; Zimmermann 2017; Hooshmand et al. 2020). Microwave techniques can be performed via batch system or a semicontinuous system.

Exposing HCW to electromagnetic radiation will cause steam to be generated from the water content in the waste and thus the types of HCW commonly treated by microwave technology will be similar to those treated by autoclaving; however, a sophisticated microwave will have lower consumption of water and be more energy-efficient than autoclaving (Zimmermann 2017). With this technology, there are minimal concerns of air, water, and solid waste contamination, compared to other methods of thermal inactivation, such as incineration and autoclaving. Compared to microwave technology, incinerators are associated with major concerns in each category, followed by autoclaving with some concerns for water contamination (Zimmermann 2017).

Exposing bacterial pathogens to this treatment method has been proven to reduce the numbers of vegetative bacteria and spores on a logarithmic scale. This includes being a simple and cost-efficient tool to inactivate *Clostridium difficile* spores (Zimmermann 2017). HCW safe for microwave treatment includes cultures and stocks, sharps, materials contaminated with blood and body fluids, isolation and surgery waste, laboratory waste and soft waste. Acceptable soft waste for microwaving consists of gauze, bandages, gowns, and bedding. Microwave techniques can even be successfully used for inactivating solid waste landfill leachate and sewage sludge (Zimmermann 2017; Hooshmand et al. 2020). Types of HCW to be excluded from microwave treatment are the following: hazardous chemical waste, volatile and semi-volatile organic compounds, chemotherapeutic waste, mercury, and radiological waste.

Microwave systems are to be fully enclosed and installed in an open area. They are usually also filtered with an air filtration system in order to prevent the release of harmful aerosols when being loaded. A typical batch cycle for microwave treatment ranges between 20 minutes to an hour, depending on the desired level of disinfection of the HCW. The system is composed of a treatment area, or chamber, which holds the waste. In this method, the chamber is designed to hold between 30 and 100 liters of HCW, depending on the manufacturer and model of the system. Once the chamber has been filled, electromagnetic radiation is directed from a microwave generator, known as a magnetron, into the HCW bin. For these systems, anywhere between two and six magnetrons are used, each having an output of about 1.2 kW (WHO 2014). Depending on manufacturer guidelines, the use of reusable and fully enclosed, microwavable containers may be necessary during treatments.

Semicontinuous microwave systems are more complex and require further precautions, but they are capable of treating up to 250 kg of HCW per hour. High-efficiency particulate air (HEPA) filters are needed to prevent the release of airborne pathogens (WHO 2014). This is because HCW will typically get loaded into a steam-filled hopper before passing through a shredder (Hooshmand et al. 2020). During these processes, HCW is broken down into smaller pieces in order to facilitate the treatment; however, this increases the risk of pathogen exposure. After being shredded, the waste will then be loaded onto an auger, also known as a conveyor screw. At this point, four to six magnetrons operate to heat the HCW to 100°C by producing more steam. If a longer exposure time is required, a holding tank may be used. If sharps require finer shredding, then a secondary shredder can be added into the system. This process can potentially allow up to 90% of waste to be safely disposed of by a landfill afterward, in which the dry-waste volume is also reduced by up to 80% (Zimmermann 2017).

2.7 CHEMICAL DISINFECTION

This method, which normally takes place on the hospital premises, uses ethylene oxide (ETO) gas, vaporized hydrogen peroxide, or liquid chemicals as the principal sterilizing agents. Microorganisms will either be eliminated, or their count reduced to acceptable levels. Since many methods exist, there is a range of associated costs and efficacies relative to the concentration of chemicals used and duration of treatment. Additionally, the targeted microorganisms must be known in order to use the most appropriate method or chemical concentration. Chemical processes typically do not sterilize, as the name implies; however, this does not mean this method should be overlooked, especially since it is a great method to disinfect liquid HCW (UNEP 2012; WHO 2014). Limitations exist to using chemical disinfection as a means to treat solid wastes and hazardous HCW. In these instances, powerful disinfectants are required, which can pose additional health risks to personnel. Material safety data sheets should be consulted and remain nearby for trained handlers to work with HCW during chemical disinfection protocols.

Since only the surface of intact solids are disinfected, shredding or milling of the solid HCW, in a closed system, is usually needed prior to chemical disinfection. Shredding has the added benefit of reducing waste volume, in addition to increasing the surface area for the disinfectant to come into contact with during treatments. Adding water to the shredding processes further facilitates the contact of the solid HCW with the chemical disinfectant. Water also protects the equipment's mechanical parts from excess wearing (WHO 2014). An adverse effect of this method is the potential generation of a toxic liquid waste, due to various factors affecting the use of the chemicals, including extent and amount of contact time to the concentration of the chemical disinfectant and the organic load of the waste. Additional variables affecting the efficacy and safety of this procedure include the operation temperature, humidity, and pH. These parameters should be optimized according to the HCW being handled and the type of chemical agent being used in the process. Compaction can then be used after disinfection in order to reduce the original waste volume by as much as 90% (WHO 2014). Chemical disinfection will often require a specialized disposal due to toxic liquid waste by-products; therefore, local guidelines should be followed.

Sterilizing commonly used instruments and equipment can be done with ETO, especially those which are temperature sensitive. However, ETO is not a cost-effective option for sterilizing HCW destined for a landfill. Much like formaldehyde, the use of ETO for HCW management should be kept to a minimum due to its carcinogenicity, flammability, and toxicity. If ETO is used, an aeration period is required in order to remove any ETO residue. Typical operational parameters are an ETO gas concentration of 450–1200 mg/L; a temperature of 37–63°C; a relative humidity of 40–80%; an exposure time of 1–6 hours; and water molecules to carry the gas (Rutala and Weber 2015). Furthermore, ETO has the ability to kill microorganisms and sterilize products during the packaging processes.

The use of a liquid agent like sodium hypochlorite (NaOCl), also known as bleach when it is in the form of a diluted solution, results in relatively mild health hazards, requires PPE for skin and eyes, and is readily available. It is widely used for disinfecting wastewater because it is bactericidal, mycobactericidal, virucidal, fungicidal, and sporicidal. Chemical disinfection using NaOCl has the shortcoming of lower biocidal activity against liquids with a high organic content, such as blood or stools. NaOCl, and many agents like it, are typically corrosive to metals at high concentrations (>550 ppm) and should be stored in plastic containers in a well-ventilated, dark room, away from acids and moisture (Rutala and Weber 2015). If there is a large quantity of leftover NaOCl solution, it is treated as a hazardous waste. Otherwise, leftover NaOCl solution must be first reduced with sodium bisulfite or sodium thiosulfate and neutralized with acids before it is let into the sewage system, because of the potential for adverse environmental and health outcomes (WHO 2014). The hypochlorite solution is typically used in concentrations of 5.25–6.15% and functions to react with organic compounds, forming toxic by-products as a result (Rutala and Weber 2015). An alternative is using chlorine dioxide, a toxic gas readily soluble and stable in water, which can also be produced onsite at the facility.

Chemical disinfectants such as peracetic acid (peroxyacetic acid) and hydrogen peroxide are advantageous because they do not cause a significant amount of irritation and no activation step is required for these disinfectants to function. However, there are concerns over their cosmetic and functional interactions with metals such as lead, brass, copper, and zinc. Hydrogen peroxide is useful in removing organic matter and organisms from surfaces. It does not coagulate blood or fix tissues to surfaces. At higher concentrations, 7.5% or above, hydrogen peroxide is capable of inactivating cryptosporidium (Rutala and Weber 2015). Ideally, hydrogen peroxide should be used at a concentration higher than 3%, since some studies, identified by Rutala and Weber, have suggested there is limited bactericidal activity at 3% levels (2015). The concentration used should match the targeted organisms and desired level of disinfection.

Vaporized hydrogen peroxide sterilizes reusable metal and nonmetal medical equipment at lower temperatures. The only by-products are water vapor and oxygen, which make it a safe option. One limitation, though, is that systems using vaporized hydrogen peroxide are unable to sterilize liquids, linens, cellulose materials, or any powders. Medical devices composed of polypropylene, polyethylene, brass, and stainless steel are all compatible with vaporized hydrogen peroxide sterilization systems. Furthermore, this system is able to sterilize items like scissors and instruments with an internal lumen diameter of 1 mm, or larger, and length of 125 mm, or shorter (Rutala and Weber 2015). Since the hydrogen peroxide is vaporized, it can infiltrate diffusion-limited spaces. Vaporized hydrogen peroxide has been shown to destroy spores, viruses, mycobacteria, bacteria, and fungi.

Peracetic acid is used for immersible instruments only and has a short cycle time of 30–45 minutes. Usually, an endoscope or medical instrument is immersed in a peracetic acid solution at a temperature of 50–55°C in a standardized cycle, in which environmentally friendly by-products are produced (Rutala and Weber 2015). The procedure is very simple and safe for users to operate, as the process is standardized to a constant dilution, perfusion of channel, temperature, and exposure. These by-products include acetic acid, oxygen, and hydrogen peroxide. Endoscopes and many other instruments may be cleaned with this method as the sterilant flow of solution through the endoscope removes residual salt, protein, and microbes. Lastly, peracetic acid is rapidly sporicidal, making it more advantageous.

Glutaraldehyde is relatively inexpensive and has a wide range of compatible materials, making it a good choice for chemical disinfection (Rutala and Weber 2015). Like many other chemical agents, glutaraldehyde is a respiratory and skin irritant associated with a strong odor. Glutaraldehyde is known to coagulate blood and fix tissues to surfaces; however, it is not inactivated in the presence of organic matter, like bleach. Other phenolic and alcohol-based disinfectants are commonly added to glutaraldehyde, because of glutaraldehyde's slow mycobactericidal activity. Before glutaraldehyde can be used, it must first be activated by making the solution alkaline, specifically to a pH of 7.5–8.5 (Rutala and Weber 2015).

It is good practice to verify if the chemical agent used will be effective against the targeted microorganism. The most resistant groups tend to be bacterial spores, mycobacteria, and hydrophilic virus. While less resistant groups are lipophilic viruses, vegetative fungi, fungal spores, and vegetative bacteria. If a chemical disinfectant is effective for a group with higher resistance, then it may be inferred effective against the lower resistant microorganisms.

2.8   GAS STERILIZATION

Plasma pyrolysis thermally degrades HCW at sub-stoichiometric air levels. This is achieved by restricting the oxygen levels, or air levels, within the combustion chamber. A highly ionized gas, also known as plasma, acts to convert electrical energy to thermal energy at temperatures of 1650°C and higher. These systems operate similar to incinerators, and in some jurisdictions are considered as such. The gas, typically argon or nitrogen, essentially passes between two electrodes and carries the energy into the HCW (UNEP 2012; WHO 2014). In other systems, a direct current plasma arc is formed between a graphite electrode and a metal in a molten bath of the HCW. The calorific value of gas ranges between 25% and 40% of that of natural gas; however, plasma requires a high-energy electrical discharge, so they need significant amounts of electricity to function. Therefore, these systems require extensive capital investment and operating costs.

When properly optimized, plasma pyrolysis produces gaseous by-products, such as carbon monoxide, carbon dioxide, hydrogen, methane, water, nitrogen, and some hydrocarbons. Other by-products include solids and liquids, which potentially have a high toxicity (UNEP 2012). Since it is very difficult to achieve a complete absence of oxygen in HCW products, there will be some oxidation taking place in the pyrolysis so that dioxins, furans, and other products of incomplete combustion are produced. These systems essentially act as incinerators and they also require APC systems in order to prevent pollutants from entering the environment.

Another method of gas sterilization uses ozone gas. With these techniques, ozone gas acts as a strong oxidant, capable of destroying microorganisms. It has the benefit of degrading into regular oxygen. Newer ozone systems have a capacity of 1,000 kg per hour and tend to be automated. They require a power of 37 kWh/h and 8 liters of water per hour to operate during peak conditions. As HCW is fed into the system, it is shredded into small particles less than 20 mm, which then fall into a treatment chamber filled with ozone–water fog. The fog is produced when ozone is mixed with aerosolized water (UNEP 2012). Once the HCW has fallen into the fog-filled chamber, it remains there for six minutes before being pushed into a transport container. After the transport container reaches capacity, the ozone gas is removed from the chamber and the transport container is detached from the chamber. The transport container can then be moved to a sanitary landfill for disposal. If any residual ozone exists, a catalytic converter and heater may be used to break down the ozone gas (UNEP 2012). Ozone treatment systems should be equipped with atmospheric ozone monitors, as well as a ventilation system. A forced-air ventilation is required for emergencies.

2.9 IRRADIATION

There are three methods to treat HCW by irradiation. These strategies include utilizing electron beams, cobalt-60, or ultraviolet sources. All of these methods are different from microwave treatments (Hooshmand et al. 2020). Since radiation is an inherent component of this treatment method, proper protective equipment in the form of shielding is essential. Lead shielding is a common safeguard to protect one against harmful doses of scatter radiation, because of its high molecular density. Electron beams can penetrate the waste bags and containers but will be impaired from fully penetrating the shielding. The level of absorbed dose by the mass of the HCW will dictate the amount of pathogen destruction. Another limitation of irradiation treatment is the shadowing effect (Hooshmand et al. 2020). In other words, HCW not in direct contact with the irradiation will be less sterile than the HCW surfaces facing the radiation. Irradiation is best served as a supplement to other treatment technologies. For example, germicidal ultraviolet radiation is only capable of destroying airborne pathogens and wastewater because it is incapable of penetrating denser materials, such as closed waste bags (WHO 2017, Hooshmand et al. 2020).

Gamma radiation for HCW management is produced from the same radioactive isotopes of cobalt used for cancer treatment. Cesium-137 is another unstable isotope commonly used to produce gamma rays. This process destroys pathogen DNA and RNA, when infectious HCW or equipment is exposed to gamma radiation, therefore it could be an effective means of sterilization. Gamma rays consist of short wavelength, high-frequency electromagnetic radiation of high energy. Gamma rays are photons, with no charge or mass, therefore they are the most penetrating type of electromagnetic radiation, which also makes them the most hazardous. These rays are able to adequately penetrate both solids and liquids.

Electron beams are particularly useful for the disinfection of wastewater and sewage sludge, since denser materials can cause attenuation of the electrons. As with gamma radiation, this method has the direct microbicidal effect of damaging genetic material, like DNA. Doing this hinders a microorganism's ability to reproduce and survive. The direct effect of electron beams is typically only responsible for treating a significant, >10% removal, number of organic compounds when the concentration of the contaminant is greater than 0.1 M (Siwek and Edgecok 2020). It also has an indirect effect to disinfect waste through water radiolysis. This indirect effect creates several reactive species, primarily these are $H\bullet$, $OH\bullet$, e_{-aq}, and hydrogen peroxide. The indirect effect and associated reactive species are the major contributor to the sterilization process.

The target pH range for the most effective sterilization using electron beams is between 4 and 10. One must take into consideration the contents of the wastewater sludge in order to choose the most suitable free radicals. The presence of heavy metals in wastewater will adversely affect the water radiolysis free radicals. Heavy metals will act as scavengers, in that they react with the free radicals produced by radiolysis and compete for them so that there are less free radicals to sterilize the wastewater. Therefore, it is recommended to filter out any potential scavengers, such as heavy metals.

Since each pathogen will have a different sensitivity to electron beams, the pathogen with the highest required decimal reduction dose should be targeted in order to determine the lowest radiation dose, and thus the most cost-effective dose. The decimal reduction is defined as the necessary dose to eliminate 90% of the targeted pathogen, or the dose required for a one-log inactivation. Decimal reduction doses range from as low as 0.06, for *Pseudomonas* species, up to 4–4.5 for *Escherichia coli* and *Cryptococcus*. At operational energies between 100 keV to 100 MeV, the penetration depth of the electron beam will be inversely proportional to the density of the material being irradiated. For a sludge water mixture, the penetration depth is said to be around 3 mm per 1 MeV. It is important to follow routine maintenance and testing in order to assure the system is operating as designed. After the sludge is treated, it does not require a landfill and it is immediately safe to be recycled as soil fertilizer.

2.10 BIOLOGICAL INACTIVATION

Biological inactivation typically allows organic matter to naturally degrade over time. It is possible to speed up the process with enzymes, which may be advantageous for biological HCW continuing pathogens. For example, kitchen waste and other undesirable organic, degradable waste, such as placentas, can successfully be dealt with via composting or vermiculture, which are both biological processes. This process also occurs after burial.

2.11 ALKALINE HYDROLYSIS

This waste management strategy, also known as alkaline digestion, serves to create a decontaminated aqueous solution from animal carcasses, human body parts, and tissues. Tissues destroyed by this method are typically pathological waste, organs, cadavers, anatomical parts, and contaminated animal carcasses. Furthermore, biological stocks, cultures, liquid blood, body fluids, and other types of infectious waste are safely treated by alkaline digestion. The process of alkaline hydrolysis has the added benefit of degrading fixative agents and other hazardous chemicals, such as formaldehyde and glutaraldehyde. Chemotherapeutic agents, like cyclophosphamide, chlorambucil, melphalan, uracil mustard, and daunomycin, are also safely broken down by the digestion process. If treated for at least six hours, prion waste is also acceptable (WHO 2014; UNEP 2012). Since a harmful by-product of hydrogen gas is formed, alkaline digestion should not be used for wastes containing metals, such as aluminum, tin, magnesium, copper, or galvanized iron. Other HCW products to avoid treating with alkaline hydrolysis include concentrated acids, flammable liquids, organohalogen compounds (especially trichloroethylene), and nitromethane and other similar nitro compounds (UNEP 2012). After treatment, the aqueous solution can then be disposed of accordingly. Alkaline digestion functions as a safe and eco-friendly alternative to incineration and burial (Tavares de Cruz, et al. 2017).

Animal and human tissue waste treated by alkaline hydrolysis is first loaded into a steam-jacketed, stainless steel basket and then into a hermetically sealed tank. Depending on the device being used, 10–4,500 kg of HCW per batch can be digested. An alkali, such as sodium or potassium hydroxide, along with water is then pumped into the vessel. The amount of alkali used is to be proportional to the quantity of tissue being digested. At this point, steam is injected into the steam jacket or vessel. The contents stirred and heated to between 100°C to 127°C, or higher. The digestion time will vary depending on factors such as the amount of alkali added, the pressure, the temperature, and mixing efficiency; however, normal ranges are between three and eight hours (UNEP 2012). If a lower-pressure system is being used, then a longer digestion time will be necessary for complete treatment. Once the digestion is complete, the machine will automatically shut off the steam, so the contents are allowed to cool overnight, down to ambient temperatures. Cooling water can also be added in some systems. After the hydrolysate has cooled, water is added to rinse the inside of the tank and dilute the aqueous solution. From here, the diluted hydrolysate is safe to be stored in a holding tank until disposal into a drain or sewer system. Disposal methods will likely vary depending on local guidelines.

In addition to an aqueous solution composed of peptide chains, amino acids, sugars, soaps, and salts, other by-products formed by alkaline hydrolysis include the biodegradable mineral constituents of bones and teeth. The mineral constituents are then crushed and recovered as sterile bone meal (Tavares de Cruz, et al. 2017). If an excess of hydroxide remains in the aqueous solution, there will be a high pH level of the liquid waste. Effluent normally has a pH of 11 and should be discharged slowly, diluted, or neutralized by bubbling carbon dioxide, depending on the local regulations. A soapy ammonia odor is usually detected in the immediate vicinity of the vessel, but this odor is generally handled well with natural ventilation (UNEP 2012).

As the digestion process creates an aqueous solution from tissues, except for calcium, metals, plastics, ceramics, or rubber in the waste, the volume of solution after digestion can range from 100 L per load for a 15 kg unit to 24,000 L per load for a large, 4,500 kg unit. Leftover solid wastes will form a slurry, which creates a hard solid when cooled. The solid wastes are sterile and may be recovered or disposed of in a dumpster to be sent to a landfill. For example, the calcium can later be used as a soil conditioner (UNEP 2012).

2.12 PROMESSION

Also known as "freeze-drying" and "composing", promession is a relatively newer technique, mainly used in the funeral industry, which allows operators to dispose of human remains and tissues by freeze-drying the organic matter to −18°C via liquid nitrogen. Then the remains are usually mechanically broken down, through vibration, and exposed to a vacuum in order to obtain the final product, a dry powder. A large magnet is typically used to separate metals from the remains, such as orthopedic implants and mercury (Tavares da Cruz, et al. 2017). The final product will not break down if it remains dry.

The organic powder can be recycled into the environment, which turns into a compost material over time. The remains should completely decompose within one year if buried in a small box made of potato or maize starch in a shallow grave, within the topsoil, where there is a higher oxygen content. This method of disposal does not require much water and it does not release any pollutants into the atmosphere, unlike incineration. The final volume of the remains is up to 20 times lower than disposal by incineration (Murigu and Mbugua 2020). Promession greatly reduces the volume of land needed for final disposal and eliminates the concern of introducing harmful substances and chemicals into the ground. All that remains are the nutrients from the body tissues, which will then be fed into nearby plants.

2.13   MECHANICAL AND FRICTIONAL HEAT TREATMENT SYSTEMS

In these systems, HCW is destroyed by grinding it into a fine power. These systems can handle a capacity of 10 kg to 500 kg per hour and are suitable for infectious wastes of cellulose, glass, plastics, metals, liquids, and pathological waste. In the machine, frictional heat is often supplemented with heat from resistance heaters, so the HCW reaches a temperature of 150°C in order to achieve sterilization, while it is turned into a fine powder (UNEP 2012).

At atmospheric pressure, the frictional heat in these systems is produced by a high-speed rotor, operating at speeds between 1,000 and 2,000 rpm, which causes a moist steam to be produced. The steam passes through heat exchangers in order to filter into a condensed steam (UNEP 2012). Newer systems may also filter this air before it enters the environment in order to reuse the condensate. This process will continue until all the moisture has evaporated, at which point the HCW is carried to superheated conditions between 135°C and 150°C for several minutes. Sterilization is achieved through multiple facets of this process. The high steam and dry temperatures combined with grinding the waste destroys cell membranes and pathogens (UNEP 2012).

Frictional heat treatment, if performed and handled correctly, will ensure that regulatory limits are followed and environmental impacts are reduced, especially if this process is done on-site. The mass of the waste will be reduced to 70% and the volume reduced to 35% relative to the starting point (UNEP 2012). This technique may also be combined with other strategies as a means to reduce the volume of HCW.

2.14   BURIAL

If no other method of treatment is available, HCW must be buried in a landfill in a manner to reduce environmental and safety hazards; untreated HCW must be contained. Encapsulation of untreated HCW is performed by adding an immobilizing material/agent into a container with the HCW and sealing the containers before burial. The containers may be made of high-density polyethylene or metal. Immobilizing agents are typically a plastic foam, bituminous sand, cement mortar, or a clay material (UNEP 2012). Taking these precautions are imperative before burial in a municipal landfill, especially because this will help prevent scavengers from gaining access to hazardous HCW. Encapsulation is best served for disposal of sharps and pharmaceutical HCW, rather than non-sharps HCW; however, a mixture of these types of HCW is also acceptable (Diaz et al. 2005). Inertization of HCW follows a similar concept, in which hazardous HCW is mixed with cement and other substances prior to disposal, in order to prevent toxic compounds leaking into water supplies. Inertization is a sufficient method to deal with pharmaceutical waste, as well as fly-ash and other by-products high in metals from incineration. Pharmaceuticals disposed of by inertization must be removed from all packaging and ground so a mixture of water, lime and cement can be added, which forms homogenous mass. The ratio targeted should be 65% pharmaceutical waste, 15% lime, 15% cement, and 5% water (UNEP 2012). Finally, the mixture may be solid or liquid as it is transported to the landfill, disposed of, and then covered by fresh municipal waste.

2.15 CONCLUSION

New strategies and technologies dealing with HCW management should always be pursued in order to preserve the environment and promote public safety. Each facility will have unique requirements and restrictions guiding them as to how HCW is managed. The WHO and UNEP provide phenomenal resources to assist waste management officers in determining the best option for the specific needs of their facility. HCW should be handled in the most cost-effective and safe manner. Much of this stems from properly educating all waste handlers on segregation practices, determining the appropriate technique to treat HCW, regularly reevaluating HCW management strategies, and routinely servicing and maintaining equipment.

REFERENCES

Adu, R. O., S. F. Gyasi, D. K. Essumang, and K. B. Otabil. 2020. "Medical Waste-Sorting and Management Practices in Five Hospitals in Ghana," *Journal of Environmental and Public Health* (March): 2934296. doi:10.1155/2020/2934296

Balushi, A. Y. M. D. A., M. M. Ullah, A. A. A. Makhamri, F. S. A. Alawi, M. Khalid, and H. M. A. Ghafri. 2018. "Knowledge, Attitude and Practice of Biomedical Waste Management among Health Care Personnel in a Secondary Care Hospital of Al Buraimi Governorate, Sultanate of Oman," *Global Journal of Health Science* 10 (3): 70–82.

Coker, A., A. Sangodoyin, M. Sridhar, C. Booth, P. Olomolaiye, and F. Hammond. 2009. "Medical Waste Management in Ibadan, Nigeria: Obstacles and Prospects," *Waste Management* 29 (2): 804–11. doi:10.1016/j.wasman.2008.06.040

Cruz, N. J. T. da, ÁG. R. Lezana, P. D. C. Freire Dos Santos, I. M. B. Santana Pinto, C. Zancan, and G. H. Silva de Souza. 2017. "Environmental Impacts Caused by Cemeteries and Crematoria, New Funeral Technologies, and Preferences of the Northeastern and Southern Brazilian Population as for the Funeral Process," *Environmental Science and Pollution Research International* 24 (31): 24121–34. doi:10.1007/s11356-017-0005-3

Dehghani, M. H., H. D. Ahrami, R. Nabizadeh, Z. Heidarinejad, and A. Zarei. 2019. "Medical Waste Generation and Management in Medical Clinics in South of Iran," *MethodsX* 6 (April): 727–33. doi:10.1016/j.mex.2019.03.029

Diaz, L. F., G. M. Savage, and L. L. Eggerth. 2005. "Alternatives for the Treatment and Disposal of Healthcare Wastes in Developing Countries," *Waste Management* 25 (6): 626–37. https://doi.org/10.1016/j.wasman.2005.01.005

Emmanuel, J. 2007. "Best Environmental Practices and Alternative Technologies for Medical Waste Management." In *Eighth International Waste Management Congress and Exhibition*. Kasane, Botswana.

Ferdowsi, A., M. Ferdosi, and M. J. Mehrani. 2013. "Incineration or Autoclave? A Comparative Study in Isfahan Hospitals Waste Management System (2010)," *Materia Socio-medica* 25 (1): 48–51. doi:10.5455/msm.2013.25.48-51

Garibaldi, B. T., M. Reimers, N. Ernst, G. Bova, E. Nowakowski, J. Bukowski, B. C. Ellis, et al. 2017. "Validation of Autoclave Protocols for Successful Decontamination of Category A Medical Waste Generated from Care of Patients with Serious Communicable Diseases," *Journal of Clinical Microbiology* 55 (2): 545–51. doi:10.1128/JCM.02161-16

Hooshmand, S., S. Kargozar, A. Ghorbani, M. Darroudi, M. Keshavarz, F. Baino, and H.W. Kim. 2020. Biomedical waste management by using nanophotocatalysts: The need for new options. *Materials* 13(16): 3511.

Hosny, G., S. Samir, and R. El-Sharkawy. 2018. "An Intervention Significantly Improves Medical Waste Handling and Management: A Consequence of Raising Knowledge and Practical Skills of Health Care Workers," *International Journal of Health Sciences* 12 (4): 56–66. https://pubmed.ncbi.nlm.nih.gov/30022905; www.ncbi.nlm.nih.gov/pmc/articles/PMC6040849/

Hossain, Md. Sohrab, Amutha Santhanam, N. A. Nik Norulaini, and A. K. Mohd Omar. 2011. "Clinical Solid Waste Management Practices and Its Impact on Human Health and Environment – A Review," *Waste Management* 31 (4): 754–66. https://doi.org/10.1016/j.wasman.2010.11.008

Kagonji, I. S., and S. V. Manyele. 2016. "Analysis of Health Workers' Perceptions on Medical Waste Management in Tanzanian Hospitals," *Engineering* 8: 445–459.

Murigu, J., and W. Mbugua. 2020. "Burying as an Interment Method and Its Impact in Kenyan Urban: A Case Study of Lang'ata Cemetery in Nairobi," *Africa Habitat Review Journal* 14 (3): 1963–1973. http://uonjournals.uonbi.ac.ke/ojs/index.php/ahr

Olaniyi, F. C., J. S. Ogola, and T. G. Tshitangano. 2019. "Efficiency of Health Care Risk Waste Management in Rural Healthcare Facilities of South Africa: An Assessment of Selected Facilities in Vhembe District, Limpopo Province," *International Journal of Environmental Research and Public Health* 16 (12): 2199. doi: 10.3390/ijerph16122199

Rutala, W. A., and D. J. Weber. 2015. "Disinfection, Sterilization, and Control of Hospital Waste." *Mandell, Douglas, and Bennett's Principles and Practice of Infectious Diseases* 2: 3294–3309.e4. doi:10.1016/B978-1-4557-4801-3.00301-5

Siwek, M., and T. Edgecock. 2020. "Application of Electron Beam Water Radiolysis for Sewage Sludge Treatment-a Review," *Environmental Science and Pollution Research International* 27 (34): 42424–48. doi:10.1007/s11356-020-10643-0

UNEP. 2012. *Compendium of Technologies for Treatment/Destruction of Healthcare Waste.* United Nations Environment Programme. Osaka, Japan.

University of Iowa. n.d. "Autoclaving Guidelines." Accessed February 20, 2021. https://ehs.research.uiowa.edu/biological/autoclaving-guidelines

WHO. 2014. *Safe Management of Wastes From-Care Activities.* World Health Organization. https://apps.who.int/iris/bitstream/handle/10665/42175/9241545259.pdf

WHO. 2017. *Safe Management of Wastes from Health-Care Activities: A Summary.* World Health Organization. https://apps.who.int/iris/handle/10665/259491

Windfeld, E. S., and M. Su-Ling Brooks. 2015. "Medical Waste Management—A Review," *Journal of Environmental Management* 163: 98–108. https://doi.org/10.1016/j.jenv man.2015.08.013

Yazie, T. D., M. G. Tebeje, and K. A. Chufa. 2019. "Healthcare Waste Management Current Status and Potential Challenges in Ethiopia: A Systematic Review," *BMC Research Notes* 12 (1): 285-019-4316-y. doi:10.1186/s13104-019-4316-y

Zimmermann, K. 2017. "Microwave as an Emerging Technology for the Treatment of Biohazardous Waste: A Mini-Review," *Waste Management & Research: The Journal of the International Solid Wastes and Public Cleansing Association, ISWA* 35 (5): 471–79. doi:10.1177/0734242X16684385

3 Management of E-waste

Rupa Rani[1] and Rajesh Ahirwar[1,]*
[1]Department of Environmental Biochemistry,
ICMR—National Institute for Research in Environmental
Health, Bhopal, India
*Corresponding author: r.ahirwar.nireh@gov.in

CONTENTS

DOI: 10.1201/9781003132349-4

3.1 INTRODUCTION

Electronic waste (e-waste) or waste electrical and electronic equipment (WEEE) refers to obsolete and discarded electronic appliances including their components, such as televisions, mobile phones, desktops, laptops, keyboards, printed board assemblies, printers, monitors, capacitors and relays, mercury switches, liquid crystal display (LCD), batteries, selenium drums (photocopier) and electrolytes, and cartridges from photocopy machines that have reached the end of their usable life (Figure 3.1). E-Waste (Management) Rules, 2016 define e-waste as "electrical and electronic equipment, whole or in part discarded as waste by the consumer or bulk consumer as well as rejects from manufacturing, refurbishment and repair processes" (Rules 2016). Similar definitions of e-waste have been adopted worldwide; for instance, the European Union defines e-waste as "electrical or electronic equipment which is waste including all components, sub-assemblies and consumables, which are part of the product at the time of discarding" (Directive 2008). In the modern era, the short life span, limited repair options, and evolving consumer demands for advanced and smart products have led to electronic gadgets to becoming waste in a comparatively shorter time (Turaga et al. 2019; Montalvo, Peck, and Rietveld 2016). Multiple sources such as domestic (e.g., television, computers, radio, cell phones, washing machines, microwave oven, electric iron, etc.), hospitals (e.g., computer, monitors, electro-encephalogram (ECG) devices, e-microscope, incubators, etc.), government sectors (e.g., computer, fax machine, keyboards, printers, photocopy machines, scanners, air-conditioner, etc.), and industries (e.g., incubators, boilers, signal generators, etc.) are contributing to the ever increasing volumes of e-waste (Ankit et al. 2021). Personalized health-care monitoring systems such as nano-biosensors and biochips, and point-of-care testing devices (POCTs) are another rapidly emerging e-waste sources (Ahirwar and Khan 2022). Thus, e-waste has become a global environmental concern for being a prominent source of environmental pollution and a human health hazard (Forti et al. 2020; Grant et al. 2013; Widmer et al. 2005).

In most countries, the e-waste generated is usually managed in one of the following ways:

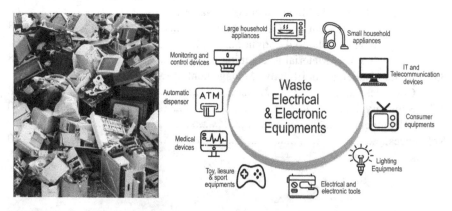

FIGURE 3.1 Representative image of e-waste and major groups of waste electrical and electronic equipment contributing to global e-waste.

(i)   Formal collection by government-recognized bodies, or by the original equipment manufacturer under the take-back initiative.
(ii) Collection outside the formal system by individual waste dealers and companies.
(iii) Dumping it to household waste bins.

E-waste collected through a formal collection is channeled to authorized recycling facilities having advanced technology, machinery, and infrastructure for efficient recycling (formal recycling). Contrarily, the e-waste collected outside the formal system or dumped into the household means they are channeled to informal recyclers who process the e-waste for extraction of valuable items under suboptimal conditions using primitive techniques that emit a large number of hazardous components to the environment. The e-waste dumped into household waste bins may also land in municipal solid waste landfills, making them a prominent source of soil and water pollution from leaching actions.

3.2   E-WASTE GENERATION

The amount of e-waste generated each successive year is increasing substantially. In comparison to previous estimates of 3–4% annual growth of e-waste, it has increased to around 6% as per recent estimates (Forti et al. 2020; Balde et al. 2017; 2015). Globally, 57.4 million metric ton (Mt) of e-waste were generated in 2021. Moreover, 53.6 Mt of e-waste were generated globally in 2019, which is projected to grow to 74.7 Mt by 2030 as per the Global E-waste Monitor estimation. Asian countries generated the largest quantity of e-waste in 2019 (24.9 Mt), followed by America (13.1 Mt), Europe (12.0 Mt), Africa (2.9 Mt), and Oceania (0.7 Mt). The total e-waste in 2019 comprised 17.4 Mt small equipment, 13.1 Mt large equipment, 10.8 Mt temperature exchange equipment, 6.7 Mt screens and monitors, 4.7 Mt small IT and telecommunication equipment, and 0.9 Mt lamps (Forti et al. 2020). Continent-wise figures of e-waste generated in the last decade, along with countries contributing the highest share of e-waste within each continent, are shown in Figure 3.2. Per capita contribution to global e-waste in 2019 among different continents was as follows: Asia, 5.6 kg; Africa, 2.5 kg; Europe, 16.2 kg; Americas, 13.3 kg; and Oceania, 16.1 kg per capita.

India is placed third after China and the United States in terms of the total e-waste generated annually. Nonetheless, most of the e-waste is still handled by the informal sector. India's total e-waste generation has increased by over 43% between 2018 and 2020. The recent COVID-19 pandemic has increased the use of electronic devices such as computers for storage and exchange of information on COVID-19 infections and vaccination. This will further contribute to an increase in e-waste production in India in the future. Also, the major share of e-waste was generated by urban areas. However, the mobile phone revolution in India is set to declare urban India as the e-waste hub. An estimate of the total amount of e-waste generated in India during the last decade is given in Table 3.1.

The definition of e-waste is very broad and covers over ten waste categories: (i) large household appliances, (ii) small household appliances, (iii) information technology and telecommunication equipment, (iv) consumer equipment and photovoltaic

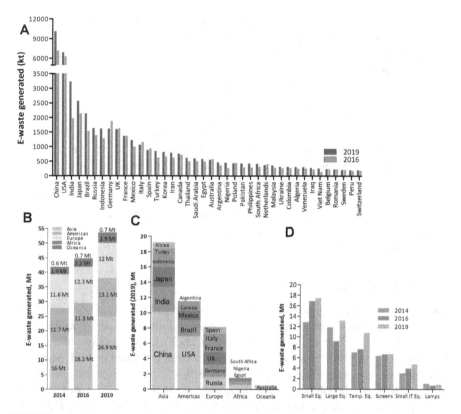

FIGURE 3.2 E-waste generation (A) Statistics of countries that generated the highest quantity of e-waste in 2016 and 2019. (B–C) Continent-wise figures of e-waste generated during 2014, 2016, and 2019, and top e-waste generating countries from each continent in 2019. (D) Quantities of e-waste for each category of e-waste. (Figure reproduced from (Ahirwar and Tripathi 2021) with permission from Elsevier).

TABLE 3.1
E-waste Generation in India

E-waste	E-waste Generated (in Million Tons/Year)				
	2012	2014	2016	2018	2020
LCD/plasma TVs, households, air conditioners, microwave oven, washing machine, refrigerators, mixture grinder, etc.	0.07	0.2	0.28	0.39	0.5
Printers, typewriters, keyboards, monitors, compact disks, central processing units, remotes, batteries, IT & communication equipment, mobile phones, chargers, headphones, semiconductors, etc.	0.35	1.04	1.48	2.09	2.65
DVDs and players, consumer electronics, remote control cars, video games, iPods, etc.	0.02	0.07	0.09	0.13	0.17

Source: Pathak, Srivastava, and Ojasvi 2017.

panels, (v) lighting equipment, (vi) electrical and electronic tools, (vii) toys, leisure, and sports equipment, (viii) medical devices, (ix) monitoring and control instruments, and (x) automatic dispensers (Table 3.2). Products within each e-waste category have broadly similar functions and average weights and compositions.

TABLE 3.2

Indicative List of Waste Electrical and Electronic Equipment (WEEE) Identified as E-wastes by the European Union

EEE Category	Indicative List of WEEE
Large household appliances	Cooling appliances (refrigerators, freezers, and other large appliances used for refrigeration, conservation and storage of food); dish washing machines, cloth washing machines, and clothes dryers; electric stoves, electric hot plates, microwaves, and appliances for cooking and processing of food; electric heating appliances (electric radiators and large appliances for heating rooms, beds, seating furniture, etc); electric fans, air conditioners, other fanning, exhaust, ventilation and conditioning equipment.
Small household appliances	Vacuum cleaners, carpet sweepers and other appliances for cleaning; appliances for sewing, knitting, weaving, and processing for textiles; appliances for ironing, mangling, and other care of clothing; toasters, fryers, grinders, coffee machines, and equipment for opening or sealing containers or packages; appliances for hair cutting, hair drying, tooth brushing, shaving, massage, and other body care appliances; clocks, watches ,and equipment to measure, indicate, or register time.
Information technology and telecommunication equipment	Centralized data processing (mainframes, minicomputers, printers); personal computing (desktop, laptop, notebook, notepad computers); printers and copying equipment; electrical and electronic typewriters; pocket and desk calculators; other equipment for collection, storage, processing, presentation, or communication of information by electronic means; user terminals and systems (facsimile machine, telex, telephones, cordless telephones, cellular telephones, answering systems, and other equipment of transmitting sound, images, or information by telecommunications).
Consumer equipment and photovoltaic panels	Radio and television sets; video cameras and video recorders; hi-fi recorders and audio amplifiers; musical instruments and other equipment for recording or reproducing sound or images, including signals or other technologies for the distribution of sound and image by telecommunications; photovoltaic panels.
Lighting equipment	Luminaires for fluorescent lamps; straight and compact fluorescent lamps; high-intensity discharge lamps, including pressure sodium lamps and metal halide lamps; low-pressure sodium lamps and other lighting equipment for spreading or controlling light, except filament bulbs.
Electrical and electronic tools	Drills, saws, and sewing machines; equipment for turning, milling, sanding, grinding, sawing, cutting, shearing, drilling, making holes, punching, folding, bending, or similar processing of wood, metal and other materials; tools for riveting, nailing, screwing, or removing rivets, nails, screws; tools for welding or soldering; equipment for spraying, spreading, dispersing, or other treatment of liquid or gaseous substances by other means; tools for mowing or other gardening activities.

(continued)

TABLE 3.2 (Continued)
Indicative List of Waste Electrical and Electronic Equipment (WEEE)
Identified as E-wastes by the European Union

EEE Category	Indicative List of WEEE
Toys, leisure and sports equipment	Electric trains or car racing sets; video games and handheld video game consoles; computers for biking, diving, running, rowing; sports equipment with electric or electronic components; coin slot machines.
Medical devices	Radiotherapy, cardiology, dialysis and nuclear medicine equipment; analyzers, freezers, and pulmonary ventilators; equipment for in-vitro diagnosis and appliances for detecting, preventing, monitoring, treating, alleviating illness, injury, or disability.
Monitoring and control instruments	Smoke detectors, heating regulators and thermostats, measuring and weighing appliances for households and laboratories; monitoring, and control instruments used in industrial installations.
Automatic dispensers	Automatic dispensers for drinks, solid products, bottles, and money; all appliances which deliver automatically all kinds of products.

Source: Directive 2003.

3.3   ENVIRONMENTAL AND HEALTH HAZARDS OF E-WASTE

E-waste contains several hazardous substances such as heavy metals including mercury (Hg), lithium (Li), barium (Ba), arsenic (As), lead (Pb), selenium (Se), cobalt (Co), copper (Cu), cadmium (Cd), and persistent organic pollutants, such as brominated flame retardants (BFRs), polycyclic aromatic hydrocarbons (PAHs), perfluoroalkyl and polyfluoroalkyl substances (PFASs), polychlorinated biphenyls (PCBs), polybrominated diphenyl ethers (PBDEs), polybrominated biphenyl (PBB), polychlorinated dibenzo-p-dioxins (PCDDs), polybrominated dioxins (PBDDs), polybrominated furans (PBDFs), and polychlorinated dibenzofurans (PCDFs), which are added to confer different properties (Table 3.3). Consequently, e-waste becomes a prominent threat to humans, animals, and the environment (Ahirwar and Tripathi 2021; Garlapati 2016). Adverse environmental and health impacts of improper e-waste management are well studied (Ahirwar and Tripathi 2021; Parvez et al. 2021). Discarded e-waste in landfills or those channeled to informal recycling facilities adversely impact the local environment and health of workers engaged in recycling activities and those residing in the vicinity of recycling sites. Improper disposal and recycling processes such as stripping of metals in open acid baths, metal recovery by burning parts and cables, and heating of printed circuit boards over a grill for removal of electronic components lead to the release of toxic pollutants into the environment through surface runoff or atmospheric precipitation, resulting in a negative effect on human health and causing water, air, and soil contamination (Khan, Besis, and Malik 2019; Youcai 2018).

TABLE 3.3
List of Toxic Substances Present in Electrical and Electronic Equipment (EEE) Components and Their Purpose of Usage

E-waste Pollutant	Component of EEE	Purpose of Usage in EEE
Arsenic (As)	Microchips, solar cells, LED, LCD, circuit boards, microprocessors, memory devices, networking, and semiconductors	Doping (n-type)
Barium (Ba)	Electronic vacuum tubes, lubricant additive, sparkplugs, fluorescent lamps, glass TV screens, and computer monitors	O_2 getters in vacuum tubes (computer monitors and TV tubes)
Beryllium (Be)	Printed circuit boards, telecommunication devices, computers, cellular phones, and x-ray equipment	Corrosion and weather resistance and fatigue-free electrical and thermal conductivity
Cadmium (Cd)	Batteries, pigments, coated metal surfaces, solder, circuit boards, and cathode ray tubes	Corrosion resistant coating
Chromium (Cr)	Coated metal surfaces (e.g., electro-galvanized steel and aluminum)	Corrosion resistant coating
Lead (Pb)	Solder, batteries, piezoelectric ceramics, sealing glass, and cathode ray tube glass	Provide conductive path in circuit elements
Lithium (Li) and Nickel (Ni)	Cellular phones, batteries, alloys, relays, and semiconductors	Provide high power, voltage in batteries
Mercury (Hg)	Barometers, thermometers, batteries, light bulbs, and medical equipment	Heat conductivity and density, liquid at ambient temperature
PCBs	Transformers, capacitors, fluorescent lighting, ceiling fans, dishwashers, and electric motors	Resistance to extreme temperature and pressure
BFRs (PBDE, TCDD, PCDD/F)	Casing, circuit boards, cables, dielectric fluids, lubricants and coolants in generators, and capacitors and transformers	Flame retardants, synergists, and smoke suppressants

Source: Ari 2016; Miliute-Plepiene and Youhanan 2019.
Note: PCBs: polychlorinated biphenyls; BFRs: brominated flame retardants; PBDEs: polybrominated diphenyl ethers; TCDD: tetrachlorodibenzo-*p*-dioxin; PCDD/F: polychlorinated dibenzo-dioxins and polychlorinated dibenzofurans.

3.3.1 Effects on Human Health

Multiple epidemiological studies have shown significantly elevated levels of heavy metals and persistent organic pollutants among individuals engaged in e-waste management and/or living in e-waste-exposed regions. In particular, children and pregnant women are found to be the most susceptible groups for the critical periods of exposure that detrimentally affect diverse biological systems and organs. Chronic exposure of pregnant women to elevated levels of pollutants at the workplace or residence negatively impacts neonatal growth and also alters hormone levels. E-waste toxicants may interact with cellular components, resulting in DNA lesions, telomere attrition, inhibited vaccine responsiveness, elevated oxidative stress, and altered thyroid, immune, lung, reproductive, and cellular functions (Ankit et al. 2021; Parvez et al. 2021; Grant et al. 2013). In a recent study, Youcai (2018) reported that e-waste present in leachates is genotoxic to mammalian cells as its absorption causes an alteration of DNA structure as well as enzymes by changing the cell pH. Xu et al. (2015c) reported that lead was observed in the blood of children, which clearly shows that the recycling of e-waste affects the immune system.

Open combustion of e-waste components such as electric cables to collect metal (Al, Cu) wires leads to production of PAHs and dioxins, which being lipophilic in nature, can enter pulmonary cells via passive diffusion upon inhalation or ingestion. Subsequent activation of cytochrome P450 (CYP) monooxygenases and other cellular enzymes to transform PAHs to their hydroxylated detoxified forms may actually produce active carcinogenic metabolites like diol-epoxides, radical cations, and reactive and redox-active o-quinones that can react with genetic material, causing mutation, abnormal gene expression, and tumorigenesis (Moorthy et al., 2015; Rani et al., 2021). A study by Eguchi et al. (2014) found that the source of perchlorate and thiocyanate in e-waste are fireworks, rocket propellants, explosives, and polyvinyl chloride (PVC) manufacturing industry. These chemicals affect the uptake of iodide by the iodide/sodium symporter and further lead to irregularities in thyroid hormone production in the thyroid gland. Long-term exposure to e-waste causes an accumulation of metals, resulting in Alzheimer's and Parkinson's disease as well as multiple sclerosis and muscular dystrophy by causing physical, neurological, and muscular degeneration (Mohod and Dhote 2013). Zhao et al. (2008) reported that PBDEs (30 ng g^{-1}), PCBs (182 ng g^{-1}), and PBBs (58 ng g^{-1}) were found in human hair from e-waste disassembly sites. An increase in concentrations of Pb and Cd were reported in children causing lower cognitive skills and respiratory problems (Zheng et al. 2008). Several health issues were reported due to an increase in concentrations of dioxins in placentas, hair, and human milk (Chan et al. 2007). Several metals present in e-waste cause adverse effects on human health such as cadmium affecting lung and kidney (Ebrahimi et al. 2020), copper affecting the liver (Danzeisen et al. 2007), and lead causing behavioral and learning changes (Sharma et al. 2020). The toxic substances present in e-waste and their harmful effect on human health are presented in Table 3.4.

3.3.2 EFFECTS ON SOIL MICROORGANISMS

Toxic substances released into the soil from informal e-waste recycling sites and landfill leachate can persist in soil for the long term, affecting the microbial community structure (Beattie et al. 2018), causing adverse effects on the functional diversity of soil microorganisms (Abdu, Abdullahi, and Abdulkadir 2017), reducing soil microbial populations, altering physical and chemical properties and enzyme activity of soil (Zou et al. 2021; de Quadros et al. 2016; Tang et al. 2013; Chaperon and Sauve 2008).

TABLE 3.4

Toxic Components Present in E-waste and Their Harmful Effect on Human Health

Study Reference	E-waste Toxicant	Health Outcomes
Growth and neurodevelopment effects		
Huo et al. (2019)	PAHs	Decrease of bodyweight, head circumference, body mass index (BMI), and decreased Apgar 1 score, associated with elevated OH-PAHs.
Li et al. (2018)	PBDEs	Neonatal BMI, Apgar 1 score, and head circumference were negatively correlated with PBDEs in the umbilical cord.
Xu et al. (2016)	Pb, Cd	Shorter neonatal length was associated with Pb in the placenta, neonatal weight and length were negatively correlated with Cd.
Xu et al. (2015a)	PBDEs	BMI, head circumference, and Apgar 1 score were negatively correlated with PBDE.
Zhang et al. (2018)	Cd	Birth weight, length, head circumference, Apgar1 and Apgar5 score were negatively associated with Cd in urine.
Xu et al. (2015b)	PAHs, Pb	Child height was negatively correlated with Pb in blood. Child height and chest circumference are negatively associated with PAH levels.
Zeng et al. (2019)	Pb	Height, weight, head circumference, and chest circumference were negatively associated with Pb in blood.
Cai et al. (2019)	Pb	Increased sensory integration difficulty scores (hearing, touch, body awareness, balance and motion, total sensory systems) was associated with elevated blood lead (>5 μg/dL).
Liu et al. (2018)	Pb, Cd	In the mediation analysis, cognitive scores and language scores were negatively correlated with Pb in blood.
Genetic and oxidative changes		
Lin et al. (2013)	Pb, Cd	Placental telomere length negatively correlated with Pb and Cd.
Xu et al. (2020)	Pb, Cd	Strong positive trend of MeCP2 promoter methylation with increasing Pb and Cd.
Xu et al. (2018)	Pb,	8-hydroxy-2'-deoxyguanosine (8-OHdG) associated with higher Pb in blood.
Lu et al. (2016)	PAH	8-OHdG significantly increased with an increase in OH-PAHs.
Respiratory, cardiovascular, and hematological changes		
Zeng et al. (2016)	Cr, Mn, Pb	Cough, wheeze, and asthma were associated with Cr, Mn, and Pb in blood, respectively.
Lu et al. (2018)	Pb	Higher Lp-PLA2, IL-6, and lower HDL were associated with elevated blood Pb.
Zheng et al. (2019)	Pb, PAHs	Higher levels of IL-6, IL12p70, IP-10, CD4+ T cell percentage, neutrophil, and monocyte counts were associated with elevated Pb in blood and PAHs in urine.
Zhang et al. (2017)	Cd	Increase in neutrophils percentage and counts were associated with Cd in blood.

Source: Data were obtained from Parvez et al. (2021).

Studies analyzing the microbial community structure found significant differences in microbiota at e-waste contaminated sites and the non-contaminated sites. Long-term exposure to the toxic pollutants can reduce the normal soil microbial biota and may favor the evolution of heavy metal resistant and organic pollutants remediating microbes. Salam and Varma (2019) analyzed bacterial 16s-RNA using denaturing gradient gel electrophoresis (DGGE) and reported that hazardous components of e-waste affected the composition and diversity of bacterial strains as well as reduced the population of predominant bacterial cells like firmicutes and proteobacteria. Meena et al. (2016) reported that metal contaminated soils may be responsible for the transfer of metals as well as metalloids into the food chain. Liu et al. (2015) reported that copper and decabromo diphenyl ether significantly altered microbial communities due to improper recycling of e-waste.

3.3.3   EFFECTS ON AQUATIC ORGANISMS

E-waste leachates can efficiently contaminate water affecting the growth and reproduction of aquatic organisms. The risk of contamination of water bodies adjacent to e-waste recycling and disposal sites with pollutants and its subsequent uptake in the aquatic food chain is particularly relevant. As shown in Figure 3.3, e-waste contaminants get accumulated in the tissues of aquatic organisms like fish via biomagnification and enter the food chain to affect human health (Steinhausen et al. 2021; Javed and Usmani 2011). A higher amount of PBDEs and PCBs were reported in aquatic birds like waterfowl in downstream areas of the Pearl River Delta in China. PCBs and PBDEs were found in concentrations of respectively 204 and 24.4 ng L^{-1} in ambient water (Luo et al. 2009). Luo, Wong, and Cai (2007b) found that PBDEs

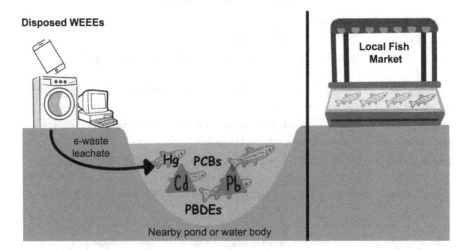

FIGURE 3.3   Schematic presentation of the entry of e-waste contaminants into the food chain via bioaccumulation in aquatic organisms. (Image reproduced from (Steinhausen et al. 2021) with permission from Elsevier).

(up to 766 ng g^{-1}) were bioaccumulated in carp fish (fresh weight) from e-waste polluting China's Nanyang river, whereas the concentration of PBDEs was maximum in sediments (up to 16,000 ng g^{-1}) (Luo, Cai, and Wong 2007a). PBDEs at a concentration of 1091, 830, 316, 490, 67.5 and 254 ng g^{-1} and PCBs at a concentration of 16,512, 12,390, 8,338, 7,052, 62.8 and 3,503 ng g^{-1} respectively were found in water snake (fresh weight), mud carp, crucian carp, northern snakehead, Chinese mystery snail, and prawns, respectively, near an e-waste recycling unit (Wu et al. 2008).

3.3.4 EFFECTS ON PLANTS

The improper disposal of e-waste in landfills or the discharge of contaminated effluents from informal e-waste recycling sites allow toxic pollutants like heavy metals, flame retardants, and other organic pollutants seep into the soil, causing the crops and plants to be enriched with pollutants. Bakare et al. (2012) reported that heavy metals present in e-waste induced cytological aberrations such as sticky chromosomes, anaphase bridge and binucleate cells, and somatic mutation in *Alliumcepa*. PBDEs and metals cause oxidative stress, affect functions of anti-oxidative enzymes such as acetyltransferase (CAT), chloramphenicol, ascorbate peroxidase (APX), superoxide dismutase (SOD), and glutathione S-transferases (GST) in algae. Furthermore, e-waste leachates adversely affect chlorophyll synthesis as well as the growth of primary producers like microalga (*Pseudokirchneriella subcapitata*) via decreasing the quantity of photosystem II (PSII) active reaction centers (Nie et al. 2015). A decrease in plant growth rate, as well as biomass, was observed in a plant grown in 100 μg L^{-1} of Cu medium (Nie et al. 2015; Bossuyt and Janssen 2004). A high concentration of copper affects plant physiological processes, inhibits PSII electron transport activity linked with photolysis of water as well as decreases the chlorophyll content and changes the energy storage capacity during photosynthesis in algae (Nie et al. 2015; Sabatini et al. 2009; Mallick and Mohn 2003). Mishra, Singh, and Arora (2017) found that heavy metals negatively affect plant growth and development by their accumulation via roots. As a consequence, these metals cause hindrance to the assimilation of essential nutrients (Djingova and Kuleff 2000), induce oxidative stress, and inhibit cytoplasmic enzymes of the plant cell (Jadia and Fulekar 2009), as well as replace the essential nutrients at the cation exchange site of the plant (Taiz and Zeiger 2002). A decrease in panicle and tiller growth in the rice plants was noticed in the presence of Hg (1 mg kg^{-1}) (Kibra 2008). Nicholls and Mal (2003) found maximum death of leaves of the plant *Lythrum salicaria* in the presence of Pb (500 mg kg^{-1}) and Cu (1000 mg kg^{-1}).

3.3.5 EFFECTS ON GROUNDWATER QUALITY

E-waste is considered one of the major sources of contamination of groundwater. The quality of groundwater is affected owing to the existence of e-waste dumping sites in the locality. Dharini et al. (2017) reported that pH, conductivity, turbidity, total dissolved solids, and chloride levels were affected by toxic substances present in e-waste, thus affecting human health. Water near the industries was contaminated due to the leakage of toxic substances from e-waste (Gupta and Nath 2020).

3.4   POLICIES AND REGULATIONS FOR E-WASTE MANAGEMENT

Considering the wide environmental and human health implications of e-waste, various countries around the globe are developing policies and legislation to deal with the growth of e-waste. These e-waste policies define actions, plans, and targets for adoption by society, institutions, or companies for efficient collection, transportation, and recycling of e-waste. Legislations and rules are enacted at the regional or national level by the government or regulators. As per recent reports, till October 2019, over 78 countries covering 71% of the world's population have a policy, legislation, or other regulating rules for e-waste management in place (Forti et al. 2020). There is a wide difference in the generation of e-waste between developed and developing countries. The richest country is producing 19.6 kg/inhabitant against the poorest country, which produces only 0.6 kg/inhabitant. Moreover, developed countries use advanced technology and management system to protect themselves from the negative effects of e-waste. However, in developing and underdeveloped countries, e-waste is either disposed of in landfill sites or fed to the informal recycling sector. Recovery of metals involves hand-picking and dismantling, nitrate/aqua-regia leaching, and throwing the residual and effluent streams in an open environment, leading to an inappropriate efficiency of resource recycling. Thus these countries are still making efforts to frame proper guidelines for effective e-waste management (SEPA 2011; Leung, Cai, and Wong 2006). A comparison of legislative policies between the developed and developing countries is listed in Table 3.5.

Developed countries have strictly implemented the legislation on the management of e-waste; several developing countries have also started to prepare and implement their specific policies. In developed countries, the national registry system, as well as the proper collection and logistics system, is very strong. Handling of e-waste in developed countries is made up of three elements: a national registry, a collection system, and logistics. The first mandated extended producer responsibility (EPR) program aimed at the avoidance of packaging waste in Germany, which imposed financial obligations on the manufacturers for the collection and reduction of packaging waste (Ongondo, Williams, and Cherrett 2011; Van Rossem, Tojo, and Lindhqvist 2006). Recently, it is adopted and extended to EEE manufacturers by Sweden, Norway, Taiwan, and Switzerland. The EU legislation restricting the use of hazardous substances in EEE and WEEE/E-waste and promoting their collection and recycling has been enacted in 2003 (Directive 2003). Presently, the legislations in China and India have implemented the EPR system but in absence of a national registry that keeps track of the produced electronics for the eventual manufacturer to take-back, as the generation of e-waste is difficult to trace (Sthiannopkao and Wong 2013). The big gray markets available for second-hand UEEE in these countries is also making the situation vulnerable while compared to the developed countries (Pathak, Srivastava, and Ojasvi 2017). In both China and India (and to a lesser extent in Pakistan), some facilities have been built, enabling the availability of proper technology for e-waste disposal. Among these countries, China at least possesses a large capacity in such things as the smelting furnaces needed for recycling nonferric metal. China is thus seen as the large-scale handler of e-waste that with the help of partnering and technology transfer has the potential for building facilities for

TABLE 3.5
An Overview of E-waste Management Norms and Rules in Various Developed and Developing Countries

Developed Country	Developing Countries
The European Union	**India**
• Waste Shipment Regulation (WSR) 1993 (amended in 2007)	• The Environmental Protection Act 1986
• Restriction of Hazardous Substances (RoHS) Directive (2002/95/EC) 2002 (revised in 2006 and 2009)	• The Ozone Depleting Substances (Regulation and Control) Rules 2000
• The Battery Directive 2006	• The Hazardous Wastes Management, Handling and Transboundary Movement Rules 2008
• WEEE Directive (2012/19/EU) 2012	• E-waste (Management and Handling) Rules 2011
Japan	• E-Waste (Management) Rules, 2016
• Law for the Promotion of Effective Utilization of Resources (LPUR) 2000	• E-waste (Management) Amendment Rules, 2018
• Law for the Recycling of Specified Kinds of Home Appliances (LRHA) 2009	**China**
• Small Electrical and Electronic Equipment Recycling Act 2013	• The Law on the Prevention and Control of Environmental Pollution by Solid Waste (2004), Technical Policy on Control of WEEE 2006
South Korea	• The Law on the Prevention and Control of Environmental Pollution by Solid Waste (2004), Technical Policy on Control of WEEE 2006
• Introduction of Waste Deposit-Refund System 1992	• The Cleaner Production Law (2002), The Ordinance on Management of Prevention and Control of Pollution from Electronic and Information Products 2007
• Guideline for Improvement of Material/Structure of Products for Stimulating Recycling 1993	• The Circular Economy Promotion law (2008), Administrative Rules on Prevention of pollution by WEEE 2008
• Extended Producer Responsibility (EPR) System 2003	
• Resource Recycling of Waste Electrical and Electronic Equipment and Vehicles, Act 2007	• The Circular Economy Promotion law (2008), Collection and Treatment Decree on Waste Electrical and Electronic Equipment 2011
Taiwan	**Pakistan**
• Waste Disposal Act (WDA) 1988 (amendment)	• Pakistan Environmental Protection Act 1997
• 4-in-1 Recycling Program (Amended WDA) 1997 (amendment) Amendment to WDA 2001	• National Environment Policy 2005
Singapore	• Import policy order, 2009.
• Environmental Protection and Management (Hazardous Substances) Act (EPMA) 1999 (revised in 2003)	**Vietnam**
	• The Law on Environmental Protection 2014
• National Voluntary Partnership for E-Waste Recycling 2016	• Commercial Law 2005 and Foreign Trade Administration Law 2014.
United States	**Nigeria**
• Resource Conservation and Recovery Act 1976	• Environmental Impact Assessment Act Cap E12
• Electronic Waste Recycling Act 2003	• Harmful Waste (Special Criminal Provisions) Act Cap HI, 1988 and updated in 2004
• National Computer Recycling Act 2009	• The National Environmental (Sanitation and Waste Control) Regulation 2009
• Responsible Electronics Recycling Act 2011	
United Kingdom	• The National Environmental (Electrical Electronic Sector) Regulations SI No 23 of 2011
• The Producer Responsibility Obligations (Packaging Waste) Regulations 2007	
• The Waste Electrical and Electronic Equipment Regulations 2013	

handling a significant amount of e-waste properly (UNEP 2009). However, the domination of informal and private players is a major bottleneck to inventory the e-waste in these countries (Abbas 2010; Jain 2010; Joseph 2007). Although these countries have lower labor costs, the large rate of e-waste generation (through both import and domestic production), the transportation cost, and above all the costlier technology for its benign disposal present hurdles despite the enactment of environmental regulations. Therefore, the informal sector keeps growing to handle the majority of e-waste in developing countries.

3.4.1 E-WASTE (MANAGEMENT) RULES, INDIA

In India, the Environmental Protection Act was firstly announced in 1986, which determines all types of environmental pollution. In 1989, the Hazardous (Management and Handling) Act was developed, which determines the management and handling of hazardous waste. In 2008, the Hazardous Act (management, handling, and transboundary movement) was developed, which introduced obligations on transboundary movement of e-waste into the rules. The transboundary movement (TBM) is defined as "an import and export of the hazardous wastes under the jurisdiction or, without jurisdiction of one country to another". The diversity in the composition of e-waste needs individual rules and regulations for collection, recycling, and disposal. Thus, in 2010, E-waste (Management and Handling) Rules were developed under Sections 6, 8, and 25 of the Environment (Protection) Act, 1986. Nevertheless, the Ministry of Environment and Forest (MoEF) received several objections/suggestions from the industries and local people. Thus the full-fledged act was passed only in May 2011, which introduced notification on e-waste legislation and commencement of e-waste legislation with EPR guidelines. EPR is defined as "setting up an effective e-waste channelization system comprising of setting up collection centers, implementing distributor take-back system (DTBS) agreements with registered dismantler or recycler either individually or collectively or through a Producer Responsibility Organization (PRO) authorized by producers". Furthermore, E-waste (Management) Amendment Rules, 2010, 2011, 2016, and 2018 were developed, which introduced fixed responsibilities and targets for refurbishers and manufacturers (Ahirwar and Tripathi 2021; Pathak, Srivastava, and Ojasvi 2017).

3.5 E-WASTE MANAGEMENT IN DEVELOPING COUNTRIES

Growing industrialization, rapid increase and innovation in electric technology, short life of electronic equipment, and increased dependence of individuals on electronic equipment and gadgets have made e-waste management a difficult task (Borthakur and Singh 2021). Despite the adoption of e-waste management rules and legislations by developing countries, they are still struggling to cover all the recommended steps of recycling and disposal due to the limited availability of infrastructure, technological access, and investments. Henceforth, implementation of the best-of-two-worlds (Bo2W) principle that provides a network and pragmatic solution for e-waste treatment can boost the emerging economies in developing countries (Nnorom and

Osibanjo 2008). Reverse logistics has been actively implemented to bring sustainability in developed countries, which is now followed by developing countries. Three major steps of the reverse logistic for e-waste recycling are (i) disassembly: selective disassembly target hazardous or valuable components for special treatment; (ii) upgrading: mechanical processing and/or metallurgical processing to increase the content and of desirable materials; and (iii) refining: purifying the recovered materials using chemical (metallurgical) processing to make them acceptable for original use (Tsydenova and Bengtsson 2011). Moreover, several factors are involved in the management of e-waste such as products and components, usage, disposal, regulations, socioeconomic, and human responsibility. Therefore, e-waste management is broadly categorized into macroscopic (products and components), microscopic (substance), and finally mesoscopic (material) scales (Gollakota, Gautam, and Shu 2020; Kumar, Holuszko, and Espinosa 2017).

3.5.1 MACRO SCALE MANAGEMENT OF E-WASTE

Several studies have reported on the regulations/legislation/management of e-waste such as (i) opinions on strengthening the prevention and control of pollution from E-waste, (ii) recycling of WEEEF, and (iii) administrative measures for the prevention and control of environmental pollution (Zhou and Xu 2012; Wang et al. 2010). Two distinguishing features were noticed; firstly, all applied and proposed laws are unevenly introduced e-waste generation, collection, recycling, and related legal liability and secondly, substantial differences were found in the e-waste regulated list in different regions of the world. Nevertheless, WEEE has properly defined the main objective of all the legislations. Thus, the applied and proposed legislation system is well established that the disposal of e-waste and its associated components has been highly fruitful in the macroscopic scale.

3.5.1.1 Role of the Government

With the increase of WEEE, the government should be highly active in guiding the nations in the right direction in minimizing environmental pollutants. In developing and developed nations, the major difference in economic growth is due to the role of the government as it plays a major role in regulations and their implementation. Moreover, due to financial limitations, several developing and underdeveloped countries drive into an e-waste black hole with an alarming level of stock-ups/imports. To overcome this, several countries have brought enactments to check the unlawful trading and reprocessing of e-waste. The governments should propose rules considering the domestic conditions as well as global measures to achieve effective management. Also, the government should arrange an innovative mechanism for storing and handling toxic and hazardous materials as well as educating the locals about the harmful effects of e-waste if left untreated (Gollakota, Gautam, and Shu 2020).

3.5.1.2 Role of the Consumers

The key to effective e-waste management is that the governing bodies should motivate and create awareness among the public regarding the toxic effects of e-waste by devising training and awareness campaigns (Borthakur and Govind 2018). In addition to this, gender, age, literacy, socio-cultural aspects, and individual perspectives have a strong impact on e-waste management programs. Successful e-waste management is only possible if people avoid excessive dependency on electronic gadgets, opt for products that are ready to recycle, efficiently use the EEE until the end of life, think ahead about the harmful effects of e-waste, and educate others regarding e-waste management (Onwosi et al. 2020). Consumer awareness regarding e-waste may make a minimal contribution toward their recycling (Gollakota, Gautam, and Shu 2020).

3.5.1.3 Extended Producer Responsibility

EPR is characterized as an ecological assurance system that enables the product manufacturer in charge of the whole life cycle of the product, particularly for the reclaim, reusing, and last transfer of the product (Lindhqvist 2000). In this manner, the producers are held responsible for products reaching the post-shopper phase of an item's life cycle (Bimbati and Rutkowski 2016). As indicated by Organization for Economic Cooperation and Development (OECD), the four salient features of EPR include source reduction (natural resource conservation/materials conservation), waste prevention, design of more environmentally compatible products, and closure of material loops to promote sustainable development (Gollakota, Gautam, and Shu 2020).

3.5.2 Micro Scale Management of E-waste

This approach involves recovery of valuable metals from e-waste via physicochemical recycling methods, which include the pretreatment of e-waste followed by extraction methods. Pretreatment of e-waste involves dismantling, shredding, and mechanical separation (Figure 3.4 and Table 3.6).

Extraction methods involve the recovery of metals by pyro-metallurgical, hydrometallurgical, or biometallurgical techniques (Kumar, Holuszko, and Espinosa 2017; Namias 2013).

3.5.2.1 Pyro-metallurgy

This method involves the formation of a mixture of desired metals containing slag by melting the materials in a high-temperature furnace. Purification is performed by using the electrorefining technique. It includes alkali smelting (plastics and ceramics may be separated at <400°C.), metal trapping (Fe and Cu), pyrolysis, roasting, and heating methods. Metals recovered by this method are gold, palladium, copper, and silver, but iron and aluminum cannot be recovered as they get oxidized (Ankit et al. 2021; Namias 2013). The advantages of this method include higher recovery rates of reactions and valuable metals, low melting point with good liquidity of materials (Ding et al. 2019). However, the need for high energy and the formation of toxic furans and dioxins adversely affect environmental and public health (Kumar, Holuszko, and Espinosa 2017).

3.5.2.2 Hydrometallurgy

This method involves the leaching of the concentrate from the pretreatment using specific solvents to dissolve the precious metals into the solution. Specific leaching agents are used for the precipitation of individual metals from the e-waste followed

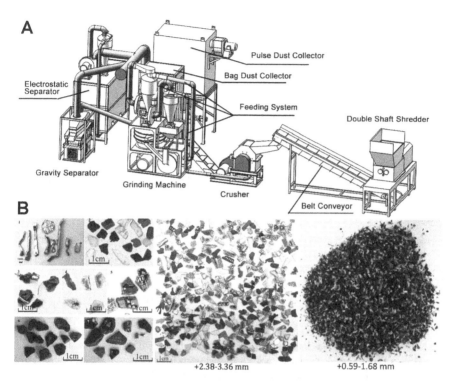

FIGURE 3.4 Steps and machinery for pre-processing of e-waste. (A) Representative image of machines used for shredding, crushing, grinding, and separating e-waste into metallic and non-metallic fractions. (B) E-waste fractions of various sizes. (Images reproduced from (Ahirwar and Tripathi 2021) with permission from Elsevier).

by purification through electro-winning. For example, leaching agents such as nitric acid, sulfuric acid or aqua regia, thiourea or cyanide, and hydrochloric acid or sodium chlorate are used for the hydrometallurgical treatment of base metals, copper, gold and silver, and palladium, respectively (Kumar, Holuszko, and Espinosa 2017). This method includes microwave-assisted leaching, liquid–liquid and solid-phase reaction, HCl+ oxidant, thiosulfate, thiourea leaching, halide leaching, cyanide leaching, iodation leaching, and chlorination leaching. The advantages of this method include lower energy consumption, more predictability and accuracy, excellent leaching efficiency, and a high recovery rate. However, high operating cost, high time consumption, need for fine grinding for efficient leaching, and requirement of more chemicals (acidic or noxious reagents), and production of a massive quantity of toxic and acidic fumes or waste solution limit its application in the long term (Ankit et al. 2021; Zhang et al. 2021; Khaliq et al. 2014; Veit, Juchneski, and Scherer 2014).

3.5.2.3 Biometallurgy

Bio-metallurgy is a biological approach that involves microorganisms to leach metal out of the e-waste followed by purification using electrowinning. Microorganisms convert metals from their insoluble solid state to water-soluble states by biochemical

TABLE 3.6
Pretreatment Techniques of E-waste

Recycling Steps	Process	Benefits	Issues
Dismantling	Remove the hazardous materials from the waste stream and then separate it manually into metal, plastics, and glass fractions	Higher grade material for end-processing, minimum dust issue, more job opportunities	Enhanced risk of public health and safety, difficult to dismantle newer complex technologies, time-consuming, maximum spending on labor and transportation cost
Shredding	Particle size reduced by crushing and grinding the electronic waste using metal shredders, hammer, and knife mills	Minimizes risk of public health and safety, faster automated systems, increased throughput, less volume for transportation	Maximum dust issue due to loss of material, decreased grade for subsequent operation
Mechanical separation	Separate various streams from the shredded material, such as magnetic separation (used to remove ferromagnetic materials such as iron, steel, and rare earth metals), density separators (air tables, air cyclones), centrifugal separators are used to recover base metals such as copper, gold, and silver from nonmetal fractions), eddy current separators (used to recover aluminum), infrared sensors (used to separate different plastics), optical sensors (used for glass)	Reduced public health and safety issues, faster automated system, increased throughput, less volume for transportation, less energy-intensive	Not suitable for small recycling businesses, an issue for removal of moisture from wet systems, and dust issue

Source: Kumar, Holuszko, and Espinosa 2017.

oxidation and reduction methods using enzymes (Narayanasamy, Dhanasekaran, and Thajuddin 2021; Pant and Dhiman 2020). Chemolithoautotrophic bacteria, acidophilic bacteria, and organic acid producing fungi are commonly used for the biological recovery of e-waste (Ankit et al. 2021). The advantages of this method include a reduction in operating cost, being eco-friendly, and minimum usage of chemicals. This method has been gaining popularity for leaching copper and gold ore. Table 3.7 represents examples of the recovery of various metals from e-waste.

TABLE 3.7
Recovery of Various Metals from E-waste Using Different Techniques

Recovered Metal	Leaching Agent	Methods	References
99% Cu	Nitric acid	Hydrometallurgy	Rao et al. (2021)
93% Cu, 100% Ni, Pd, Zn		Pyrolysis + roasting	Panda et al. (2020)
97% Au	Potassium chlorate and hydrochloric acid	Hydrometallurgy + selective adsorption on glutaraldehyde cross-linked chitosan	Bui, Jeon, and Lee (2021)
75% Au, 94% Cu	*Frankia casuarinae* DDNSF-02	Bioleaching	Marappa et al. (2020)
98% Au, 78% Cu	*Chromobacterium Violaceum*	Bioleaching + ion exchange using AmberjetTM 4200 and Amberlite IRC-86 resins	Choi et al. (2020)
96% Cu, 73% Ni, 85% Zn, 93% Co	*Leptospirillum ferriphilum, Sulfobacillus benefaciens*	Bioleaching in a two-staged continuous bioreactor	Hubau et al. (2020)
97.7%	Cyanogenic bacteria isolated from e-waste landfill	Bioleaching	Arab et al. (2020)
97% Cu	*Acidithiobacillus*	Bioleaching + Direct current	Wei et al. (2020)
90%, Cu, 89% Ni	*Penicillium simplicissimum*	Bioleaching	Arshadi, Nili, and Yaghmaei (2019)
98% Au, 100% Ag	Thiourea	Hydrometallurgy	Lee, Molstad, and Mishra (2018)
90% Au	Ammonium Thiosulfate, *Lactobacillus acidophilus*	Hydrometallurgy + biosorption	Sheel and Pant (2018)
75% Au	Sodium, ammonium Thiosulfate	Hydrometallurgy + electrowinning	Kasper and Veit (2018)
92% Zn, 64% Pb, 81% Ni	*Acidithiobacillus ferrooxidans*	Bioleaching	Priya and Hait (2018a)
68.5% Au, 33.8% Ag,	*Pseudomonas balearica* SAE1	Bioleaching	Kumar, Saini, and Kumar (2018)
90% Au	Aqua regia	Hydrometallurgy + reduction of gold by polyaniline coated cotton fibers	Wu et al. (2017)
79% Cu, 39% Ni, 29% Zn 10% Pb, 94% Cu	*Acidiphilium acidophilum* (NCIM 5344; ATCC 27807)	Bioleaching	Priya and Hait (2018b)
63% Cu	*Acinetobacter sp.*	Sequential Bioleaching	Jagannath, Shetty, and Saidutta (2017)

3.5.3   Meso Scale Management of E-waste

This approach includes material compatibility, material fatigue, and material reclaiming for the design stage, consumption stage, and end-of-life (EOL) stage (Gollakota, Gautam, and Shu 2020).

3.5.3.1   Material Compatibility

The chemical composition of electronic products is a critical factor to identify the compatibility of material, during the production, chemical storage, delivery, and EOL. However, only material-based analysis is not sufficient to estimate the performance of the processing system.

3.5.3.2   Material Fatigue

In the case of the utilization of electronics, the maximum number of recurrent applications is due to the early failure of materials. As a result of rapid advancement in technology and higher demand for a new generation, the duration of old electronics is reported to be very short of the material fatigue cycle.

3.5.3.3   Material Reclaiming

Recovery of precious metals from e-waste is the major aim during the EOL of the product. While e-waste is a handful of valuable metals present in electronic components like printed circuit boards, graphic and memory cards, hard drives, and cables and connectors, it is also a source of a variety of toxic chemicals such as heavy metals (Pb, Hg, As, Cd) and organic pollutants (PCBs, dioxins and furans, PAHs, etc.).

3.6   MAJOR CHALLENGES IN EFFECTIVE MANAGEMENT OF E-WASTE IN DEVELOPING COUNTRIES

In developing countries, ineffective legislation, lack of infrastructure, and informal recycling of e-waste are emerging as a new environmental challenge for the 21st century. Environmental groups such as the Greenpeace, Korea Zero Waste Movement Network (KZWMN), Silicon Valley Toxicity Coalition (SVTC), Basel Action Network (BAN), and Toxic Links have established through their investigations that a large amount of highly toxic e-waste is still illegally dumped into developing countries and that home-grown recycling activities are wreaking an environmental havoc (Osibanjo and Nnorom 2007). Several environmentalists have criticized the dumping of e-waste at various forums around the world, thus indicating that pouring of e-waste into developing countries by developed countries flouts the Basel Convention in the transboundary movement of hazardous wastes as well as allows electronic manufacturers to escape from their duties over the ultimate fate of the products they put out in the market (Toxic Dispatch 2004). The main challenge before effective e-waste management is the massively high volume and complexity of the generated e-waste. The generation of e-waste is increasing

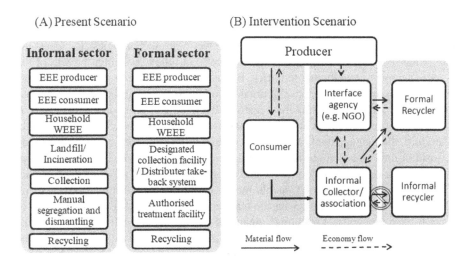

FIGURE 3.5 (A) Schematic illustration of the existing e-waste recycling system at most developing countries, and (B) an informal-formal sector partnerships model for pollution-free management of e-waste.

constantly at an estimated rate of 2 Mt annually. In addition to the huge amount of e-waste, improper treatment methods will also be responsible for potential environmental disasters involving e-waste in developing countries. In China, most of the e-waste recycling and disposal operations such as open burning of plastic waste, exposure to toxic solders, river dumping of acids, and widespread general dumping are quite polluting and likely to be very damaging to the ecology and human health (ECOFLASH 2003). Partnering of the informal sector and the formal sector (Figure 3.5) can be a big step toward pollution-free and health hazard-free management of e-waste.

Along with this, advancing technologies and innovations are adding highly advanced but complex electronic gadgets into the market. Both the tendency of individuals to dispose of a variety of e-waste together and the difficulty in effective dismantling of e-waste components from complex electronic gadgets that are not designed for effective recycling (i.e., some components may be bolted, others screwed, snapped, glued or soldered together) make effective segregation of e-waste components a difficult task. Notwithstanding, the enacted laws and rules on e-waste management and individuals' lack of knowledge on e-waste disposal methods are another hurdle in effective management of e-waste. Mostly, such conditions arise when governments and authorized agencies do not take many efforts to popularize e-waste collection and recycling policies. Limited option of e-waste collection is yet another stumbling block. It promotes informal setups to grow. The lack of infrastructure and advanced technologies enforce informal recyclers to extract only limited quantities of precious metals, releasing them along with toxic materials into the environment from primary, secondary, and tertiary emissions.

3.7 SUMMARY

Rapid advancement and development in electronic technologies to meet the increasing demand for electronic devices have led to the exponential growth of the e-waste sector in developing countries. E-waste contains various toxic components such as heavy metals and persistent organic pollutants, which cause serious environmental pollution. E-waste imparts negative effects on soil microorganisms (such as bacteria and fungi), aquatic organisms, plants, groundwater as well as human health. Therefore, the management of e-waste is crucial to tackling the problem of growing e-waste in developing countries like India and China. The most important concern about e-waste is its improper disposal and informal processing, which involve manual dismantling and recovery of valuable metals (gold, silver, platinum, copper) using homemade equipment in a very unsafe manner. Efficiently harnessing the benefits of e-waste recycling without affecting human and environmental health require (i) robust e-waste collection approaches and channels, (ii) advanced and high-capacity formal recycling facilities, (iii) assisting informal recyclers to upgrade to formal recyclers, and (iv) spreading awareness on e-waste collection methods and environmental and human health hazards of e-waste. While the management of e-waste is well tackled by developed countries through stringent law enactments and establishing proper recycling facilities, developing countries still lack the statutes, have gaps in policy-making, face socio-economic-cultural barriers, unable to deploy technology, and lack appropriate treatment facilities. Therefore, timely amendments in e-waste legislation and its strict compliance is also very important as effective implementation of legislation can contribute to the proper handling of toxic wastes as well as the recycling and recovery of precious metals.

ACKNOWLEDGMENT

Research in the laboratory of the corresponding author is supported by funds from the Indian Council of Medical Research (Grant No. 65/2/AKT/NIREH/2018-NCD-II) and Science and Engineering Research Board (Grant no ECR/2017/003179). Authors declare no conflict of interest.

REFERENCES

Abbas, Z. 2010. *E-waste Management in Pakistan. Regional Workshop on E-waste/WEEE Management. Osaka, Japan.* 6–9 July.

Abdu, N., A. A. Abdullahi, and A. Abdulkadir. 2017. "Heavy metals and soil microbes." *Environmental Chemistry Letters* 15(1): 65–84. doi: 10.1007/s10311-016-0587-x

Ahirwar, R., and A. K. Tripathi. 2021. "E-waste management: a review of recycling process, environmental and occupational health hazards, and potential solutions." *Environmental Nanotechnology, Monitoring & Management* 15: 100409. doi: 10.1016/j.enmm.2020.100409

Ahirwar, R., and N. Khan. 2022. "Smart Wireless Nanosensor Systems for Human Healthcare". In *Nanosensors for Futuristic Smart and Intelligent Healthcare Systems.* Kaushik S, V. Soni, and E. Skotti (Ed.) CRC Press, pp. 265–292.

Ankit, Saha, L., V. Kumar, J. Tiwari, S. Rawat, J. Singh, and K. Bauddh. 2021. "Electronic waste and their leachates impact on human health and environment: Global ecological threat and management." *Environmental Technology & Innovation* 24: 102049. doi: 10.1016/j.eti.2021.102049

Arab, B., F. Hassanpour, M. Arshadi, S. Yaghmaei, and J. Hamedi. 2020. "Optimized bioleaching of copper by indigenous cyanogenic bacteria isolated from the landfill of e-waste." *Journal of Environmental Management* 261: 110124. doi: 10.1016/j.jenvman.2020.110124

Ari, V. 2016. "A review of technology of metal recovery from electronic waste." *E-Waste in Transition—From Pollution to Resource, InTech* 121–158. doi: 10.5772/61569

Arshadi, M., S. Nili, and S. Yaghmaei. 2019. "Ni and Cu recovery by bioleaching from the printed circuit boards of mobile phones in non-conventional medium." *Journal of Environmental Management* 250: 109502. doi: 10.1016/j.jenvman.2019.109502

Bakare, A. A., A. O. Adeyemi, A. Adeyemi, O. A. Alabi, and O. Osibanjo. 2012. "Cytogenotoxic effects of electronic waste leachate in *Allium cepa*." *Caryologia* 65(2): 94–100. doi: 10.1080/00087114.2012.709786

Balde, C. P., V. Forti, V. Gray, R. Kuehr, and P. Stegmann. 2017. *The Global E-waste Monitor 2017: Quantities, Flows and Resources*. United Nations University, International Telecommunication Union, and International Solid Waste Association.Balde, C.P., F. Wang, R. Kuehr, and J. Huisman. 2015. *The Global E-waste Monitor—2014*. United Nations University, IAS–SCYCLE, Bonn, Germany. ISBN 978-92-808-4555-6.

Beattie, R. E., W. Henke, M. F. Campa, T. C. Hazen, L. R. McAliley, and J. H. Campbell. 2018. "Variation in microbial community structure correlates with heavy-metal contamination in soils decades after mining ceased." *Soil Biology and Biochemistry* 126: 57–63. doi: 10.1016/j.soilbio.2018.08.011

Bimbati, T., and E. W. Rutkowski. 2016. "A Responsabilidade compartilkada e seus instrumentos na promocao da reciclagem." *X Simpósio Internacional de Qualidade Ambiental*. Porto Alegre: PUCRS. 19: 11.

Borthakur, A., and M. Govind. 2018. "Management of the challenges of electronic waste in India: an analysis." *Proceedings of the Institution of Civil Engineers-Waste and Resource Management* 171(1): 14–20.

Borthakur, A., and P. Singh. 2021. "The journey from products to waste: a pilot study on perception and discarding of electronic waste in contemporary urban India." *Environmental Science and Pollution Research* 28(19): 24511–24520. doi: 10.1007/s11356-020-09030-6

Bossuyt, B. T., and C. R. Janssen. 2004. "Long-term acclimation of *Pseudokirchneriella subcapitata* (Korshikov) Hindak to different copper concentrations: changes in tolerance and physiology." *Aquatic Toxicology* 68(1): 61–74. doi: 10.1016/j.aquatox.2004.02.005

Bui, T. H., S. Jeon, and Y. Lee. 2021. "Facile recovery of gold from e-waste by integrating chlorate leaching and selective adsorption using chitosan-based bioadsorbent." *Journal of Environmental Chemical Engineering* 9(1): 104661. doi: 10.1016/j.jece.2020.104661

Cai, H., X. Xu, Y. Zhang, X. Cong, X. Lu, and X. Huo. 2019. "Elevated lead levels from e-waste exposure are linked to sensory integration difficulties in preschool children." *Neurotoxicology* 71: 150–158. doi: 10.1016/j.neuro.2019.01.004

Chan, J. K., G. H. Xing, Y. Xu, Y. Liang, L. X. Chen, S. C. Wu, C. K. Wong, C. K. Leung, and M. H. Wong. 2007. "Body loadings and health risk assessment of polychlorinated dibenzo-p-dioxins and dibenzofurans at an intensive electronic waste recycling site in China." *Environmental Science & Technology* 41(22): 7668–7674. doi: 10.1021/es071492j

Chaperon, S., and S. Sauvé. 2008. "Toxicity interactions of cadmium, copper, and lead on soil urease and dehydrogenase activity in relation to chemical speciation." *Ecotoxicology and Environmental Safety* 70(1): 1–9. doi: 10.1016/j.ecoenv.2007.10.026

Choi, J. W., M. H. Song, J. K. Bediako, and Y. S. Yun. 2020. "Sequential recovery of gold and copper from bioleached wastewater using ion exchange resins." *Environmental Pollution* 266: 115167. doi: 10.1016/j.envpol.2020.115167

Danzeisen, R., M. Araya, B. Harrison, C. Keen, M. Solioz, D. Thiele, and H. J. McArdle. 2007. "How reliable and robust are current biomarkers for copper status?" *British Journal of Nutrition* 98(4): 676–683. doi: 10.1017/S0007114507798951

de Quadros, P. D., K. Zhalnina, A. G. Davis-Richardson, J. C. Drew, F. B. Menezes, A. D. O. Flávio, and E. W. Triplett. 2016. "Coal mining practices reduce the microbial biomass, richness and diversity of soil." *Applied Soil Ecology* 98: 195–203. doi: 10.1016/j.apsoil.2015.10.016

Dharini, K., J. B. Cynthia, B. Kamalambikai, J. A. S. Celestina, and D. Muthu. 2017. "Hazardous e-waste and its impact on soil structure." *IOP Conference Series: Earth and Environmental Science* 80(1): 012057.

Ding, Y., S. Zhang, B. Liu, H. Zheng, C. C. Chang, and C. Ekberg. 2019. "Recovery of precious metals from electronic waste and spent catalysts: a review." *Resources, Conservation and Recycling* 141: 284–298. doi: 10.1016/j.resconrec.2018.10.041

Directive. 2003. "Directive 2002/96/EC of the European parliament and of the council of 27 January 2003 on waste electrical and electronic equipments (WEEE)." *Union, T.E.P.a.t.C.p.t.E (Ed.), Official Journal of the European Union* 24–38.

Directive. 2008. "Directive 2008/98/EC of the European parliament and of the council of 19 November 2008 on waste and repealing certain directives." *Union, T.E.P.a.t. C.p.t.E (Ed.), Official Journal of the European Union, Official Journal of the European Union* 3–30.

Djingova, R., and I. Kuleff. 2000. "Instrumental techniques for trace analysis." In: B. Markert and K. Friese (eds) *Trace Elements—Their Distribution and Effects in the Environment.* Elsevier, Amsterdam, 137–185.

Ebrahimi, M., N. Khalili, S. Razi, M. Keshavarz-Fathi, N. Khalili, and N. Rezaei. 2020. "Effects of lead and cadmium on the immune system and cancer progression." *Journal of Environmental Health Science and Engineering* 18(1): 335–343. doi: 10.1007/s40201-020-00455-2

Ecoflash. 2003. "Current situation of e-waste in China." In: *M. Menant & Y. Ping (Eds), Delegation of German Industry and Commerce Shanghai.* ECOFLASH, 10–13.

Eguchi, A., T. Kunisue, Q. Wu, P. T. K. Trang, P. H. Viet, K. Kannan, and S. Tanabe. 2014. "Occurrence of perchlorate and thiocyanate in human serum from e-waste recycling and reference sites in Vietnam: association with thyroid hormone and iodide levels." *Archives of Environmental Contamination and Toxicology* 67(1): 29–41. doi: 10.1007/s00244-014-0021-y

Forti, V., C. P. Baldé, R. Kuehr, and G. Bel. 2020. *The Global E-waste Monitor 2020: Quantities, Flows and the Circular Economy Potential.* United Nations University (UNU)/United Nations Institute for Training and Research (UNITAR)-co-hosted SCYCLE Programme. International Telecommunication Union (ITU) & International Solid Waste Association (ISWA), Bonn/Geneva/Rotterdam.

Garlapati, V. K. 2016. "E-waste in India and developed countries: management, recycling, business and biotechnological initiatives." *Renewable and Sustainable Energy Reviews* 54: 874–881. doi: 10.1007/s00244-014-0021-y

Gollakota, A. R., S. Gautam, and C. M. Shu. 2020. "Inconsistencies of e-waste management in developing nations—facts and plausible solutions." *Journal of Environmental Management* 261: 110234. doi: 10.1016/j.jenvman.2020.110234

Grant, K., F. C. Goldizen, P. D. Sly, M. N. Brune, M. Neira, M. van den Berg, and R. E. Norman. 2013. "Health consequences of exposure to e-waste: a systematic review." *Lancet Global Health* 1(6): e350–e361. doi: 10.1016/S2214-109X(13)70101-3

Gupta, N., and M. Nath. 2020. "Groundwater contamination by e-waste and its remedial measure—a literature review." *Journal of Physics: Conference Series* 1531(1): 012023.

Hubau, A., M. Minier, A. Chagnes, C. Joulian, C. Silvente, and A. G. Guezennec. 2020. "Recovery of metals in a double-stage continuous bioreactor for acidic bioleaching of printed circuit boards (PCBs)." *Separation and Purification Technology* 238: 116481. doi: 10.1016/j.seppur.2019.116481

Huo, X., Y. Wu, L. Xu, X. Zeng, Q. Qin, and X. Xu. 2019. "Maternal urinary metabolites of PAHs and its association with adverse birth outcomes in an intensive e-waste recycling area." *Environmental Pollution* 245: 453–461. doi: 10.1016/j.envpol.2018.10.098

Jadia, C. D., and M. H. Fulekar. 2009. "Phytoremediation of heavy metals: recent techniques." *African Journal of Biotechnology* 8(6): 921–928.

Jagannath, A., V. Shetty, and M. B. Saidutta. 2017. "Bioleaching of copper from electronic waste using *Acinetobacter* sp. Cr B2 in a pulsed plate column operated in batch and sequential batch mode." *Journal of Environmental Chemical Engineering* 5(2): 1599–1607. doi: 10.1016/j.jece.2017.02.023

Jain, A. 2010. *E-waste Management in India: Current Status, Emerging Drivers and Challenges.* Regional Workshop on E-waste/WEEE Management, Osaka, Japan, 6–9.

Javed, M., and N. Usmani. 2011. "Accumulation of heavy metals in fishes: a human health concern." *International Journal of Environmental Sciences* 2(2): 659–670.

Joseph, K. 2007. "Electronic waste management in India—issues and strategies." *In Eleventh International Waste Management and Landfill Symposium, Sardinia.*

Kasper, A. C., and H. M. Veit. 2018. "Gold recovery from printed circuit boards of mobile phones scraps using a leaching solution alternative to cyanide." *Brazilian Journal of Chemical Engineering* 35: 931–942. doi: 10.1590/0104-6632.20180353s20170291

Khaliq, A., M. A. Rhamdhani, G. Brooks, and S. Masood. 2014. "Metal extraction processes for electronic waste and existing industrial routes: a review and Australian perspective." *Resources* 3: 152–179. doi: 10.3390/resources3010152

Khan, M. U., A. Besis, and R. N. Malik. 2019. "Environmental and health effects: Exposure to e-waste pollution." In: M. Hashmi and A. Varma (eds) *Electronic Waste Pollution, Soil Biology*, vol 57. Springer, Cham. 111–137. https://doi.org/10.1007/978-3-030-26615-8_8

Kibra, M. G. 2008. "Effects of mercury on some growth parameters of rice (*Oryza sativa* L.)." *Soil & Environment* 27(1): 23–28.

Kumar, A., M. Holuszko, and D. C. R. Espinosa. 2017. "E-waste: an overview on generation, collection, legislation and recycling practices." *Resources, Conservation and Recycling* 122: 32–42. doi: 10.1016/j.resconrec.2017.01.018

Kumar, A., H. S. Saini, and S. Kumar. 2018. "Bioleaching of gold and silver from waste printed circuit boards by *Pseudomonas balearica* SAE1 isolated from an e-waste recycling facility." *Current Microbiology* 75(2): 194–201. doi: 10.1007/s00284-017-1365-0

Lee, H., E. Molstad, and B. Mishra. 2018. "Recovery of gold and silver from secondary sources of electronic waste processing by thiourea leaching." *JOM* 70(8): 1616–1621. doi: 10.1007/S11837-018-2965-2

Leung, A., Z. W. Cai, and M. H. Wong. 2006. "Environmental contamination from electronic waste recycling at Guiyu, southeast China." *Journal of Material Cycles and Waste Management* 8(1): 21–33. doi: 10.1007/s10163-005-0141-6

Li, M., X. Huo, Y. Pan, H. Cai, Y. Dai, and X. Xu. 2018. "Proteomic evaluation of human umbilical cord tissue exposed to polybrominated diphenyl ethers in an e-waste recycling area." *Environment International* 111: 362–371. doi: 10.1016/j.envint.2017.09.016

Lin, S., X. Huo, Q. Zhang, X. Fan, L. Du, X. Xu, S. Qiu, Y. Zhang, Y. Wang, and J. Gu. 2013. "Short placental telomere was associated with cadmium pollution in an electronic waste recycling town in China." *PloS one* 8(4): 60815. doi: 10.1371/journal.pone.0060815

Lindhqvist, T. 2000. *Extended Producer Responsibility in Cleaner Production: Policy Principle to Promote Environmental Improvements of Product Systems.* Lund University.

Liu, J., X. X. He, X. R. Lin, W. C. Chen, Q. X. Zhou, W. S. Shu, and L. N. Huang. 2015. "Ecological effects of combined pollution associated with e-waste recycling on the composition and diversity of soil microbial communities." *Environmental Science & Technology* 49(11): 6438–6447. doi: 10.1021/es5049804

Liu, L., B. Zhang, K. Lin, Y. Zhang, X. Xu, and X. Huo. 2018. "Thyroid disruption and reduced mental development in children from an informal e-waste recycling area: a mediation analysis." *Chemosphere* 193: 498–505. doi: 10.1016/j.chemosphere.2017.11.059

Lu, S. Y., Y. X. Li, J. Q. Zhang, T. Zhang, G. H. Liu, M. Z. Huang, X. Li, J. J. Ruan, K. Kannan, and R. L. Qiu. 2016. "Associations between polycyclic aromatic hydrocarbon (PAH) exposure and oxidative stress in people living near e-waste recycling facilities in China." *Environment International* 94: 161–169. doi: 10.1016/j.envint.2016.05.021

Lu, X., X. Xu, Y. Zhang, Y. Zhang, C. Wang, and X. Huo. 2018. "Elevated inflammatory Lp-PLA2 and IL-6 link e-waste Pb toxicity to cardiovascular risk factors in preschool children." *Environmental Pollution* 234: 601–609. doi: 10.1016/j.envpol.2017.11.094

Luo, Q., Z. W. Cai, and M. H. Wong. 2007a. "Polybrominated diphenyl ethers in fish and sediment from river polluted by electronic waste." *Science of the Total Environment* 383(1–3): 115–127. doi: 10.1016/j.scitotenv.2007.05.009

Luo, Q., M. Wong, and Z. Cai. 2007b. "Determination of polybrominated diphenyl ethers in freshwater fishes from a river polluted by e-wastes." *Talanta* 72(5): 1644–1649. doi: 10.1016/j.talanta.2007.03.012

Luo, X. J., X. L. Zhang, J. Liu, J. P. Wu, Y. Luo, S. J. Chen, B. X. Mai, and Z. Y. Yang. 2009. "Persistent halogenated compounds in waterbirds from an e-waste recycling region in South China." *Environmental Science & Technology* 43(2): 306–311. doi: 10.1021/es8018644

Mallick, N., and F. H. Mohn, 2003. "Use of chlorophyll fluorescence in metal-stress research: a case study with the green microalga Scenedesmus." *Ecotoxicology and Environmental Safety* 55(1): 64–69. doi: 10.1016/s0147-6513(02)00122-7

Marappa, N., L. Ramachandran, D. Dharumadurai, and T. Nooruddin. 2020. "Recovery of gold and other precious metal resources from environmental polluted e-waste printed circuit board by bioleaching frankia." *International Journal of Environmental Research* 14(5): 165–176. doi: 10.1007/s41742-020-00254-5

Meena, R., S. P. Datta, D. Golui, B. S. Dwivedi, and M. C. Meena. 2016. "Long-term impact of sewage irrigation on soil properties and assessing risk in relation to transfer of metals to human food chain." *Environmental Science and Pollution Research* 23(14): 14269–14283. doi: 10.1007/s11356-016-6556-x

Miliute-Plepiene, J., and L. Youhanan. 2019. *E-Waste and Raw Materials: From Environmental Issues to Business Models.* IVL Swedish Environmental Research Institute.

Mishra, J., R. Singh, and N. K. Arora, 2017. "Alleviation of heavy metal stress in plants and remediation of soil by rhizosphere microorganisms." *Frontiers in Microbiology* 8: 1706. doi: 10.3389/fmicb.2017.01706

Mohod, C. V., and J. Dhote. 2013. "Review of heavy metals in drinking water and their effect on human health." *International Journal of Innovative Research in Science, Engineering and Technology* 2(7): 2992–2996.

Montalvo, C., D. Peck, and E. Rietveld. 2016. *A Longer Lifetime for Products: Benefits for Consumers and Companies.* European Parliament, Directorate General for Internal Policies, Brussels.

Moorthy, B., C. Chu, and D. J. Carlin. 2015. "Polycyclic aromatic hydrocarbons: from metabolism to lung cancer." *Toxicological Sciences* 145: 5–15.

Namias, J. 2013. *The Future of Electronic Waste Recycling in the United States: Obstacles and Domestic Solutions.* Earth Resources Engineering Department of Earth and Environmental Engineering Columbia University.

Narayanasamy, M., D. Dhanasekaran, and N. Thajuddin. 2021. "Bioremediation of noxious metals from e-waste printed circuit boards by Frankia." *Microbiological Research* 245: 126707. doi: 10.1016/j.micres.2021.126707

Nicholls, A. M., and T. K. Mal. 2003. "Effects of lead and copper exposure on growth of an invasive weed, *Lythrum salicaria* L.(Purple Loosestrife)." *Ohio Journal of Science* 103(5): 129–133.

Nie, X., C. Fan, Z. Wang, T. Su, X. Liu, and T. An. 2015. "Toxic assessment of the leachates of paddy soils and river sediments from e-waste dismantling sites to microalga, *Pseudokirchneriella subcapitata.*" *Ecotoxicology and Environmental Safety* 111 : 168–176. doi: 10.1016/j.ecoenv.2014.10.012

Nnorom, I. C., and O. Osibanjo. 2008. "Overview of electronic waste (e-waste) management practices and legislations, and their poor applications in the developing countries." *Resources, Conservation and Recycling* 52(6): 843–858. doi: 10.1016/j.resconrec.2008.01.004

Ongondo, F. O., I. D. Williams, and T. J. Cherrett. 2011. "How are WEEE doing? A global review of the management of electrical and electronic wastes." *Waste Management* 31(4): 714–730. doi: 10.1016/j.wasman.2010.10.023

Onwosi, C. O., V. C. Igbokwe, T. N. Nwagu, J. N. Odimba, and C. O. Nwuche. 2020. "E-waste management from macroscopic to microscopic scale." In: A. Khan, Inamuddin, and A. Asiri (eds) *E-waste Recycling and Management. Environmental Chemistry for a Sustainable World*, vol. 33. Springer, Cham. 143–157. doi: 10.1007/978-3-030-14184-4_8

Osibanjo, O., and I. C. Nnorom. 2007. "The challenge of electronic waste (e-waste) management in developing countries." *Waste Management & Research* 25(6): 489–501. doi: 10.1177/0734242X07082028

Panda, R., P. R. Jadhao, K. K. Pant, S. N. Naik, and T. Bhaskar. 2020. "Eco-friendly recovery of metals from waste mobile printed circuit boards using low temperature roasting." *Journal of Hazardous Materials* 395: 122642. doi: 10.1016/j.jhazmat.2020.122642

Pant, D., and V. Dhiman. 2020. "An overview on environmental pollution caused by heavy metals released from e-waste and their bioleaching." *Advances in Environmental Pollution Management: Wastewater Impacts and Treatment Technologies* 1: 41–53. doi: 10.26832/aesa-2020-aepm-04

Parvez, S. M., F. Jahan, M. N. Brune, J. F. Gorman, M. J. Rahman, D. Carpenter, Z. Islam, M. Rahman, N. Aich, L. D. Knibbs, and P. D. Sly. 2021. "Health consequences of exposure to e-waste: an updated systematic review." *Lancet Planetary Health* 5(12): e905–e920. doi: 10.1016/S2542-5196(21)00263-1

Pathak, P., R. R. Srivastava, and Ojasvi. 2017. "Assessment of legislation and practices for the sustainable management of waste electrical and electronic equipment in India." *Renewable and Sustainable Energy Reviews* 78: 220–232. doi: 10.1016/j.rser.2017.04.062

Priya, A., and S. Hait. 2018a. "Extraction of metals from high grade waste printed circuit board by conventional and hybrid bioleaching using *Acidithiobacillus ferrooxidans.*" *Hydrometallurgy* 177: 132–139. doi: 10.1016/j.hydromet.2018.03.005

Priya, A., and S. Hait. 2018b. "Feasibility of bioleaching of selected metals from electronic waste by *Acidiphilium acidophilum.*" *Waste and Biomass Valorization* 9(5): 871–877. doi: 10.1007/s12649-017-9833-0

Rani, R., A. Kela, G. Dhaniya, K. Arya, A. K. Tripathi, and R. Ahirwar. 2021. "Circulating microRNAs as biomarkers of environmental exposure to polycyclic aromatic hydrocarbons: potential and prospects." *Environ Sci Pollut Res* 28: 54282–54298.

Rao, M. D., K. K. Singh, C. A. Morrison, and J. B. Love. 2021. "Recycling copper and gold from e-waste by a two-stage leaching and solvent extraction process." *Separation and Purification Technology* 263: 118400. doi: 10.1016/j.seppur.2021.118400

Rules. 2016. *E-Waste (Management) Rules 2016.* Ministry of Environment, Forest and Climate Change, the Gazette of India: Extraordinary [PART II—SEC. 3(i)].

Sabatini, S. E., Á. B. Juárez, M. R. Eppis, L. Bianchi, C. M. Luquet, and M. D. C. R. de Molina. 2009. "Oxidative stress and antioxidant defenses in two green microalgae exposed to copper." *Ecotoxicology and Environmental Safety* 72(4): 1200–1206. doi: 10.1016/j.ecoenv.2009.01.003

Salam, M., and A. Varma. 2019. "Bacterial community structure in soils contaminated with electronic waste pollutants from Delhi NCR, India." *Electronic Journal of Biotechnology* 41: 72–80.

SEPA. 2011. *Recycling and Disposal of Electronic Waste: Health Hazards and Environmental Impacts.* Stockholm, Elsevier Amsterdam.

Sharma, S., S. Wakode, A. Sharma, N. Nair, M. Dhobi, M. A. Wani, and F. H. Pottoo. 2020. "Effect of environmental toxicants on neuronal functions." *Environmental Science and Pollution Research* 27(36): 44906–44921. doi: 10.1007/s11356-020-10950-6

Sheel, A., and D. Pant. 2018. "Recovery of gold from electronic waste using chemical assisted microbial biosorption (hybrid) technique." *Bioresource Technology* 247: 1189–1192. doi: 10.1016/j.biortech.2017.08.212

Steinhausen S. L., N. Agyeman, P. Turrero, A. Ardura, E. Garcia-Vazquez. 2021. "Heavy metals in fish nearby electronic waste may threaten consumer's health. Examples from Accra, Ghana." *Marine Pollution Bulletin* 175: 113162. ISSN 0025-326X.

Sthiannopkao, S., and M. H. Wong. 2013. "Handling e-waste in developed and developing countries: Initiatives, practices, and consequences." *Science of the Total Environment* 463: 1147–1153. doi: 10.1016/j.scitotenv.2012.06.088

Taiz, L., and E. Zeiger. 2002. *Plant Physiology. Sinauer Associates Inc Publishers, Sunderland,* 690.

Tang, X., J. Qiao, C. Chen, L. Chen, C. Yu, C. Shen, and Y. Chen. 2013. "Bacterial communities of polychlorinated biphenyls polluted soil around an e-waste recycling workshop." *Soil and Sediment Contamination: An International Journal* 22(5): 562–573. doi: 10.1080/15320383.2013.750269

Toxic Dispatch. 2004. "Environmentalists denounce toxic waste dumping in Asia." *A Newsletter from Toxic Links* 23: 1–2.

Tsydenova, O. and M. Bengtsson. 2011. "Chemical hazards associated with treatment of waste electrical and electronic equipment." *Waste Management* 31(1): 45–58. doi: 10.1016/j.wasman.2010.08.014

Turaga, R. M. R., K. Bhaskar, S. Sinha, D. Hinchliffe, M. Hemkhaus, R. Arora, S. Chatterjee, D. S. Khetriwal, V. Radulovic, P. Singhal, and H. Sharma. 2019. "E-waste management in India: issues and strategies." *Vikalpa* 44(3): 127–162. doi: 10.1177/0256090919880655

UNEP (United Nations Environment Programme). 2009. *Recycling—From E-waste to Resources.* UNEP Division of Technology, Industry and Economics, Sustainable Consumption and Production Branch.

Van Rossem, C., N. Tojo, and T. Lindhqvist. 2006. *Extended Producer Responsibility: An Examination of Its Impact on Innovation and Greening Products.* Greenpeace International www.greenpeace.org/international/press/reports/epr

Veit, H. M., N. C. D. F. Juchneski, and J. Scherer. 2014. "Use of gravity separation in metals concentration from printed circuit board scraps." *Rem: Revista Escola de Minas* 67: 73–79. doi: 10.1590/S0370-44672014000100011

Wang, Y., Y. Ru, A. Veenstra, R. Wang, and Y. Wang. 2010. "Recent developments in waste electrical and electronics equipment legislation in China." *The International Journal of Advanced Manufacturing Technology* 47(5): 437–448. doi: 10.1007/s00170-009-2339-6

Wei, X., D. Liu, W. Huang, W. Huang, and Z. Lei. 2020. "Simultaneously enhanced Cu bioleaching from E-wastes and recovered Cu ions by direct current electric field in a bioelectrical reactor." *Bioresource Technology* 298: 122566. doi: 10.1016/j.biortech.2019.122566

Widmer, R., H. Oswald-Krapf, D. Sinha-Khetriwal, M. Schnellmann, and H. Böni. 2005. "Global perspectives on e-waste." *Environmental Impact Assessment Review* 25(5): 436–458. doi: 10.1016/j.eiar.2005.04.001

Wu, J. P., X. J. Luo, Y. Zhang, Y. Luo, S. J. Chen, B. X. Mai, and Z. Y. Yang. 2008. "Bioaccumulation of polybrominated diphenyl ethers (PBDEs) and polychlorinated biphenyls (PCBs) in wild aquatic species from an electronic waste (e-waste) recycling site in South China." *Environment International* 34(8): 1109–1113. doi: 10.1016/j.envint.2008.04.001

Wu, Y., Q. Fang, X. Yi, G. Liu, and R. W. Li. 2017. "Recovery of gold from hydrometallurgical leaching solution of electronic waste via spontaneous reduction by polyaniline." *Progress in Natural Science: Materials International* 27(4): 514–519. doi: 10.1016/j.pnsc.2017.06.009

Xu, L., J. Ge, X. Huo, Y. Zhang, A. T. Lau, and X. Xu. 2016. "Differential proteomic expression of human placenta and fetal development following e-waste lead and cadmium exposure in utero." *Science of the Total Environment* 550: 1163–1170. doi: 10.1016/j.scitotenv.2015.11.084

Xu, L., X. Huo, Y. Liu, Y. Zhang, Q. Qin, and X. Xu. 2020. "Hearing loss risk and DNA methylation signatures in preschool children following lead and cadmium exposure from an electronic waste recycling area." *Chemosphere* 246: 125829. doi: 10.1016/j.chemosphere.2020.125829

Xu, L., X. Huo, Y. Zhang, W. Li, J. Zhang, and X. Xu. 2015a. "Polybrominated diphenyl ethers in human placenta associated with neonatal physiological development at a typical e-waste recycling area in China." *Environmental Pollution* 196: 414–422. doi: 10.1016/j.envpol.2014.11.002

Xu, X., J. Liu, C. Huang, F. Lu, Y. M. Chiung, and X. Huo. 2015b. "Association of polycyclic aromatic hydrocarbons (PAHs) and lead co-exposure with child physical growth and development in an e-waste recycling town." *Chemosphere* 139: 295–302. doi: 10.1016/j.chemosphere.2015.05.080

Xu, X., X. Chen, J. Zhang, P. Guo, T. Fu, Y. Dai, S. L. Lin, and X. Huo. 2015c. "Decreased blood hepatitis B surface antibody levels linked to e-waste lead exposure in preschool children." *Journal of Hazardous Materials* 298: 122–128. doi: 10.1016/j.jhazmat.2015.05.020

Xu, X., W. Liao, Y. Lin, Y. Dai, Z. Shi, and X. Huo. 2018. "Blood concentrations of lead, cadmium, mercury and their association with biomarkers of DNA oxidative damage in preschool children living in an e-waste recycling area." *Environmental Geochemistry and Health* 40(4): 1481–1494. doi: 10.1007/s10653-017-9997-3

Youcal, Z. 2018. *Pollution Control Technology for Leachate from Municipal Solid Waste: Landfills, Incineration Plants, and Transfer Stations*. Butterworth-Heinemann, United Kingdom. ISBN: 9780128158135.

Zeng, X., X. Xu, Q. Qin, K. Ye, W. Wu, and X. Huo. 2019. "Heavy metal exposure has adverse effects on the growth and development of preschool children." *Environmental Geochemistry and Health* 41(1): 309–321. doi: 10.1007/s10653-018-0114-z

Zeng, X., X. Xu, X. Zheng, T. Reponen, A. Chen, and X. Huo. 2016. "Heavy metals in PM2. 5 and in blood, and children's respiratory symptoms and asthma from an e-waste recycling area." *Environmental Pollution* 210: 346–353. doi: 10.1016/j.envpol.2016.01.025

Zhang, K., L. Qiu, J. Tao, X. Zhong, Z. Lin, R. Wang, and Z. Liu. 2021. "Recovery of gallium from leach solutions of zinc refinery residues by stepwise solvent extraction with N235 and Cyanex 272." *Hydrometallurgy* 205: 105722. doi: 10.1016/j.hydromet.2021.105722

Zhang, Y., X. Xu, A. Chen, C. B. Davuljigari, X. Zheng, S. S. Kim, K. N. Dietrich, S. M. Ho, T. Reponen, and X. Huo. 2018. "Maternal urinary cadmium levels during pregnancy associated with risk of sex-dependent birth outcomes from an e-waste pollution site in China." *Reproductive Toxicology* 75: 49–55. doi: 10.1016/j.reprotox.2017.11.003

Zhang, Y., X. Xu, D. Sun, J. Cao, Y. Zhang, and X. Huo. 2017. "Alteration of the number and percentage of innate immune cells in preschool children from an e-waste recycling area." *Ecotoxicology and Environmental Safety* 145: 615–622. doi: 10.1016/j.ecoenv.2017.07.059

Zhao, G., Z. Wang, M. H. Dong, K. Rao, J. Luo, D. Wang, J. Zha, S. Huang, Y. Xu, and M. Ma. 2008. "PBBs, PBDEs, and PCBs levels in hair of residents around e-waste disassembly sites in Zhejiang Province, China, and their potential sources." *Science of the Total Environment* 397(1–3): 46–57. doi: 10.1016/j.scitotenv.2008.03.010

Zheng, L., K. Wu, Y. Li, Z. Qi, D. Han, B. Zhang, C. Gu, G. Chen, J. Liu, S. Chen, and X. Xu. 2008. "Blood lead and cadmium levels and relevant factors among children from an e-waste recycling town in China." *Environmental Research* 108(1): 15–20. doi: 10.1016/j.envres.2008.04.002

Zheng, X., X. Huo, Y. Zhang, Q. Wang, Y. Zhang, and X. Xu. 2019. "Cardiovascular endothelial inflammation by chronic coexposure to lead (Pb) and polycyclic aromatic hydrocarbons from preschool children in an e-waste recycling area." *Environmental Pollution* 246: 587–596. doi: 10.1016/j.envpol.2018.12.055

Zhou, L., and Z. Xu. 2012. "Response to waste electrical and electronic equipments in China: legislation, recycling system, and advanced integrated process." *Environmental Science & Technology* 46(9): 4713–4724. doi: 10.1021/es203771m

Zou, L., Y. Lu, Y. Dai, M. I. Khan, W. Gustave, J. Nie, Y. Liao, X. Tang, J. Shi, and J. Xu. 2021. "Spatial variation in microbial community in response to as and Pb contamination in paddy soils near a Pb-Zn mining site." *Frontiers in Environmental Science* 9: 138. doi: 10.3389/fenvs.2021.630668

4 Management of Marine Plastic Debris and Microplastics

G. G. N. Thushari[1,2], J. D. M. Senevirathna[1,2], Naveen M. Wijesena[3], and H. K. S. De Zoysa[4,5]*
[1]Department of Animal Science, Faculty of Animal Science and Export Agriculture, Uva Wellassa University, Badulla, Sri Lanka
[2]Department of Aquatic Bioscience, Faculty of Agricultural and Life Sciences, The University of Tokyo, Tokyo, Japan
[3]Department of Biology, University of Bergen, Bergen, Norway
[4]Department of Bioprocess Technology, Faculty of Technology, Rajarata University Sri Lanka, Mihintale, Sri Lanka
[5]Department of Biology, University of Naples Federico II, Naples, Italy
*Corresponding author: dezoysahks@yahoo.com, ksdezoys@tec.rjt.ac.lk

CONTENTS

DOI: 10.1201/9781003132349-5

79

4.1   INTRODUCTION

The global ocean is suffering from plastic pollution due to a constant and unprecedented accumulation of plastic, which is emerging as one of the environmental problems. This is mainly due to the adoption of plastics as a substitute for traditional materials since the 1950s (Gallo et al. 2018; Kandziora et al. 2019; Welden 2019) and the problem emerged in the marine environment in the 1970s (Xanthos and Walker 2017). Among marine debris, plastic debris is reported as the most abundant with up to 80% and more persisting in the ocean with its long-life durability (Ockelford, Cundy, and Ebdon 2020). This is comprised of diverse materials with sizes ranging from several nanometers to several meters. These materials are introduced into the ocean by a variety of processes and are subjected to fragmentation (Gallo et al. 2018). This fragmentation process leads to large plastic items breaking into small pieces including microplastics and nanoplastics (Napper and Thompson 2019). However, this has become a widespread emerging issue as the end point for two-thirds of the global plastic production is at the seabed. Half of the remaining plastics are floating on or under the surface, and the rest washes into the beaches and sink with sediments in both shallow and deep sea and in arctic sea ice (Napper and Thompson 2019; Welden 2019). Plastic debris has become one of the most studied areas in the world due to its severe ecological, economic, and social impact (Niaounakis 2017). As marine plastic debris and microplastics have already become a global issue, more international collaboration and effective solutions are needed to overcome the problem, especially in developing regions of the world. Therefore, this chapter focuses on the problem of plastics in the ocean using new strategies, legislations, and technological methods with potential reduction measures to mitigate this emerging problem in developing countries that contribute most to the accumulation of marine plastics debris in the ocean.

4.2 TYPES OF PLASTICS

Plastics are considered as synthetic or semisynthetic, highly durable, lightweight, inexpensive, corrosion-resistant, and strong organic polymers. Most of the plastics are composed of derivatives of hydrocarbons and a wide variety of other additives (Hahladakis, Iacovidou, and Gerassimidou 2020; Napper and Thompson 2019). Based on previous studies, plastics in the ocean can be categorized as shown in Table 4.1 (Carney Almroth and Eggert 2019; Zalasiewicz, Gabbott and Waters, 2019).

TABLE 4.1
Main Types of Plastics and Their Densities Found Among Marine Debris
(Seawater Density = 1,027 kg m^{-3})

Types of Plastic	Polymer Type		Density (kg m^{-3})	Debris Items	Reference
Non-fiber plastics	PE	High density (HDPE) Low density (LDPE) Linear low density (LLDPE)	890–980	Plastic bags, storage containers, fishing floats, preproduction pellets, bottles for milk and dishwashing liquids, buckets, and craters	(Birch et al. 2020; Mountford and Morales Maqueda 2019; Eriksen, Thiel, and Lebreton 2016; Driedger et al. 2015; Kershaw 2016)
	PP		900–1050	Rope, bottle caps, strapping, fishing gear, detergent bottles, combs, toothbrushes, preproduction pellets	
	PVC		1150–1410	Pipe, film, containers, soft vinyl toys, shampoo bottles	
	PS	*Expanded PS (EPS)*	1010–1050	Cool boxes, floats, cups	
		Extruded PS (XPS)	1040–1090	Utensils, bottles, containers	
	Polycarbonate (PC)		1200–1220	Optical disks	
	PET		1300–1410	Bottles, strapping, textiles	
	PUR		871–1420	Durable foams (insulation materials)	(Mountford and Morales Maqueda 2019)
Fiber plastics	Polyester (PES) + Glass fiber		>1350	Boats, textiles	(Kershaw 2016; Birch et al. 2020; Driedger et al. 2015)
	Polyamide (PA)		1130–1160	Rope, fishing nets, textiles, toothbrush bristles	
	Acrylic (PMMA)		1160–1200	Optical lenses, paint, shatterproof windows	
	Cellulose Acetate (CA)		1220–1300	Cigarette filters	(Kershaw 2016; Birch et al. 2020; Eriksen, Thiel, and Lebreton 2016; Driedger et al. 2015)
	Polytetrafluoroethylene (PTEF)		2100–2300	Wires, cables, bearings, gears	

Source: Kershaw 2016.

TABLE 4.2
Marine Plastic Categorization Based on Size

Type		Size (mm)	Reference
Nanoplastics		< 0.0001 (0.1 µm)	(Eriksen et al. 2014; Napper
Microplastics	Small	0.33–1.00	and Thompson 2019; Welden
	Large	1.01–4.75	2019; Birch et al. 2020)
Mesoplastics		4.76–200	
Macroplastics		>200	

Plastics debris in the ocean can be defined in various ways using size, origin, shape, polymer type, color, and original usage. There are four main categories of marine plastic debris according to their size (Table 4.2). However, macroplastics and mesoplastics can easily be distinguished due to their visibility compared to microplastics and nanoplastics. However, nanoplastics and microplastics ranging from 1 to 100 nm (0.001–0.1 µm) and 0.1 to 5,000 µm, respectively, are difficult to distinguish from each other in aquatic environments.

These plastics can be further categorized into primary and secondary microplastics based on their origin. Primary microplastics are fibers, capsules, pellets, and microbeads that directly enter the environment whereas secondary microplastics are formed through the fragmentation of larger plastics (macroplastics and mesoplastics) and further fragmentation of microplastic debris into the nanoplastics (Welden 2019; Birch et al. 2020). However, this is facilitated by several factors such as sunlight (UV radiation) and temperature, the density of plastics, physical and mechanical action, hydrolysis, weathering, biodegradation, photooxidation, erosion, and embrittlement (Welden 2019; Van Sebille et al. 2020).

4.3   SOURCES AND PATHWAYS OF MARINE PLASTICS

The production of global plastics is increasing every day. It represents the exponential trend that started in the 1950s and is projected to reach over 2.5 million tons in 2050 (Lebreton, Egger, and Slat 2019). In addition, ocean-based sources such as commercial fishery activities, aquaculture practices, coastal tourism, private ships, and commercial cargo or cruise ships are also responsible for this emerging problem, in a smaller proportion compared to land-based inputs (Gallo et al. 2018; Ajith et al. 2020; Goodman et al. 2020). Around 98% of the total amount of microplastics in the ocean are derived from land-based sources (Table 4.3), whereas 2% is from ocean-based sources (Boucher and Friot 2017).

Most of the plastics from land-based sources and other marine debris are mainly brought into the ocean by the rivers. This can be seen in most of the developing countries in the world. Those countries contribute to main plastic inputs sent to the ocean by rivers, though most of the plastics producers are from developed countries (Table 4.4) (Lebreton et al. 2017; Purba et al. 2019; Adam et al. 2020). Moreover, among the rivers in the world, the topmost polluted 122 rivers are in Asia and Africa.

TABLE 4.3
Land-Based Sources of Marine Plastic Litter and Pathways

Source	Loss (%)	Way of Losses	Origin of Release	Associated Pathway
Synthetic textiles	35	Abrasion during laundry	land-based losses and releases	Wastewater pathway
Tires	28	Abrasion while driving		Road runoff pathway, wind pathway
City dust	24	Spills, weathering & abrasion		
Road markings	7	Weathering and abrasion by vehicles		
Marine coatings	3.7	Weathering, application, and maintenance	Sea-based losses and releases	Ocean pathway
Personal care products	2	Loss during use	Land-based losses and releases	Wastewater pathway
Plastic pellets	0.3	Manufacturing, transport, and recycling	Land-based losses and releases, sea-based losses, and releases	Road runoff pathway, wastewater pathway, ocean pathway

Source: Boucher and Friot 2017.

TABLE 4.4
The Global Release of Plastics into the Ocean

Region(s)	Plastic Waste (Metric Tons/Year)	Primary Microplastics (Metric Tons/Year)
South Asia and Southeast Asia	5.50	0.75
Africa and the Middle East	1.53	0.13
South America	0.58	0.14
Europe and Central Asia	0.28	0.24
North America	0.07	0.26

Source: Boucher and Friot 2017.

Furthermore, research shows that most of these inputs of plastics are seasonal with regional variations (Akenji et al. 2020; Lebreton et al. 2017).

The main reasons behind the large contribution by developing countries with regards to marine plastic pollution are rapid industrialization, changes in patterns of consumption and production, improper waste management plans, and expansion of the middle-income population (Akenji et al. 2020; Alpizar et al. 2020). Therefore, there is a pressing need for monitoring and mitigating marine debris as rapid economic growth occurs in developing countries.

Floating debris, including marine plastics, move with ocean currents (Gallo et al. 2018) and coupled with the durability of plastics, this contributes to the wide dispersion of plastics in the ocean worldwide (Napper and Thompson 2019). Most of the suspended plastics can be transported to remote areas from the point of release and could accumulate at ocean gyres (Welden and Lusher 2017). Furthermore, distribution and transferring of plastics in the open ocean and ocean floor are determined by several notable factors such as human activities, material characters, ocean dynamics associated with wind, hydrodynamics at the open and semi-closed areas, ocean topography, shoreline features, and biochemical interactions. Marine plastics are distributed in the ocean predominantly by horizontal transportation, especially in the case of floating plastics. In addition, the distribution of plastics is driven by the mass movement of water and the movement of conveyers and gyres. Plastics are packed into convergence zones of the gyres and are known as garbage patches (Welden 2019) (Figure 4.1).

A study conducted in 2014 reported that an estimated 269,000 tons of plastic debris is floating in all the gyres in the open ocean worldwide (Xanthos and Walker 2017). There are four main ways in which microplastics and nanoplastics are distributed vertically in the oceans: (a) sinking of marine snow aggregates, (b) fecal pellet incorporated sinking,(c) plastic pump by plankton, giant larvaceans, and mesopelagic fishes, and (d) aggregate sinking with suspended inorganic particles. In addition,

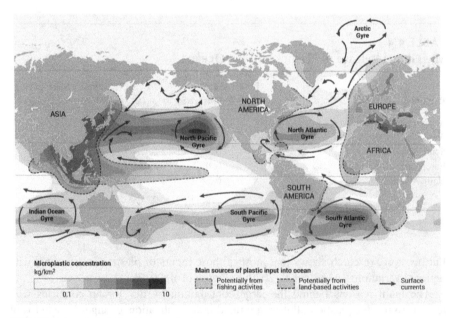

FIGURE 4.1   Global distribution of microplastics in gyres and open ocean with their potential origin.

Source: wwwgrida.no/resources/13339 (Creator credit: Riccardo Pravettoni and Philippe Rekacewicz).

other processes such as biofouling and hyperpycnal flows generated by flash floods or deep-water cascading events are also responsible (Van Sebille et al. 2020). Further, those sunk into the ocean bottom could be remobilized due to sedimentary gravity processes, and catastrophic events like earthquakes, floods, hurricanes, tsunamis, and storms (Maximenko et al. 2019; Van Sebille et al. 2020).

4.4 IMPACTS AND FATES OF MICROPLASTICS

Microplastics in marine ecosystems can be found both floating on the surfaces and accumulating in sediments (Auta et al. 2017). While most microplastics that are lighter than seawater float on the sea surface upon entering marine ecosystems (Ryan et al. 2009), some of them can sink to the bottom and get accumulated in sediments due to fouling by marine organisms and through binding to denser particles (Lobelle and Cunliffe 2011). A recent study has shown that the level of accumulated microplastics in the form of fibers in deep-sea sediments was four orders of magnitude higher than that of surface waters (Woodall et al. 2014), suggesting that deep-sea sediments may act as sinks for some types of microplastics. The microplastic concentration of sediments is considerably higher in affected marine zones of Asia, Africa, America, and Europe. Mangrove ecosystems, which are considered as enhancers of sedimentation (Valiela and Cole 2002), have also become heavily impacted by microplastic pollution. Studies carried out in mangrove ecosystems in different regions of the world have shown that these habitats were acting as sinks to a variety of microplastics containing polypropylene, polyvinyl chloride, nylon, polyethylene (Mohamed Nor and Obbard 2014), polystyrene foams, and plastic pellets (Jayanthi et al. 2014).

4.4.1 ECOLOGICAL AND ENVIRONMENTAL IMPACT OF MICROPLASTICS

Increasing levels of microplastics in marine environments have resulted in an increase in interactions between microplastics and marine organisms. The nature of these interactions depends on the various properties of microplastics such as their density, color, shape, size, and charge (Van Cauwenberghe et al. 2015). Most marine organisms ingest microplastics as food particles and these include invertebrates such as mussels (Avio et al. 2015), oysters, barnacles, sea cucumbers, amphipods, zooplankton (Cole et al. 2013; Goldstein and Goodwin 2013; Rehse, Kloas, and Zarfl 2016; Wijethunga et al. 2019) and vertebrates such as fishes, turtles, birds, and mammals (Ferreira et al. 2016; Batel et al. 2016; Fossi et al. 2016; Caron et al. 2016). When ingested, these particles can cause either physical damage by adhering to the surface of the digestive tract and blocking the passage of food or chemical damage in the forms of inflammation and hepatic stress (Setälä, Norkko, and Lehtiniemi 2016). Microplastics also act as carriers of organic pollutants that are added during their production or absorbed from the environment (Bakir, Rowland, and Thompson 2014). Given their high surface area to volume ratio, microplastics have a great affinity to adsorb water-borne contaminants such as persistent organic pollutants, metals (Cole et al. 2011), and endocrine disrupters (Ng and Obbard 2006). Marine microorganisms metabolize these pollutants as seen in amphipod *Allorchestes compresa,* which has been shown to assimilate polybrominated diphenyl ethers (Chua et al. 2014). In turn, these pollutants get assimilated in the tissues of fish feeding on these microorganisms (Wardrop et al. 2016) and travel up in the food chain. Most of these pollutants cause a wide variety of harmful effects including cancer, endocrine disruption, birth defects, immune system defects, and developmental defects (Setälä, Norkko, and Lehtiniemi 2016), resulting in long-lasting detrimental effects on the environment.

4.4.2 Impact of Microplastics on Marine Organisms

Microplastics ingested by organisms at lower trophic levels such as zooplankton gets concentrated along food chains when these organisms get eaten by predators at higher trophic levels (Hollman et al. 2013). Studies carried out on European flat oyster (*Ostrea edulis*) has shown that exposure to microplastics resulted in elevated levels of stress in oysters. Benthic organisms such as periwinkles (*Littorina sp*), isopods (*Idotea balthica*), and peppery shell clams (*Scrobicularia plana*) showed reduced reproductive output and increased mortality rates (Green 2016). Laboratory experiments have shown that when exposed to polystyrene microbeads, marine copepod *Calanus helgolandicus* showed a significant reduction in the rate of ingesting algal cells, resulting in a 40% reduction of biomass. Long-term exposure to these microbeads resulted in a reduction in fecundity and hatching rates and higher mortality (Cole et al. 2016). Bioaccumulation of microplastics and organic pollutants associated with them in marine food webs act as a severe ecological threat to many marine environments (Auta, Emenike, and Fauziah 2017). This issue has received a lot of attention in fish, given its commercial importance in addition to its ecological importance. Laboratory studies on zebrafish have shown that ingested microplastics are retained within epithelial cells and intestinal villi (Batel et al. 2016). Ingestion and bioaccumulation of microplastics and associated toxic chemicals result in pathological and oxidative stress and inflammation of the liver in *Oryzias latipes* (Lu et al. 2016).

4.4.3 Socioeconomic Impact of Marine Microplastics

Any disruptions to ecosystem services provided by marine ecosystems can result in a significant negative impact on human life due to loss of food security, livelihoods, and health. The negative impacts of microplastics on marine organisms could significantly affect marine ecosystem services. However, the scarcity of data on the holistic effects of marine microplastics on ecosystem services has been a hindrance to understanding the ecological, social, and economic impacts of marine plastic pollution (Beaumont et al. 2019). Marine plastics, including microplastics, can significantly reduce the efficiency and productivity of commercial fisheries, resulting in catastrophic effects on global food production (Beaumont et al. 2019). Even though there is limited data on the effects of microplastics and associated harmful chemical ingestion, the perceived risk of consuming contaminated seafood will have long-term detrimental impacts on marine fisheries (Beaumont et al. 2019). Negative impacts of microplastic pollution in charismatic marine species such as marine mammals and sea birds could have indirect effects on the well-being of humans (Beaumont et al. 2019). Marine microplastic pollution also directly affects the recreational industry. Apart from these immediate socioeconomic impacts, microplastic pollution has the potential to shift the ecology of marine environments

(Galloway, Cole, and Lewis 2017). Together with other environmental stressors, plastic pollution can intensify the effects of other pollutants, changing ocean temperatures, ocean acidification, and overexploitation of marine resources (Beaumont et al. 2019).

4.5 MONITORING, GOVERNANCE, AND POLICY DEVELOPMENT FRAMEWORKS AND SELECTION OF MEASURES FOR DEVELOPING COUNTRIES

4.5.1 POLICY FORMULATION

Policy formulation is the basis for plastic pollution management in developing countries. It involves converting identified policy goals into appropriate and concrete strategies for dealing with related policy issues. Further, it involves the efficacy and the sustainability or suitability of proposed activities (UN 2021; Kershaw 2016). Moreover, policymaking plays a significant role in overcoming emerging environmental problems. Accordingly, pathways of plastic pollution into marine ecosystems are to be explored for the identification of goals and policy toolbox for developing countries (Alpizar et al. 2020). The impact pathway analysis process provides an overview of interconnections of different anthropogenic activities and pollution sources by emphasizing the need for planning and implementing a package of policies. In low- and middle-income countries, land-based sources are responsible for more than 75% of marine plastic pollution (Premakumara and Onogawa, 2019). Accordingly, policy goals have been identified for major pollution sources and impact pathways in developing countries (Alpizar et al. 2020; Premakumara and Onogawa, 2019) as illustrated in the Figure 4.2.

For low- and middle-income countries, policy framework for a plastic waste management regime needs to be focused on plastic production, consumption, and waste management, considering six approaches in an integrated or single manner: (i) prevention of disposal of plastic litter, (ii) reduction of load of plastic wastes, (iii) removal, (iv) recovery, recycling, and reusing of plastics, (v) ecological rights and responsibilities, and (vi) an eco-friendly lifestyle and behavioral change (Premakumara and Onogawa, 2019; Chen 2015). For developing countries, combined policy instruments will be more effective than individual policy objectives (Walls 2013). A policy toolbox with incentives act as a powerful instrument for behavioral, monitoring, and informational tools during waste collection, sorting, recycling, and reusing processes (Kirakozian 2016).

4.5.2 MAJOR POLICY INSTRUMENTATIONS

Developing countries can apply diverse policy networks (Figure 4.2) in addressing marine plastic pollution issues at different levels (Abbott and Sumaila 2019) as shown below.

- Banning or setting standard benchmarks, prohibiting the consumption of single-use plastic food service products and the introduction of standard levels for recyclable materials and/or polymer composition for the plastic products (Ocean Conservancy 2018).

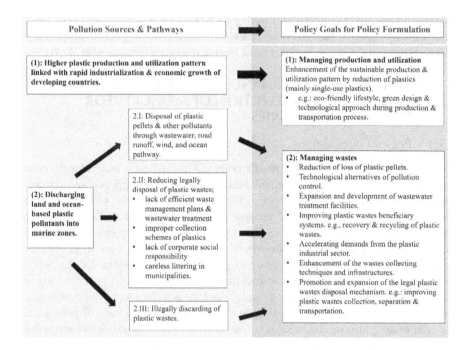

FIGURE 4.2 Major plastic pollution sources, pathways, and appropriate policy goals for developing countries.

- Economic instrumentation acts as a price-based policy with a direct impact on plastic products processing and disposal sector. Fixed penalty fees, highly flexible price-based systems, and a two-part disposal policy are key price-based tools (Fullerton and Wolverton 2000).
- Behavioral intervention acts as one of the most powerful policy instrumentations. Further, eco-friendly behaviors are expected to influence social norms via different modes (Dorothea Kübler 2001; Thaler and Sunstein 2009). Financial incentives and payment for disposable plastic items are common practices of behavioral interventions.
- Rights-based policy instrumentation focused mainly on plastic producers and consumers. A deposit refund scheme and extended producer responsibility (EPR) are major rights-based policy tools.

Policy instrumentation depends on social, psychological, economic, and environmental factors.

4.5.3 Governance and Policy

Governance (a set of actions or processes in governing or managing the control and direction of a country/nation) is an essential component of controlling marine plastic pollution. Sound public governance plays a major role in respective policy formulation, designing, implementation and evaluation. International level governance has a direct intervention on controlling marine plastic pollution globally. At the regional level, governance is proposed with a framework/toolbox connected at multiple levels (Kershaw 2016). Governance related to plastic pollution is always associated with the waste management process. Assemblage of task forces of different agencies in the UN system will act as a powerful framework to conduct sustainable solid waste (including marine plastics) management programs for developing countries (UN 2021). Any government needs to play a leading role as a facilitator of all aspects of plastic pollution control measures with systematic coordination at both horizontal and the vertical dimensions while creating a sound communication network among all the stakeholders as a strategy for effective waste management. Provision of incentives for the development of practical tools, sound decision-making to shift away from plastic pollution-related issues at the environment and society levels, and building a more favorable environment for dialogue among stakeholders are major requirements needed for effective governance (Akenji et al. 2020). The governance system needs to be focused on the industrial and business sectors by promoting public–private partnerships (3Ps). In Asia, governments face challenges, when unorganized small to medium enterprises (SME) engage with waste management practices at a large scale. The adaptive management process referred to as "learning by doing" acts as a novel approach that can be incorporated into governance (Akenji et al. 2020) in developing countries. Key factors and challenges affecting the effective plastic policy and governance (Akenji et al. 2020) are identified as follows:

- lack of proper databases and knowledge
- inadequate integration and coordination among national, regional, and ministerial stakeholders
- insufficient guidance and tools for facilitating activities of stakeholders
- ambiguity and controversy of regulations, mandates, roles, and responsibilities on plastic pollution
- absence of an inclusive framework for policy instruments and tools
- fragmented governance

These gaps and challenges are to be addressed by developing countries to implement management measures effectively in overcoming the plastic pollution issue.

4.5.4 Policy Framework at the Regional and National Level

The policy framework is intended to govern the reviewing, making alterations, developing, establishment of policies by ensuring authority, relevance, and consistency with legislation. Additionally, identification of roles, responsibilities, duties, and authorities with respect to reviewing, amending, developing, and establishment of policies are expected from the policy framework. Existing regional and national level policies applicable for developing countries are summarized in Table 4.5.

With respect to the Asian region, East Asia Summit (EAS) conference takes a regional approach to overcome environmental issues related to marine debris

TABLE 4.5
Major Policies Applicable for Developing Countries

Type of Policy	Policy Interventions	Reference
1. Plastic bag policy	• Price-based (tax system for the selling of specific plastic products by shops/stores) payment scheme for lightweight bags by customers/consumers • Legislative (complete or partial bans of lightweight bags) measures	Xanthos and Walker 2017
2. Microbead policy	• Legal restriction of sale and utilization of microbeads for controlling plastic pollution	Xanthos and Walker 2017
3. Modified EU industrial policy	• Sustainable and green approach in a circular economy: EU plastic strategic action plan	Iverson 2019

accumulation. This conference has proposed to take concrete actions such as expanding research, knowledge and awareness of marine plastic debris, developing cooperation in law enforcement and policy reformation, enhancing sound environmental management of plastic wastes and resources efficiency, strengthening international and regional cooperation to overcome marine plastic pollution, and establishing policies that promote the end-user and the business sector to combat marine plastic pollution. In 2017, the ASEAN conference was conducted to manage marine debris in the Asian region and the members of this respective meeting identified the transboundary nature of the marine plastic pollution issue. Accordingly, they emphasized the urgent need for implementing coordinated and collective actions among member states for the reduction of marine litter accumulation (including marine plastic pollution) regionally (Akenji et al. 2020). The suggested actions in this meeting to combat the marine debris issue including plastic pollution are still applicable to be incorporated into the policy toolbox of developing countries at the national and regional level. In general, developing countries need to consider the technological, regulatory, scientific, voluntary, and economic interventions when developing a policy toolbox for management of plastic wastes at the national level (Akenji et al. 2020; Premakumara & Onogawa, 2019; UNEP, 2018) as shown in Figure 4.3.

In summary, both upstream and downstream waste management approaches are to be considered as regional level initiatives. At the national level, both developing and developed countries have applied policy instrumentation for waste management and control. The existing national-level pollution control mechanism needs to be improved and expanded when considering case studies from most of developing countries. For developing countries, it is recommended to develop environmental governance with pollution control approaches considering biological and ecological settings.

The following best practices are suggested for developing countries to develop a roadmap to control plastic pollution toward a plastic-free environment based on Figures 4.2 and 4.3.

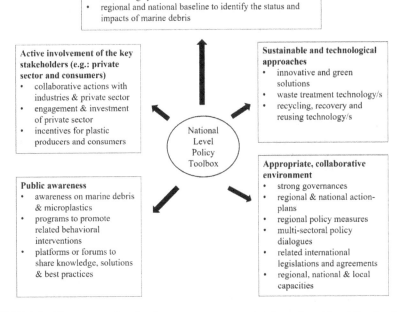

Scientific approach, monitoring and baseline identification
- integration & applicability of scientific knowledge
- monitoring system
- regional and national baseline to identify the status and impacts of marine debris

Active involvement of the key stakeholders (e.g.: private sector and consumers)
- collaborative actions with industries & private sector
- engagement & investment of private sector
- incentives for plastic producers and consumers

Sustainable and technological approaches
- innovative and green solutions
- waste treatment technology/s
- recycling, recovery and reusing technology/s

National Level Policy Toolbox

Public awareness
- awareness on marine debris & microplastics
- programs to promote related behavioral interventions
- platforms or forums to share knowledge, solutions & best practices

Appropriate, collaborative environment
- strong governances
- regional & national action-plans
- regional policy measures
- multi-sectoral policy dialogues
- related international legislations and agreements
- regional, national & local capacities

FIGURE 4.3 Key aspects for plastic waste management at the national level for developing countries.

- Strengthening and amending the existing national policy framework and the formulation of an integrated policy toolbox while promoting policy dialogue at the regional level by gradual fade away from plastic production and consumption
- Strong, efficient, transparent, and properly established legal framework
- Financial, technical, and intellectual capacity building
- Sustainable, green, user-friendly, and cost-effective technologies
- Enhancing related scientific knowledge, research, and innovations by shifting toward biodegradable alternatives
- Promoting engagement of private sector and/or industrial associations with the public
- Implementation of sound marine plastic litter monitoring programs at the national level
- Development of cooperation and participation for regional and international initiatives
- Outreach actions and community awareness (mode of changing attitudes and behaviors)
- Multisectoral collaboration and coordination among different organizations and parties

4.6 MONITORING

Qualitative and quantitative scientific information acts as an indicator data for effective coastal and marine debris management programs. Water, sediments, and biological indicators are commonly used in the environmental monitoring process. Three major nonexhaustive techniques are employed in monitoring plastic litter: numerical modeling (plastic tracking modeling), in-situ and visual monitoring, and satellite/aerial monitoring (Copernicus 2021). Modeling is a novel application used in plastic litter monitoring. The plastic tracking model is also termed as a drift model that considers hydrodynamic factors (Copernicus 2021). An in-situ survey is the monitoring of on-site plastic samples and qualitative and quantitative assessment over different geographical locations at a stipulated time. Vessel (boat/ship) and beach or shoreline surveys are well-known in-situ measurements. The satellite/aerial technique is another novel trend in the monitoring of plastic debris in coastal and marine environments (Corcoran, Biesinger, and Grifi 2009). However, the application of satellite techniques for marine debris assessment is still at a developing stage (Copernicus 2021). Moreover, monitoring of sediments in the benthic environment is another indicator to qualitative and quantitative assessment of a load of plastic budget at the bottom of the ocean. Bio-indicators play a key role in the evaluation of plastic pollution levels in respective marine and coastal ecosystems. The direct effect and fate of plastic pollution can be understood using biological assessments.

Many countries have launched plastic litter monitoring programs at the national level. However, these existing monitoring activities are not effective for building a global database on marine plastic litter. Hence, systematic monitoring mechanisms have to be developed at the regional and global level for creating a holistic database to address this environmental issue in a sustainable manner. For developing countries, following regional level marine litter monitoring measures can be used as effective management directives.

- Guidelines for surveying and monitoring of marine litter by regional sea program of the United Nations Environment Programme (UNEP) with the Intergovernmental Oceanographic Commission (IOC) of UNESCO (UNEP/IOC) (Cheshire et al. 2009).
- Guidelines for Harmonizing Ocean Surface Microplastic Monitoring Methods publicized by the Ministry of the Environment Japan (MOEJ) (Osaka Blue Ocean Vision 2021).
- Standard guidance for monitoring of marine litter in European Seas (European Commission 2013).
- Training program to lay a foundation on monitoring and assessing marine plastic litter under the guidance of Coordinating Body on Seas of East Asia (COBSEA), Global Partnership on Marine Litter (GPML), and Global Program of Action on East Asia (COBSEA 2019).
- Marine debris monitoring and evaluation programs at the regional level Helsinki Commission(HELCOM), Northwest Pacific Action Plan (NOWPAP), Mediterranean Action Plan (MAP) Oslo, and Paris Conventions(OSPAR) under the Regional Seas Conventions and Action Plans.

4.7 LEGAL ASPECTS ON CONTROLLING MARINE PLASTIC POLLUTION

The legal framework is a kind of enforcement strategy for preventing and controlling plastic pollution in the marine environment. In general, law enforcement is based on three major aspects: disposal, consumption, and production of plastics. With respect to the regulation of plastic waste, the legal framework needs to be focused mainly on harmful residuals and illegal dumping of garbage. Mandatory recycling laws act as effective tools in increasing the recycling rate of plastic waste compared to the announced recycling aims (Viscusi, Huber, and Bell 2012). The legal framework of plastic consumption is related to the market regime of the consumer. At the national level, several countries and states have made restrictions to the manufacture, supply, import, trade, and consumption of disposable bags either completely and/or partially. These kinds of legal actions are to be linked with the replacement and promotion of low-cost, green products at the time of implementation. Regulations can be focused on the plastic industry, including the production sector (Alpizar et al. 2020). With the legal aspect, the composition of the raw materials in the product is to be modified and improved by incorporating fewer chemical additives and polymers during the production phase. Most developing countries have not adequately regulated plastic industrial sector through their national legal framework. Hence, national legislations of developing countries are to be focused on the plastic production process (simplification of the composition of plastic products toward a green approach), recycling, solid waste management, and wastewater treatment of the plastic industrial sector.

4.8 MITIGATION AND MANAGEMENT APPROACHES

4.8.1 EXTENDED PRODUCER RESPONSIBILITY (EPR)

EPR is a principle of environmental policy focused on the responsibilities and duties of stakeholders at the post-consumer phase of product life (OECD and Ministry of the Environment 2014). EPR approach is illustrated in the Figure 4.4.

During the EPR process, plastic producers act as a key stakeholder group with the responsibility for properly collecting or taking back, final discarding, and treatment (wastes recycling) of the product after consumption. According to the Organization for Economic Co-operation and Development (OECD), the EPR policy addresses the following two approaches:

(1) provision of incentives and facilities to manufacturers for developing the green design of products by incorporating environmental concerns.
(2) empowering the plastic producers with management, financial, and/or physical responsibilities partially and/or completely without handling by the government or the public sector.

As a rights-based policy tool, the key objective of EPR is to promote plastic recycling and to reduce the usage of virgin materials and waste levels (Brouillat and Oltra 2012). The main feature of EPR is the mandatory take-back option by producer or

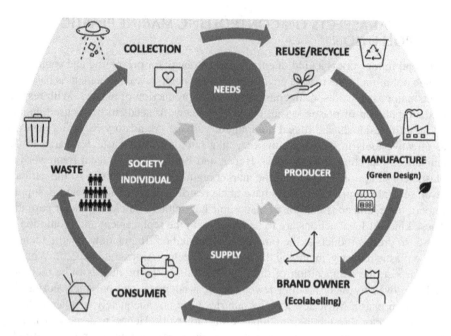

FIGURE 4.4 EPR system as a sustainable waste management measure.

plastic industries, which enhances the recycling rate of plastics by providing required facilities and incentives (Tibbetts 2015). The EPR program is implemented as a collective approach toward the producer responsible organizations, due to high transaction cost by individual responsibility (Leal Filho et al. 2019). Therefore, the EPR acts as a suitable low-cost approach for governments of developing countries. Developing countries and regions also need to focus on sustaining this environmental policy for other plastic products such as carpets, agro-plastics, pharmaceutical packaging, furniture, printer cartridges, and plastics related to the construction sector at a large scale by addressing the issue of marine plastic pollution at the national level (Leal Filho et al. 2019).

4.8.2 PUBLIC–PRIVATE PARTNERSHIP (PPPS, 3PS)

Public–private partnership (PPPs, 3Ps) is the organization between private and public institutions in which duties and responsibilities of the public sector are partially assigned to the private sector in the medium or long term (World Bank 2011) as illustrated in Figure 4.5.

The 3Ps allows to combine the responsibilities of the public sector with the private sector through a decision-making system and governance structure (Ahmed and Ali 2006). Citizens act as the third party of this approach. The 3Ps is considered as one of the common and popular approaches for waste management (including plastics), especially in developing countries (Kershaw 2016). The European Commission

FIGURE 4.5 Public–private partnership (PPPs, 3Ps) for an advanced and powerful solid waste management program.

has identified four types of 3Ps: Build Own Operate (BOT), Design Build Finance Operate (DBFO), and Contracting (European Commission 2003). The approaches of BOT and DBFO promote the engagement with private partners for design, construction, operation, and financial investment for services during a given time. Developing countries can adopt the aforesaid two approaches when national governments face financial difficulties in investment for implementing marine plastic pollution control projects. The contracting approach does not rely on private money (European Commission 2013), and it may not act as an effective model for developing countries having critical monetary challenges. The approach of concession promotes the "polluter pay principle" with the revenue from another party (e.g., local people) of the 3Ps. In most developing countries, the poverty level of the local community may hinder the expected outcome through the concession and is likely to create conflicts among different stakeholders. However, the 3Ps have several advantages for the effective solid waste management approach in developing countries. Major positive aspects of this approach are the accessibility to advanced techniques, potential engagement with the informal and unofficial waste removal activities, easy resource availability, and strong capacity for financial investment (e.g., in Sri Lanka, a private partner "Eco Spindles" makes sustainable initiatives through eco-awareness campaigns and recycling infrastructure facilities and also the John Keells Group has launched a social entrepreneurship project "Plasticcycle" to reduce plastic pollution). Moreover, 3Ps act as a one of the cost-effective approaches in the operational and management regime with a facilities delivery system to achieve the expected outcomes of waste management. Also, the availability of various human resources and experts in the private sector further ensures the flexibility and effectiveness of the services under the 3Ps approach (UNESCAP 2011, World Bank 2011). The 3Ps will be a sustainable approach to introduce and encourage output-focused contracts to control plastic pollution. Powerful 3Ps can make interventions for all stakeholders toward effective waste management regimes. Different limitations and constraints have been identified for the successful implementation of the 3Ps as given below (Ahmed and Ali 2006, UNESCAP 2011).

- inadequate capacity
- emerging risks for technical aspects
- financial and transaction expenses
- lack of skills, accessibility, and financial capacity of the NGOs and private sector
- conventional attitudes of municipalities

For cost-effective and protective approach of 3Ps, developing countries need to consider national and local factors individually and overcome the above challenges.

4.8.3 ECONOMIC INSTRUMENTS

Economic instruments are adopted based on price-based policies. This strategy is applied in different phases of the product life cycle such as disposal, purchasing, selling, consumption, and production. Moreover, price-based tools are implemented focusing on plastic manufacturers and companies. At the disposal stage, the fee is assigned considering the concepts of the "polluter pay" or "pay as you throw" (Dijkgraaf and Gradus 2004). The waste disposal fee is calculated using the volumetric price of the garbage. However, effective monitoring and enforcement tools, strong cultural norms, and powerful official enforcements are required to achieve the expected goals of this waste management system (Kim, Chang, and Kelleher 2008). These provisions act as barriers to implement this tool successfully for most developing countries. On the other hand, economic instrumentation is also used to encourage responsible behavior related to waste collection, disposal, and recycling.

The advanced disposal fee approach of economic instruments is assessed by the consideration of consumers and manufacturers at the marketing phase. This approach will also be useful in reducing pollution sources and encouraging green designs. City of Toronto, the United States, and Ireland have successfully adopted this system for disposable plastic bags at the selling place (Rivers, Shenstone-Harris and Young 2017; Wagner 2017). On the other hand, two-part instrument price-based system under economic instruments links with the following two standards (Fullerton and Wolverton 2000):

(1) advanced disposal payment mechanisms
(2) incentives to companies and consumers as encouragement for proper disposal of plastic products and packaging items

The traditional deposit refund system becomes ineffective due to different challenges such as administrative charges for the refund during the implementation phase (Abbott and Sumaila 2019). Commercial and recreational/tourism-based tax systems can be introduced to reduce plastic pollution in coastal and marine zones. The tax is charged from the plastic producers and the amount of tax is calculated by considering the degree of plastic emission into the ocean. However, the above tax-based system is not effective for non-point and/or indirect pollution sources. A producer-based tax system can also be assessed considering recyclability and ecological damage of respective plastic products (Alpizar et al. 2020). An interconnected price-based mechanism allows the producers to allocate the eco/environmental tax of the product to the consumer either partially or completely. Accordingly, this tax stimulates the consumers to reduce the consumption of plastic products with additional expenses. However, economic-based tools are not very effective for developing countries due to the expensiveness of these measures and financial constraints. Developing countries need to adopt this tool by allocating reasonable and acceptable charges to consumers and producers upon the agreement of all stakeholders.

4.8.4 Awareness and Capacity Building Campaigns

Legal enforcement-related to waste management is always sustained through nonregulatory measures such as public attitude, perception, and actions (Garcia, Fang, and Lin 2019). Accordingly, environmental education and awareness are powerful tools to make significant changes in human attitudes, and this consequently influences eco-friendly, sustainable behavioral interventions for developing countries. Public awareness is also an effective solution to prevent local misconceptions on plastic waste management in regions with high consumption rates of plastic products (Garcia, Fang, and Lin 2019). The younger generation is the most influential citizen group on any kind of action, project, or management measures in any nation. Therefore, the younger generation can be enriched with environmental knowledge at early stages of life through the education system. Consumers should also become aware of the damage caused by plastics in marine ecosystems and other issues caused directly and indirectly. Ultimately, improved attitudes would result in behavioral changes that can improve recycling and proper disposal of plastic trash in a responsible way. However, these awareness programs can become successful only through the active participation of all the stakeholders. Citizen science further acts as a novel concept of this approach.

4.8.5 Citizen Science

This is one of the most effective new approaches that could be used with the support of volunteers who are untrained citizens able to contribute to data collection, information, and even for sampling for scientific research studies (Emmerik and Schwarz 2020; Hidalgo-Ruz and Thiel 2015). Through participation in such projects, those volunteers will become more aware of the impacts of marine plastics. Ultimately, they can contribute to improve awareness among their close friends. This will initiate a region-wise awareness program with the incorporation of new approaches like the development of mobile apps and other interactive ways. Educational projects within schools and volunteer programs guided by professional scientists with the participation of environmental organizations are conducted using the citizen science approach. Further, this approach can cover a wide area from local to international levels for short-term to long-term periods with big data sets by large numbers of participants from the public (Emmerik and Schwarz 2020; Hidalgo-Ruz and Thiel 2015).

4.8.6 Reduce, Reuse, and Recycling Methods

The reduce, reuse, and recycle (3Rs) concept could be used as a long-term sustainable solution in an effective way to cut down on plastics (Kumar 2019). There are many other ways of reducing the use of plastics-based products in the human lifestyle by practicing the use of alternatives such as purchase of items with cardboard containers instead of PS containers, avoiding the use of plastics associated with personal care products and minimizing the use of single-use plastics (Letcher 2020; Bano et al. 2020). Reuse would delay the plastic's entry into the environment through consumers or even avoid new entry of plastics into the environment. This could be achieved with combined recycling and reuse of plastics (Visvanathan and Norbu 2006). However, marine plastic debris that have persisted for a long time in the ocean cannot be used directly due to their downgraded properties. Therefore, it is only possible to reuse them with combined applications with other methods of recycling such as preparation of artificial stones, soil cement, road construction materials, bricks, and construction of artificial reefs and habitats platforms (Kumar 2019; Niaounakis 2017). Recycling of plastics depends on the type of plastic as the end-product should consist of only pure plastic materials without any non-plastic materials upon recycling. Recycling of plastics found in marine and associated aquatic ecosystems can be carried out using different steps and some could be used in a combined manner using both conventional and new treatment methods (Table 4.6). However, most of the time, true recycling is rare, and most often they are only hydrolyzed and turned into monomers (Letcher 2020; Niaounakis 2017).

Recently, 3Rs approach has been developed into the expanded Rs approach for controlling and preventing plastic pollution effectively. A new concept of the "6Rs Golden Rule" was developed with the approach of zero pollution (El-Haggar 2007). Then, 6Rs Golden Rule was shifted toward the 7Rs management system by incorporating another "R", which represents the enforcement approach, "Regulation". The 7Rs approach covers all key elements required for sustainable waste management (El-Haggar 2007). Refusing, recruiting, re-Gifting, rethinking, repurposing, repairing, rotting, recovering, renovation, rechoosing, regulation, renew, redesign, and retrieve

---

TABLE 4.6
Conventional Methods of Removal and Treatment of Plastics

	Methods	Reference
Collection and removal	Cleaning shorelines and beaches Removal of floating plastic debris Removal of submerged plastic debris	(Niaounakis 2017)
Mechanical and/ or physical treatments	Categorization Cleaning Volume reduction [grinding, compression, thermal volume reduction and melt, dissolution] Separation	(Niaounakis 2017; Bano et al. 2020)
Disposal	Landfill	(Buekens and Letcher 2020; Niaounakis 2017)

are expanded best practices beyond typical 3Rs. The 7Rs acts as an industrial eco-logical hierarchy and a sustainable waste management approach for zero pollution (El-Haggar 2007). This mechanism is successfully adopted for large-scale industrial sectors and community-based projects toward the zero-waste approach. Eventually, eco-friendly 7Rs approaches act as best practices for developing countries to make sustainable efforts to control marine plastic pollution in an effective way. In addition to the Rs, another long-term sustainable plan is a circular economy, which is newly implemented instead of a linear economy. This has been implemented to reduce waste and impacts while adapting sustainable production and consumption patterns (Abbott and Sumaila 2019).

4.8.7 New Approaches and Methods of Treatments and Removal of Marine Plastics

Treatment, degradation, and removal of marine plastic debris using new technological experiments and innovative applications are limited. Therefore, the development of successful new methods and new applications are also worthy solutions, mainly in developing countries to overcome negative impacts of plastics in the oceans. There are many promising new approaches and technological applications (Table 4.7), which are eco-friendly solutions to handle the plastic problem while providing affordability (Ganesh Kumar et al. 2020; Onda et al. 2020; Padervand et al. 2020). The natural, common method of removal of marine debris and plastics is biodegrad-ation, which can happen in different ways (Table 4.7) with microbial biofilm for-mation, biodeterioration, bio fragmentation, and mineralization. These degradations also depend on different factors such as polymer characteristics and environmental conditions (Sheth et al. 2019; Ganesh Kumar et al. 2020; Onda et al. 2020).

Further, new approaches for removal and treatment of marine debris including plastics are mentioned below.

4.8.7.1 Adsorption on Algae and Ingestion by Other Marine Organisms

Several studies have reported the possibility of removal of microplastics in the ocean using algae. Microplastics with a large surface area/volume ratio makes it easy for adsorption into the surface of tissues using their gelatinous nature. Positively charged microplastics can be adsorbed by negatively charged anionic polysaccharide algal cell structure. Further, this could be achieved using other marine organisms like bac-teria (*Bacillus gottheilii, B. cereus*), and clams (*Tridacna maxima*) (Padervand et al. 2020; Sheth et al. 2019).

4.8.7.2 Membrane Technology

Diatomite platforms, coated polymers, and membrane bioreactor approaches can be used as efficient ways to remove microplastics (Padervand et al. 2020). However, this is more useful for wastewater treatment from municipal areas and effective for the removal of microplastics. In addition, with the membrane bioreactor method, it has been shown that the capability of removing small size plastic particles varied from 20 to 100 μm by the combination of rapid sand filters, dissolved air floating, and disk filters.

TABLE 4.7
Some New Methods and Approaches of Removal and Treatments of Plastics

	New Methods		Reference
Organic and biodegradation	Biodegradation by specific microorganisms (enzymatic decomposition)	Biodegradation under controlled conditions Biodegradation in uncontrollable conditions	(Crippa et al. 2019)
	Physicochemical-biological	Ozonation pretreatment by Penicillium variable Thermal pretreatment by *Klebsiella pneumoniae*	(Ganesh Kumar et al. 2020)
Chemical treatment (also called chemical or tertiary recycling)	*Monomerization and/or oligomerization*		(Niaounakis 2017; Buekens and Letcher 2020)
	Thermal decomposition (also called thermolysis)	Incineration (or combustion) Energy recovery Pyrolysis Fuel Fiber-reinforced plastics Other thermal technologies [Steam or catalytic degradation (cracking), Liquefaction, gasification, hydrogenation, as a reducing agent in the blast furnace, degradative extrusion or viscosity breaking, thermal plasma]	
	Solvent-based purification and depolymerization technologies		(Crippa et al. 2019)
	Feedstock recycling technologies		
Treatment of marine waste on board of a ship	Treatment of recovered marine plastic waste on board of a ship		(Niaounakis 2017)
	Treatment of plastic waste generated on board of a ship	Containment Pressure Heating Extrusion Thermocompression Incineration	
	Treatment of plastic-containing waste generated inland on board of a ship		

4.8.7.3   Protein Engineering Technology

Enzymatic degradation of marine plastics can be triggered by specific enzymatic activities with protein engineering. Modified bacteria with their modified hydrolases have been shown to be effective for some polymer types such as PET and fibers. In addition, studies have identified the possibility of degrading the internal structure of polyurethane using engineered cutinases and polyurethanes (Ganesh Kumar et al. 2020; Sheth et al. 2019).

4.8.7.4 New Strain of Microorganisms

Modified organisms with plastic degradable enzymes have been used for the formation of biofilms. Generally, wild types of organisms are used for organism modification with ribosomal DNA (rDNA) techniques. This is an eco-friendly approach and an effective way to treat marine plastic debris using this modified engineered strain.

4.8.7.5 Metagenomics Approach

This is a widely used emerging technique with both terrestrial and marine microorganism communities to remove microplastics. This approach has been confirmed as an advanced and innovative approach for the recognition of specific enzymes, which can treat marine plastic debris polymers without unculturable microorganisms. The widely used method is a shot gun metagenomic sequencing of biofilm fouling microorganism with marine plastics (Ganesh Kumar et al. 2020).

4.8.7.6 In-silico Application

This is based on computational technologies with the connection of genomics, transcriptomics, and metabolomics data, which would be an efficient, novel, innovative approach for marine plastic bioremediation with in-silico genome mining. Many researchers have used this approach on the PET with the expression of PETase using marine sponge-associated strain *Streptomyces* sp. SM14. This kind of genome mining approach by focusing on biodegradation would be an efficient and effective way to manage marine debris in the ocean (Ganesh Kumar et al. 2020).

4.8.7.7 The Marine Plastic Footprint

This is a new concept to measure the inventory of plastic leakage step by step with the life cycle approach, which offers valuable data by identified sources with leakage of marine plastics. This also could be used as a management approach for the reduction of entry of plastics into the ocean and even as a decision-making tool. Therefore, this approach is named a marine plastic footprint (Boucher et al. 2020). However, this life cycle approach investigates the generation of plastic waste and losses due to usage, maintenances, and transport of the products, and associated environmental impacts. Furthermore, this can be evaluated at the industry level, region level, country level, and even globally. There are a few clear objectives on which this method is focused on, namely, less leakage of more materiality, more circularity, and no trade-offs (Boucher et al. 2019; Boucher et al. 2020).

4.9 SUMMARY

This chapter critically evaluated the current scenario of marine plastic pollution with respect to the origin of marine plastics, distribution pattern, fate, negative environmental, social, ecological, biological, and health consequences, and management approaches. Primary and secondary microplastics with long residence periods have adverse health effects on biological compartments representing different trophic levels. Hence, this environmental issue needs to be addressed through a combined approach using international, regional, and national level initiatives. The plastic management regime needs to be focused mainly on global plastic production, consumption, waste management, and disposal against the entry points of plastic pollution. The plastic litter management measures emphasize six main aspects: (i) prevention of disposal of plastic litter, (ii) reduction of load of plastic waste, (iii) removal, (iv) recovery, recycling, and reusing of plastics, (v) ecological rights and responsibilities, and (vi) eco-friendly lifestyle and behavioral changing approach. This chapter identified potential gaps and challenges to implementing plastic waste management actions effectively in developing countries and further proposed sustainable, applicable best practices toward a "plastic-free environment". The identified different aspects of this emerging threat and recommended green solutions will also contribute to adopting better management programs and a shift toward the zero-emission of plastics into the world oceans.

REFERENCES

Abbott, Joshua K., and U. Rashid Sumaila. 2019. "Reducing Marine Plastic Pollution: Policy Insights from Economics." *Review of Environmental Economics and Policy* 13 (2): 327–36. https://doi.org/10.1093/reep/rez007.

Adam, Issahaku, Tony R. Walker, Joana Carlos Bezerra, and Andrea Clayton. 2020. "Policies to Reduce Single-Use Plastic Marine Pollution in West Africa." *Marine Policy* 116: 1–10. https://doi.org/10.1016/j.marpol.2020.103928

Ahmed, Shafiul Azam, and Syed Mansoor Ali. 2006. "People as Partners: Facilitating People's Participation in Public-Private Partnerships for Solid Waste Management." *Habitat International* 30 (4): 781–96. https://doi.org/10.1016/j.habitatint.2005.09.004

Ajith, Nithin, Sundaramanickam Arumugam, Surya Parthasarathy, et al. 2020. "Global Distribution of Microplastics and Its Impact on Marine Environment—A Review. *Environmental Science and Pollution Research* 27: 25970–86. https://doi.org/10.1007/s11356-020-09015-5

Akenji, Lewis, Magnus Bengtsson, Yasuhiko Hotta, Mizuki Kato, and Matthew Hengesbaugh. 2020. "Policy Responses to Plastic Pollution in Asia." In *Plastic Waste and Recycling*, edited by Trevor M. Letcher, 531–67. Elsevier. https://doi.org/10.1016/b978-0-12-817880-5.00021-9

Alpizar, F., F. Carlsson, G. Lanza, B. Carney, R.C. Daniels, M. Jaime, T. Ho, et al. 2020. "A Framework for Selecting and Designing Policies to Reduce Marine Plastic Pollution in Developing Countries." *Environmental Science & Policy* 109: 25–35. https://doi.org/10.1016/j.envsci.2020.04.007

Auta, H.S., C.U. Emenike, and S.H. Fauziah. 2017. "Distribution and Importance of Microplastics in the Marine Environment: A Review of the Sources, Fate, Effects, and Potential Solutions." *Environment International* 102: 165–76. https://doi.org/10.1016/j.envint.2017.02.013

Avio, Carlo Giacomo, Stefania Gorbi, Massimo Milan, Maura Benedetti, Daniele Fattorini, Giuseppe D'Errico, Marianna Pauletto, Luca Bargelloni, and Francesco Regoli. 2015. "Pollutants Bioavailability and Toxicological Risk from Microplastics to Marine Mussels." *Environmental Pollution* 198: 211–22.

Bakir, Adil, Steven J. Rowland, and Richard C. Thompson. 2014. "Transport of Persistent Organic Pollutants by Microplastics in Estuarine Conditions." *Estuarine, Coastal and Shelf Science* 140: 14–21. https://doi.org/10.1016/j.ecss.2014.01.004

Bano, Nasreen, Tanzila Younas, Fabiha Shoaib, Dania Rashid, and Naqi Jaffri. 2020. "Plastic: Reduce, Recycle, and Environment." In *Environmentally-Benign Energy Solutions. Green Energy and Technology*, edited by Ibrahim Dincer, Can Colpan, and Mehmet Ezan, 191–208. Cham: Springer. https://doi.org/10.1007/978-3-030-20637-6_10

Barasarathi, Jayanthi, P. Agamuthu, C.U. Emenike, and S.H. Fauziah. 2014. "Microplastic Abundance in Selected Mangrove Forests in Malaysia." *Proceeding of the ASEAN Conference on Science and Technology 2014.*

Batel, Annika, Frederic Linti, Martina Scherer, Lothar Erdinger, and Thomas Braunbeck. 2016. "Transfer of Benzo[a]Pyrene from Microplastics to Artemia Nauplii and Further to Zebrafish via a Trophic Food Web Experiment: CYP1A Induction and Visual Tracking of Persistent Organic Pollutants." *Environmental Toxicology and Chemistry* 35 (7): 1656–66. https://doi.org/10.1002/etc.3361

Beaumont, Nicola J., Margrethe Aanesen, Melanie C. Austen, Tobias Börger, James R. Clark, Matthew Cole, Tara Hooper, Penelope K. Lindeque, Christine Pascoe, and Kayleigh J. Wyles. 2019. "Global Ecological, Social and Economic Impacts of Marine Plastic." *Marine Pollution Bulletin* 142 (March): 189–95. https://doi.org/10.1016/j.marpolbul.2019.03.022

Birch, Quinn T., Phillip M. Potter, Patricio X. Pinto, Dionysios D. Dionysiou, and Souhail R. Al-Abed. 2020. "Sources, Transport, Measurement and Impact of Nano and Microplastics in Urban Watersheds." *Reviews in Environmental Science and Biotechnology* 19: 1–62. https://doi.org/10.1007/s11157-020-09529-x

Boucher, Julien, Guillaume Billard, Eleonora Simeone, and Joao Sousa. 2020. *The Marine Plastic Footprint*. Gland, Switzerland: International Union for Conservation of Nature (IUCN). https://doi.org/10.2305/iucn.ch.2020.01.en

Boucher, Julien, Carole Dubois, Anna Kounina, and Philippe Puydarrieux. 2019. *Review of Plastic Footprint Methodologies*. Gland, Switzerland: International Union for Conservation of Nature (IUCN). https://doi.org/10.2305/IUCN.CH.2019.10.en

Boucher, Julien, and Damien Friot. 2017. *Primary Microplastics in the Oceans: A Global Evaluation of Sources*. Gland, Switzerland: International Union for Conservation of Nature (IUCN). https://doi.org/10.2305/iucn.ch.2017.01.en

Brouillat, Eric, and Vanessa Oltra. 2012. "Extended Producer Responsibility Instruments and Innovation in Eco-Design: An Exploration through a Simulation Model." *Ecological Economics* 83: 236–45. https://doi.org/10.1016/j.ecolecon.2012.07.007

Buekens, Alfons, and Trevor M. Letcher. 2020. "The Treatment of Plastic in Automobile Shredder Residue." In *Plastic Waste and Recycling*, edited by Trevor M. Letcher, 401–14. Elsevier. https://doi.org/10.1016/b978-0-12-817880-5.00015-3

Carney Almroth, Bethanie, and Håkan Eggert. 2019. "Marine Plastic Pollution: Sources, Impacts, and Policy Issues." *Review of Environmental Economics and Policy* 13 (2): 317–26. https://doi.org/10.1093/reep/rez012

Caron, Alexandra G.M., Colette R. Thomas, Ellen Ariel, Kathryn L.E. Berry, Steven Boyle, Cherie A. Motti, and Jon E. Brodie. 2016. "Extraction and Identification of Microplastics from Sea Turtles: Method Development and Preliminary Results." Tropical Water Report No. 15/52. Centre for Tropical Water & Aquatic Ecosystem Research (TropWATER) Publication 15/52, James Cook University, Townsville.

Chen, Chung Ling. 2015. "Regulation and Management of Marine Litter". In *Marine Anthropogenic Litter*, edited by Melanie Bergmann, Lars Gutow, and Michael Klages, 395–428. Springer. https://doi.org/10.1007/978-3-319-16510-3_15

Cheshire, Anthony, Ellik Adler, Julian Barbière, Yuval Cohen, Sverker Evans, Srisuda Jarayabhand, Ljubomir Jeftic, et al. 2009. *UNEP/IOC Guidelines on Survey and Monitoring of Marine Litter*. Regional Seas Reports and Studies No. 186 IOC Technical Series No. 83. UNEP/IOC.

Chua, Evan M., Jeff Shimeta, Dayanthi Nugegoda, Paul D. Morrison, and Bradley O. Clarke. 2014. "Assimilation of Polybrominated Diphenyl Ethers from Microplastics by the Marine Amphipod, *Allorchestes compressa*." *Environmental Science and Technology* 48 (14): 8127–34. https://doi.org/10.1021/es405717z

COBSEA. 2019. *East Asian Seas Countries Welcome Harmonization of Marine Litter Monitoring Efforts at Regional Training*. Coordinating Body on the Seas of East Asia (COBSEA). www.unep.org/cobsea/news/story/east-asian-seas-countries-welcome-harmonization-marine-litter-monitoring-efforts

Cole, Matthew, Penelope K. Lindeque, Elaine Fileman, James Clark, Ceri Lewis, Claudia Halsband, and Tamara S. Galloway. 2016. "Microplastics Alter the Properties and Sinking Rates of Zooplankton Faecal Pellets." *Environmental Science and Technology* 50 (6): 3239–46. https://doi.org/10.1021/acs.est.5b05905

Cole, Matthew, Pennie Lindeque, Elaine Fileman, Claudia Halsband, Rhys Goodhead, Julian Moger, and Tamara S. Galloway. 2013. "Microplastic Ingestion by Zooplankton." *Environmental Science and Technology* 47 (12): 6646–55. https://doi.org/10.1021/es400663f

Cole, Matthew, Pennie Lindeque, Claudia Halsband, and Tamara S. Galloway. 2011. "Microplastics as Contaminants in the Marine Environment: A Review." *Marine Pollution Bulletin* 62 (12): 2588–97. https://doi.org/10.1016/j.marpolbul.2011.09.025

Copernicus. 2021. "Detecting Plastic Pollution | CMEMS." 2021. https://marine.copernicus.eu/services/plastic-pollution

Corcoran, Patricia L., Mark C. Biesinger, and Meriem Grifi. 2009. "Plastics and Beaches: A Degrading Relationship." *Marine Pollution Bulletin* 58 (1): 80–84. https://doi.org/10.1016/j.marpolbul.2008.08.022.

Crippa, Maurizio., Bruno De Wilde, Rudy Koopmans, Jan Leyssens, Jane Muncke, Anne-Christine Ritschkoff, Karine Van Doorsselaer, Costas Velis, and Martin Wagner. 2019. *A Circular Economy for Plastics: Insights from Research and Innovation to Inform Policy and Funding Decisions*. Luxembourg: Publications Office of the European Union.

Dijkgraaf, E., and Raymond H.J.M. Gradus. 2004. "Cost Savings in Unit-Based Pricing of Household Waste. The Case of the Netherlands." *Resource and Energy Economics* 26 (4): 353–71. https://doi.org/10.1016/j.reseneeco.2004.01.001

Driedger, Alexander G.J., Hans H. Dürr, Kristen Mitchell, and Philippe Van Cappellen. 2015. "Plastic Debris in the Laurentian Great Lakes: A Review." *Journal of Great Lakes Research* 41 (1): 9–19. https://doi.org/10.1016/j.jglr.2014.12.020

El-Haggar, Salah M. 2007. *Sustainable Industrial Design and Waste Management: Cradle-to-Cradle for Sustainable Development*. Amsterdam: Elsevier.

Emmerik, Tim, and Anna Schwarz. 2020. "Plastic Debris in Rivers." *WIREs Water* 7 (1): 1–24. https://doi.org/10.1002/wat2.1398

Eriksen, Marcus, Laurent C.M. Lebreton, Henry S. Carson, Martin Thiel, Charles J. Moore, Jose C. Borerro, Francois Galgani, Peter G. Ryan, and Julia Reisser. 2014. "Plastic Pollution in the World's Oceans: More than 5 Trillion Plastic Pieces Weighing over 250,000 Tons Afloat at Sea." *PLoS ONE* 9 (12): 1–15. https://doi.org/10.1371/journal.pone.0111913

Eriksen, Marcus, Martin Thiel, and Laurent Lebreton. 2016. "Nature of Plastic Marine Pollution in the Subtropical Gyres." In *Hazardous Chemicals Associated with Plastics in the Marine Environment*, edited by Hideshige Takada and Hrissi K. Karapanagioti, vol. 78, 135–62. Cham: Springer Verlag. https://doi.org/10.1007/698_2016_123

European Commission. 2003. *Regional Policy Guidelines for Successful Public Private Partnerships*. Brussels: European Commission.

European Commission. 2013. *MSDF Guidance on Monitoring Marine Litter*. https://doi.org/10.2788/99475

Ferreira, Pedro, Elsa Fonte, M. Elisa Soares, Felix Carvalho, and Lúcia Guilhermino. 2016. "Effects of Multi-Stressors on Juveniles of the Marine Fish *Pomatoschistus microps*: Gold Nanoparticles, Microplastics and Temperature." *Aquatic Toxicology* 170: 89–103. https://doi.org/10.1016/j.aquatox.2015.11.011

Fossi, Maria Cristina, Letizia Marsili, Matteo Baini, Matteo Giannetti, Daniele Coppola, Cristiana Guerranti, Ilaria Caliani, et al. 2016. "Fin Whales and Microplastics: The Mediterranean Sea and the Sea of Cortez Scenarios." *Environmental Pollution* 209: 68–78. https://doi.org/10.1016/j.envpol.2015.11.022

Fullerton, Don, and Ann Wolverton. 2000. "Two Generations of a Deposit-Refund System." *NBER Working Paper Series* 53: 1689–1699.

Gallo, Frederic, Cristina Fossi, Roland Weber, David Santillo, Joao Sousa, Imogen Ingram, Angel Nadal, and Dolores Romano. 2018. "Marine Litter Plastics and Microplastics and Their Toxic Chemicals Components: The Need for Urgent Preventive Measures." *Environmental Sciences Europe* 30 (1): 1–14. https://doi.org/10.1186/s12302-018-0139-z

Galloway, Tamara S., Matthew Cole, and Ceri Lewis. 2017. "Interactions of Microplastic Debris throughout the Marine Ecosystem." *Nature Ecology and Evolution* 1 (5): 1–8. https://doi.org/10.1038/s41559-017-0116

Ganesh Kumar, A., K. Anjana, M. Hinduja, K. Sujitha, and G. Dharani. 2020. "Review on Plastic Wastes in Marine Environment—Biodegradation and Biotechnological Solutions." *Marine Pollution Bulletin* 150 (January): 1–8. https://doi.org/10.1016/j.marpolbul.2019.110733

Garcia, Beatriz, Mandy Meng Fang, and Jolene Lin. 2019. "Marine Plastic Pollution in Asia: All Hands on Deck!" *Chinese Journal of Environmental Law* 3 (1): 11–46. https://doi.org/10.1163/24686042-12340034

Goldstein, M.C., and D.S. Goodwin. 2013. "Gooseneck Barnacles (*Lepas* spp.) Ingest Microplastic Debris in the North Pacific Subtropical Gyre." *PeerJ* 1: e184. http://dx.doi.org/10.7717/peerj.184.

Goodman, Alexa J., Tony R. Walker, Craig J. Brown, Brittany R. Wilson, Vicki Gazzola, and Jessica A. Sameoto. 2020. "Benthic Marine Debris in the Bay of Fundy, Eastern Canada: Spatial Distribution and Categorization Using Seafloor Video Footage." *Marine Pollution Bulletin* 150: 1–6. https://doi.org/10.1016/j.marpolbul.2019.110722

Green, Dannielle Senga. 2016. "Effects of Microplastics on European Flat Oysters, Ostrea Edulis and Their Associated Benthic Communities." *Environmental Pollution* 216: 95–103. https://doi.org/10.1016/j.envpol.2016.05.043

Hahladakis, John N., Eleni Iacovidou, and Spyridoula Gerassimidou. 2020. "Plastic Waste in a Circular Economy." In *Plastic Waste and Recycling*, edited by Trevor M. Letcher, 481–512. Elsevier. https://doi.org/10.1016/b978-0-12-817880-5.00019-0

Hidalgo-Ruz, Valeria, and Martin Thiel. 2015. "The Contribution of Citizen Scientists to the Monitoring of Marine Litter." In *Marine Anthropogenic Litter*, edited by Melanie Bergmann, Lars Gutow, and Michael Klages, 429–47. Springer International Publishing. https://doi.org/10.1007/978-3-319-16510-3_16

Hollman, P.C.H., H. Bouwmeester, and R.J.B. Peters. 2013. *Microplastics in the Aquatic Food Chain: Sources, Measurements, Occurrence and Potential Health Risks*. RIKILT Report 2013.003. Wageningen: RIKILT Wageningen UR (University & Research Centre).

Iverson, Autumn R. 2019. "The United States Requires Effective Federal Policy to Reduce Marine Plastic Pollution." *Conservation Science and Practice* 1 (6): e45. https://doi.org/10.1111/csp2.45

Kandziora, J. H., N. van Toulon, P. Sobral, H.L. Taylor, A.J. Ribbink, J.R. Jambeck, and S. Werner. 2019. "The Important Role of Marine Debris Networks to Prevent and Reduce Ocean Plastic Pollution." *Marine Pollution Bulletin* 141 (April): 657–62. https://doi.org/10.1016/j.marpolbul.2019.01.034

Kershaw, Peter J. 2016. *Marine Plastic Debris and Microplastics—Global Lessons and Research to Inspire Action and Guide Policy Change*. Nairobi: United Nations Environment Programme. https://ec.europa.eu/environment/marine/good-environmental-status/descriptor-10/pdf/Marine_plastic_debris_and_microplastic_technical_report_advance_copy.pdf

Kim, Geum Soo, Young Jae Chang, and David Kelleher. 2008. "Unit Pricing of Municipal Solid Waste and Illegal Dumping: An Empirical Analysis of Korean Experience." *Environmental Economics and Policy Studies* 9 (3): 167–76. https://doi.org/10.1007/BF03353988

Kirakozian, Ankinee. 2016. "One Without the Other? Behavioural and Incentive Policies for Household Waste Management." *Journal of Economic Surveys* 30 (3): 526–51. https://doi.org/https://doi.org/10.1111/joes.12159

Kübler, Dorothea. 2001. "On the Regulation of Social Norms." *Journal of Law, Economics, and Organization* 17 (2): 449–76. https://EconPapers.repec.org/RePEc:oup:jleorg:v:17:y:2001:i:2:p:449-76

Kumar, Prashant. 2019. "Designing Zero Plastic Policy and Its Implementation: Major Role of Law to Protect Environment, Plastic Everywhere." *National Journal of Environmental Law* 2 (2): 45–55.

Leal Filho, Walter, Ulla Saari, Mariia Fedoruk, Arvo Iital, Harri Moora, Marija Klöga, and Viktoria Voronova. 2019. "An Overview of the Problems Posed by Plastic Products and the Role of Extended Producer Responsibility in Europe." *Journal of Cleaner Production* 214: 550–58. https://doi.org/10.1016/j.jclepro.2018.12.256

Lebreton, Laurent, Matthias Egger, and Boyan Slat. 2019. "A Global Mass Budget for Positively Buoyant Macroplastic Debris in the Ocean." *Scientific Reports* 9 (1): 1–10. https://doi.org/10.1038/s41598-019-49413-5

Lebreton, Laurent C.M., Joost Van Der Zwet, Jan Willem Damsteeg, Boyan Slat, Anthony Andrady, and Julia Reisser. 2017. "River Plastic Emissions to the World's Oceans." *Nature Communications* 8 (1): 1–10. https://doi.org/10.1038/ncomms15611

Letcher, Trevor M. 2020. "Introduction to Plastic Waste and Recycling." In *Plastic Waste and Recycling*, edited by Trevor M. Letcher, 3–12. Elsevier. https://doi.org/10.1016/b978-0-12-817880-5.00001-3

Lobelle, Delphine, and Michael Cunliffe. 2011. "Early Microbial Biofilm Formation on Marine Plastic Debris." *Marine Pollution Bulletine* 62 (1): 197–200. doi: 10.1016/j.marpolbul.2010.10.013

Lu, Yifeng, Yan Zhang, Yongfeng Deng, Wei Jiang, Yanping Zhao, Jinju Geng, Lili Ding, and Hongqiang Ren. 2016. "Uptake and Accumulation of Polystyrene Microplastics in Zebrafish (*Danio rerio*) and Toxic Effects in Liver." *Environmental Science and Technology* 50 (7): 4054–60. https://doi.org/10.1021/acs.est.6b00183

Maximenko, Nikolai, Paolo Corradi, Kara L. Law, Erik Van Sebille, Shungudzemwoyo P. Garaba, Richard S. Lampitt, Francois Galgani, et al. 2019. "Towards the Integrated

Marine Debris Observing System." *Frontiers in Marine Science* 6: 1–25. https://doi.org/10.3389/fmars.2019.00447

Mohamed Nor, Nur Hazimah, and Jeffrey Philip Obbard. 2014. "Microplastics in Singapore's Coastal Mangrove Ecosystems." *Marine Pollution Bulletin* 79 (1–2): 278–83. https://doi.org/10.1016/j.marpolbul.2013.11.025

Mountford, A.S., and M.A. Morales Maqueda. 2019. "Eulerian Modeling of the Three-Dimensional Distribution of Seven Popular Microplastic Types in the Global Ocean." *Journal of Geophysical Research: Oceans* 124 (12): 8558–73. https://doi.org/10.1029/2019JC015050

Napper, Imogen E., and Richard C. Thompson. 2019. "Marine Plastic Pollution: Other Than Microplastic." In *Waste*, edited by Trevor M. Letcher and Daniel A. Vallero, 425–42. Elsevier. https://doi.org/10.1016/b978-0-12-815060-3.00022-0

Ng, K.L., and J.P. Obbard. 2006. "Prevalence of Microplastics in Singapore's Coastal Marine Environment." *Marine Pollution Bulletin* 52 (7): 761–67. https://doi.org/10.1016/j.marpolbul.2005.11.017

Niaounakis, Michael. 2017. *Management of Marine Plastic Debris. Management of Marine Plastic Debris.* 1st ed. Elsevier.

Ocean Conservancy. 2018. "Building a Clean Swell—Report," 28. https://oceanconservancy.org/wp-content/uploads/2018/07/Building-A-Clean-Swell.pdf

Ockelford, Annie, Andy Cundy, and James E. Ebdon. 2020. "Storm Response of Fluvial Sedimentary Microplastics." *Scientific Reports* 10 (1): 1–10. https://doi.org/10.1038/s41598-020-58765-2

OECD and Ministry of the Environment. June 2014. The State of Play on Extended Producer Responsibility (EPR): Opportunities and Challenges." *Ministry of Environment*, 17. www.oecd.org/environment/waste/Global%20Forum%20Tokyo%20Issues%20Paper%2030-5-2014.pdf

Onda, Deo Florence L., Norchel Corcia F. Gomez, Daniel John E. Purganan, Mark Paulo S. Tolentino, Justine Marey S. Bitalac, Jahannah Victoria M. Calpito, Jose Nickolo O. Perez, and Alvin Claine A. Viernes. 2020. "Marine Microbes and Plastic Debris: Research Status and Opportunities in the Philippines." *Philippine Journal of Science* 149 (1): 89–100.

Osaka Blue Ocean Vision. 2021. *G20 Workshop on Harmonized Monitoring and Data Compilation of Marine Plastic Litter.* Towards Osaka Blue Ocean Vision. https://g20mpl.org/archives/893

Padervand, Mohsen, Eric Lichtfouse, Didier Robert, and Chuanyi Wang. 2020. "Removal of Microplastics from the Environment. A Review." *Environmental Chemistry Letters* 18 (3): 807–28. https://doi.org/10.1007/s10311-020-00983-1

Premakumara, J.D.G., and Onogawa, K. 2019. *Strategies to Reduce Marine Plastic Pollution from Land-based Sources in Low and Middle—Income Countries.* United Nations Environment Programme.

Purba, Noir P., Dannisa I.W. Handyman, Tri D. Pribadi, Agung D. Syakti, Widodo S. Pranowo, Andrew Harvey, and Yudi N. Ihsan. 2019. "Marine Debris in Indonesia: A Review of Research and Status." *Marine Pollution Bulletin* 146: 134–44. https://doi.org/10.1016/j.marpolbul.2019.05.057

Rehse, Saskia, Werner Kloas, and Christiane Zarfl. 2016. "Short-Term Exposure with High Concentrations of Pristine Microplastic Particles Leads to Immobilisation of *Daphnia magna*." *Chemosphere* 153: 91–99. https://doi.org/10.1016/j.chemosphere.2016.02.133

Rivers, Nicholas, Sarah Shenstone-Harris, and Nathan Young. 2017. "Using Nudges to Reduce Waste? The Case of Toronto's Plastic Bag Levy." *Journal of Environmental Management* 188: 153–62. https://doi.org/10.1016/j.jenvman.2016.12.009

Ryan, Peter G., Charles J. Moore, Jan A. Van Franeker, and Coleen L. Moloney. 2009. "Monitoring the Abundance of Plastic Debris in the Marine Environment." *Philosophical Transactions of the Royal Society B: Biological Sciences* 364 (1526): 1999–2012. https://doi.org/10.1098/rstb.2008.0207

Sebille, Erik Van, Stefano Aliani, Kara Lavender Law, Nikolai Maximenko, José M. Alsina, Andrei Bagaev, Melanie Bergmann, et al. 2020. "The Physical Oceanography of the Transport of Floating Marine Debris." *Environmental Research Letters* 15 (2): 1–32. https://doi.org/10.1088/1748-9326/ab6d7d

Setälä, Outi, Joanna Norkko, and Maiju Lehtiniemi. 2016. "Feeding Type Affects Microplastic Ingestion in a Coastal Invertebrate Community." *Marine Pollution Bulletin* 102 (1): 95–101. https://doi.org/10.1016/j.marpolbul.2015.11.053

Sheth, Maya U., Sarah K. Kwartler, Emma R. Schmaltz, Sarah M. Hoskinson, E.J. Martz, Meagan M. Dunphy-Daly, Thomas F. Schultz, Andrew J. Read, William C. Eward, and Jason A. Somarelli. 2019. "Bioengineering a Future Free of Marine Plastic Waste." *Frontiers in Marine Science* 6: 1–10. https://doi.org/10.3389/fmars.2019.00624

Thaler, Richard H., and Cass R. Sunstein. 2009. *NUDGE: Improving Decisions About Health, Wealth, and Happiness*, Vol. 53. New York: Penguin Books.

Tibbetts, John H. 2015. "Managing Marine Plastic Pollution: Policy Initiatives to Address Wayward Waste." *Environmental Health Perspectives* 123 (4): A90–93. https://doi.org/10.1289/ehp.123-A90

UN. 2021. "Addis Ababa Action Agenda." Sustainable Development Knowledge Platform. https://sustainabledevelopment.un.org/index.php?page=view&type=400&nr=2051&menu=35

UNEP. 2018. *Single-Use Plastics: A Roadmap for Sustainability*, 1–104. Rev. ed. United Nations Environment Programme. www.unep.org/ietc/resources/publication/single-use-plastics-roadmap-sustainability

UNESCAP. 2011. *A Guidebook on Public-Private Partnership in Infrastructure*. ESCAP. www.unescap.org/resources/guidebook-public-private-partnership-infrastructure

Valiela, Ivan, and Marci L. Cole. 2002. "Comparative Evidence that Salt Marshes and Mangroves May Protect Seagrass Meadows from Land-Derived Nitrogen Loads." *Ecosystems* 5 (1): 92–102. https://doi.org/10.1007/s10021-001-0058-4

Van Cauwenberghe, Lisbeth, Michiel Claessens, Michiel B. Vandegehuchte, and Colin R. Janssen. 2015. "Microplastics Are Taken Up by Mussels (Mytilus Edulis) and Lugworms (Arenicola Marina) Living in Natural Habitats." *Environmental Pollution* 199: 10–17. https://doi.org/10.1016/j.envpol.2015.01.008

Van Sebille, Erik, Stefano Aliani, Kara Lavender Law, Nikolai Maximenko, Jose M. Alsina, Andrei Bagaev, Melanie Bergmann, et al. 2020. "The Physical Oceanography of the Transport of Floating Marine Debris." *Environmental Research Letters* 15 (2): 1–32. doi: 10.1088/1748-9326/ab6d7d.

Viscusi, W. Kip, Joel Huber, and Jason Bell. 2012. "Alternative Policies to Increase Recycling of Plastic Water Bottles in the United States." *Review of Environmental Economics and Policy* 6 (2): 190–211. https://doi.org/10.1093/reep/res006

Visvanathan, C., and T. Norbu. 2006. "Reduce, Reuse, and Recycle: The 3Rs in South Asia Frontiers Topic: Methane: A Bioresource for Fuel and Biomolecules View Project Thermophilic MBR (Membrane Bioreactor) View Project." *3 R South Asia Expert Workshop* 1–61.

Wagner, Travis P. 2017. "Reducing Single-Use Plastic Shopping Bags in the USA." *Waste Management* 70: 3–12. https://doi.org/10.1016/j.wasman.2017.09.003

Walls, Margaret 2013. "Deposit-Refund Systems in Practice and Theory." *Encyclopedia of Energy, Natural Resource, and Environmental Economics* 3: 133–37. https://doi.org/10.1016/B978-0-12-375067-9.00035-8

Wardrop, Peter, Jeff Shimeta, Dayanthi Nugegoda, Paul D. Morrison, Ana Miranda, Min Tang, and Bradley O. Clarke. 2016. "Chemical Pollutants Sorbed to Ingested Microbeads from Personal Care Products Accumulate in Fish." *Environmental Science and Technology* 50 (7): 4037–44. https://doi.org/10.1021/acs.est.5b06280

Welden, Natalie. 2019. "Microplastics: Emerging Contaminants Requiring Multilevel Management." In *Waste*, edited by Trevor M. Letcher and Daniel A. Vallero, 405–24. Elsevier. https://doi.org/10.1016/b978-0-12-815060-3.00021-9

Welden, Natalie A.C., and Amy L. Lusher. 2017. "Impacts of Changing Ocean Circulation on the Distribution of Marine Microplastic Litter." *Integrated Environmental Assessment and Management* 13 (3): 483–87. https://doi.org/10.1002/ieam.1911

Wijethunga, H.N. Sampath, Anuradha M.G.A.D. Athawuda, P. Charith B. Dias, Ayantha P. Abeygunawardana, J.D. Mahesh Senevirathna, G.G. Nadeeka Thushari, Nuwan P.P. Liyanage, and Sepalika C. Jayamanne. 2019. "Screening the Effects of Microplastics on Selected Invertebrates along Southern Coastal Belt in Sri Lanka: A Preliminary Approach to Coastal Pollution Control." Proceedings of International Research Conference of Uva Wellassa University 2019.

Woodall, Lucy C., Anna Sanchez-Vidal, Miquel Canals, Gordon L.J. Paterson, Rachel Coppock, Victoria Sleight, Antonio Calafat, Alex D. Rogers, Bhavani E. Narayanaswamy, and Richard C. Thompson. 2014. "The Deep Sea Is a Major Sink for Microplastic Debris." *Royal Society Open Science* 1 (4): 1–8. https://doi.org/10.1098/rsos.140317.

World Bank. 2011. *What Are Public Private Partnerships?* Public Private Partnership. Washington, DC: World Bank.

Xanthos, Dirk, and Tony R. Walker. 2017. "International Policies to Reduce Plastic Marine Pollution from Single-Use Plastics (Plastic Bags and Microbeads): A Review." *Marine Pollution Bulletin* 118 (1–2): 17–26. https://doi.org/10.1016/j.marpolbul.2017.02.048

Zalasiewicz, Jan, Sarah Gabbott, and Colin N. Waters. 2019. "Plastic Waste: How Plastics Have Become Part of the Earth's Geological Cycle." In *Waste*, edited by Trevor M. Letcher and Daniel A. Vallero, 443–52. Elsevier. https://doi.org/10.1016/b978-0-12-815060-3.00023-2

5 Nanowaste Management

Jagath Illanagsinghe[1*]
[1]Lyceum International School, Sri Lanka
*Corresponding author: lakmal21@yahoo.com

CONTENTS

DOI: 10.1201/9781003132349-6

5.1 INTRODUCTION

One billionth (10^{-9}) in meter (m) is defined as one nanometer (1 nm). In general nanoscale refers to a length from the atomic level of around 0.1–100 nm. Nanoscience is the study of the phenomena and manipulation of material at the nanoscale. Nanotechnology is the design, characterization, production, and application of structures, devices, and systems by controlling shape and size at the nanoscale.

Nanotechnology can be considered as an interdisciplinary subject. Therefore, there are various physical, chemical, biological, and hybrid techniques available to synthesize nanomaterials as described by Kulkarni (2015) (Figure 5.1). However, the production methods should be selected considering several factors such as the nature of nanomaterials (NMs) with various degrees of quality, speed, and cost. Basically, the production techniques can be divided into two fundamental categories: (1) bottom-up approach and (2) top-down assembly (Berger 2008; Dowling et al. 2004; Pokropivny et al. 2007; Khan et al. 2017).

According to Younis et al. (2018), the specific properties of NMs such as size distribution, symmetry, and purity can be optimized using different methodologies associated with the use of hazardous chemicals, which release nanowastes to the environment with a potential risk for life (Pokropivny et al. 2007; Berger 2008;

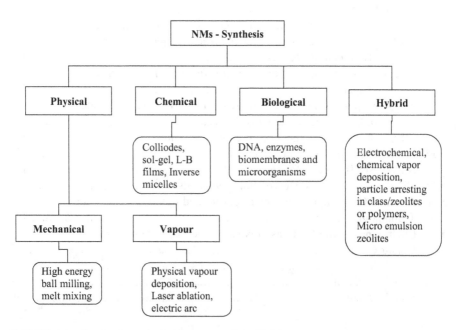

FIGURE 5.1 Synthesis methods of nanomaterials (NMs).

Bystrzejewska-Piotrowska et al. 2009; Adams and Barbante 2013; Schulte et al. 2013; Bhatia 2016; Kwak and An 2016; Al-Mubaddel et al. 2017; Purohit et al. 2017, Younis et al. 2018). Therefore, it is recommended that there is a need to move toward eco-friendly biological methods such as microorganisms, enzymes, and plants to synthesize green nanostructure materials (Kharissova et al. 2013; Lu and Ozcan 2015; Purohit et al. 2017).

5.2   WHAT IS NANOWASTE?

Nanotechnology has been widely recognized and accepted as the fourth industrial revolution which triggered the production of 58,000 tons of nanomaterials in 2020. Nanowaste is closely associated with the nanomaterials. The dramatic growth is there in NM production because nanotechnology has novel and revolutionary characteristics. It is reported that the production of fabricated NMs increased so rapidly from 2,600 types in 2015 to more than 4,000 types on the market in 2017.

There are more than 6,970 nano products on the market produced by 52 countries worldwide. Furthermore, it has been reported by Mrowiec in 2016 that the worldwide production rate of manufactured NMs increased from 268,000 to 318,000 metric tons at a rate of 25% per annum (Mirowiec 2016). As a consequence, the rapid unregularized production growth rate of NMs has created technical and regulatory challenges, leading to new issues.

Due to the uniqueness of properties of nanomaterials, there is a significant increase in nanomaterial products and applications in the last two decades, such as consumer products, personal care products, electronics, textiles, pharmaceuticals, and energy and environmental applications (Khan et al. 2013).

As we identify, nanowaste is a new group of waste (waste containing nanoparticles), which enters the environment as a by-product of nano-engineered products. Nanowaste may be defined as waste stream(s) containing nanomaterials; nanoscale synthetic by-products produced during production, storage, or distribution; end-of-life (EOL) nanotechnological materials or products, and materials contaminated with nanomaterials such as pipes and protective clothing (Musee 2011).

There are small quantities of nanomaterials in certain household EOL products, such as household waste, which is simply identified as "nanowaste". Further, it is not clearly mentioned as collectable and separable nanowaste in the definition proposed above, it could be simply recognized that all waste flows in society are potentially nanowaste (Boldrin et al. 2014).

According to the definition, nanowaste can be of four forms:

1. Pure nanomaterials produced at the production point.
2. Surfaces or substances contaminated by nanomaterials (containers, personal protective clothing, etc.).
3. Liquid suspensions containing nanomaterials.
4. Solids containing nanomaterials.

Some researchers have recently found that nanowaste can be a serious health hazard in many ways (Musee 2011; Piotrowska et al. 2009). The risk factor with nanoparticles is that nanoparticles or nanoparticle containing waste matter can last for a significant duration in landfills, waste treatment plants, and waste incineration plants. It is a clear fact that waste treatment methods and techniques for NMs are not well understood (Gottschalk et al. 2010; Keller and Lazareva 2014).

5.3   SOURCES OF NANOMATERIALS/NANOWASTE

There are a number of nanowaste sources that exist but identification of those have been neglected for some time due to the unawareness of nanowastes. The combustion process, which triggered the industrial revolution, is one of the main sources for emitting nanowaste to the environment (Biswas et al. 2005).

In addition to combustion, mobile sources, industrial settings, atmospheric reactions, and some other sources of nucleation of vapor precursors due to radioactive decay have been recognized as other sources of nanoparticle formation (Biswas et al. 2005; Kulmala et al. 2004).

Fine particle emissions from stationary combustion systems have been identified, using nanometer-sized particulate matter characterization technology, as a potential source of nanowaste.

The massive industrialization over the past century has rapidly increased the emission content, creating a more hazardous impact for the environment and ecological systems.

In addition to the conventional combustion systems used in industry for power generation, boilers and incinerators, new stationary emissions sources such as refineries and other chemical production methods have been significant emission sources of nanowaste.

It has been noted that catalytic nanomaterials used in process improvement purposes can lead to unintentional direct or indirect emission of nanowaste to the environment. Due to lack of appropriate treatment for waste-handling, the release to the environment is possible despite the direct emission being controlled and prevented (Saleem et al. 2018).

The nature of emissions including peak particle size, number of particles, and chemistry differ largely from source to source depending on the composition of the base process. Maguhn reported that a peak particle size of ultrafine particles in the flue gas and the municipal waste incinerator stack was 90 nm at 700°C (within the combustor) (Maguhn et al. 2003). However, the processes such as the condensation, agglomeration, nucleation, and physicochemical sorption of gaseous chemicals onto the nucleated particulate matter increase the downstreamed particle size. Furthermore, the peak size emissions are affected by the fuel composition, which is used for combustion. The air dilution ratio and temperature (operating condition) determine the end particle size distribution for various fuels (Table 5.1) according to Chang (Chang et al. 2004).

There are some outdoor activities that act as a stationary source of nanoparticle emissions. For example, cigarette smoke (Anderson et al. 1989) and cooking (especially frying and sautéing) (Li et al. 1992) can be identified as the prime sources of ultrafine particulate matter within buildings. Furthermore, Wallace et al. in 2004 reported that 90% particles originating from cooking, on a number basis, are assembled

TABLE 5.1
Peak Particle Size from Various Emission Sources

Source	Peak Particle Size (nm)
Municipal waste incinerator	90
Backup oil burners	30
Medium sulfur bituminous coal	40–50
No. 6 fuel oil	70–100
Natural gas	15–25

into ultrafine strata with a peak at 60 nm. The biofuels, used for cooking, have been a major source of nanoparticle emission, especially in India and Asian countries. In addition, mosquito coils, incense, candles, coal combustors, and fireplaces are not only strong contributors to indoor nanoparticle emissions but also these strong emission levels could cause significant climatic impact in the South Asian region (Venkataraman 2005).

Out of the other mobile sources of nanowaste emission, the diesel engine is identified as the major mobile source of nanoparticle emissions. Biswas and Wu have shown that the peak particle size occurs at 50 nm on a number concentration basis, while the mass distribution is characterized by 50–1,000 nm size (Biswas et al. 2005). The main constituents of diesel exhaust are sulfate particles, lubricating oil, and unburnt hydrocarbons (soluble organic fraction). The facilitating of the condensation of organic species is due to sulfuric acid, which is present in small amounts in the exhaust (Tobias et al. 2001).

The formation of downstream nuclei mode and the mass concentration can be reduced by introducing the catalytic converters to vehicle engines as a modem device, which enhances the oxidation of organic compounds.

The CMD value of a two-stroke engine is greater in the emission with a catalytic converter than without a catalytic converter for various fuel types (Alander et al. 2005) (Table 5.2). In addition to the nanoparticle emission in engines, due to combustion of fuels and lubricating oils, the catalytic converters and the metallic components of engines can also contribute to ultrafine metal particle emission in the engine exhausts (Rühle et al. 1997). Moreover, some research studies have reported the emission of platinum nanocrystals affixed to alumina particles (Artelt et al. 1999) and heavy metals such as palladium (Pd), rhodium (Rh), and other similar metals also have the potential to be emitted as nanoparticles (Artelt et al. 1999).

5.4 ATMOSPHERIC CONVERSION

The existence and formation of nanoparticles by nature is a common phenomenon. It has been identified that ultrafine particles are formed in the atmosphere by a process known as nucleation. Interestingly, nanoparticles in the range of 20–50 nm were first observed by Aiken in 1884 and today, these nuclei mode particles are called Aiken

TABLE 5.2
Count Median Diameter for Various Engine Types

Engine Type	Count Median Diameter (CMD) (nm)	Number Concentration (cm^3)
Diesel	50	N.A.
Spark-ignition	20–40	N.A.
Unleaded gasoline	45	8.0×10^4
Leaded gasoline	45	1.8×10^6
LPG	60	2.6×10^5
Two-stroke (with catalyst)	55-75	$2.6–3.0 \times 10^{15}$ (kWh^{-1})
Two-stroke(without catalyst)	75–120	$1.2–2.3 \times 10^{15}$ (kWh^{-1})

Source: Ålander et al. 2005; Harris et al. 2001; Rickeard et al. 1996; Shi et al. 2001; Ristovski et al. 1998; Faiz et al. 1996.

mode particles (Saleem 2018). It has been identified that sulfuric and nitric acid along with organic gases act as some of the main precursors for the formation of nuclei mode particles in the atmosphere (Seinfeld and Pandis 1998).

As a consequence of advanced detection instrumentation, ultrafine particles and nanoparticles on the order of a few nanometers have been observed in the atmosphere of some of the urban areas in different regions.

It has been suggested that the formation of nucleation mode particles in the free troposphere cloud outflows, leading to a significant contribution to the global particle production, resulting in global climate change by promoting atmospheric optical effect (Covert et al. 1996b; Wiedensohler et al. 1996; Clarke 1993). The main concern of these particles is the potential to be transported over a long distance; however, the life span of these ultrafine particles is very short, for instance, a 10 nm particle can be lasting only 15 min.

Interestingly, the emission patterns differ in different geographical locations along with the anthropogenic atmospheric disturbances (i.e., surface emissions/removal of gas-phase species). The formation of ammonium sulfate nanoparticles in the marine atmosphere, which is close to penguin colonies, are suggested to be due to ammonia and sulfuric acids in the atmosphere (Weber et al. 1998).

The photochemicals formed such as photo-oxidized diiodomethane (CH2I2) can cause the ultrafine particle formation in some of the coastal areas (O'Dowd et al. 2002; Jimenez 2003). During winter, there is evidence of identification of ultrafine particles, which consist mainly of organic components and iron with traces of sulfate (Hughes et al. 1998), in Los Angeles, California.

Moreover, Chung et al. reported that the location can be a factor for the composition of particles formed including water-soluble cations and anions, for example, 20% calcium (Ca^{+2}), 11% nitrate, and 5% sulfates in Bakersfield, California, during winter (Chung et al. 2001). Numerous studies conducted in different locations suggested that the local conditions play a key role for the formation of ultrafine particles with varying compositions (Cabada et al. 2004; Fine et al. 2004).

5.5   INDUSTRIAL SETTING

The industrial setting has a huge impact on the net production of nanoparticles/ engineered nanomaterials (ENMs). Having analyzed the information sources available for engineered nanomaterials (ENMs), it has been observed that there is a significant growth in the demand for ENMs and subsequently their production. A report has mentioned that in 2010, 268,00–318,000 mt of ENMs were produced and the year-on-year increase has been 25% (Future Markets Inc. 2012).

To understand how the industrial process releases the ENMs, there is a need for a method using which the sources can be classified appropriately. Vicent and Clement's classification is considered as one of the effective classifications. According to Vincent and Clement's classification, sources of these ENMs can be identified as follows (Vincent and Clement 2000):

1. Exhaust from hot processes (i.e., refining, smelting, welding, etc.)
2. Combustion exhausts (incomplete) (i.e., transportation)
3. Bioaerosols (i.e., endotoxins, viruses, etc.)

The availability of adequate amounts of vaporizable materials in high and sufficient temperature to vaporize the material followed by rapid cooling for aerosol formation triggers the formation of ENMs (Vincent and Clement 2000).

There are different types of industrial processes with high emission potential to emit ultrafine nanoparticles. The powder coating, plasma processing, printing, milling, copying, cooking, and baking are classified under the category of high-emission potential process in industrial settings, according to Vincent and Clement (2000).

The synthesis of ENMs is carried out in a closed system avoiding unnecessary emission to the environment. Hence, unintentional exposure to ENMs while being synthesized is expected to be highly impossible. However, system failure or startup and shutdown of continuous processes can make the accidental release to the environment possible (Luther 2004). Generally, human exposure to ENMs can be due to post-manufacture treatment, handling, or transportation of ENM products. Furthermore, we can consider that isolation, separation, and collection of ENMs are other high-risk processes, especially if the nature of ENMs being handled is dry. Therefore, it is suggested that extra precautions such as ventilation and filtration systems are necessary in handling ENMs as aerosols (Biswas et al. 2005).

Another critical stage in the production of ENMs, with a high risk of releasing, is cleaning operations, which vary from process to process, making different modes of releases. Both simple physical removal methods such as brushes, tissues, or sponges, and strong removal techniques based on chemicals (i.e., chemical/solvent etching) make room for ENM penetration into the solid and liquid waste streams, resulting in solid waste and wastewater a source of ENM in the environment.

One of the characteristic physical properties of NMs is their size, which is insensitive to gravitational effects, therefore leading to a prolonged suspension in the air (Lecoanet and Wiesner 2004). As a result of that, health risks to humans can increase along with the potential explosion risks of pyrophoric material ENMs.

In addition to the possible accessible methods of ENMs to the environment, the solid and liquid waste streams release the consumer products containing the ENMs into the environment. It has been seen that there is a sharp increase in demand for ENMs containing consumer products, especially in the health, fitness, and home and garden sectors. Moreover, the direct link of these sectors with the environment has created a difficult situation in controlling and containment of nanowaste. Well-known consumer products containing ENMs such as zinc oxide, fumed silica, carbon nanotubes, carbon fibers, silica, nanosized alumina, and titanium oxide are part of products consumed daily (Saleem et al. 2018).

5.6 DISTRIBUTION OF NANOMATERIALS IN WASTE STREAMS

In order to design and implement appropriate treatment methods for nanowaste, understanding how the nanowastes enter and their distribution in various waste streams is very important. Currently, there is a lack of scientific work, which suggests possible approaches for effective handling and managing of ENMs in waste streams. Unawareness and negligence of nanowaste management have created an uncontrolled release of NMs into the air, soil, and water streams. Hence, such a release can result in unhealthy contamination of soil, air, and water. In case of water, the contamination can be both in surface water and groundwater. Moreover, Biswas and Wu showed that the geographical scale of NM dispersion, lack of monitoring and remediation technologies, and economic feasibility are the main factors making remediation of the contaminated environment a difficult task (Biswaset et al. 2005).

The huge demand for NMs, as catalytic and performance enhancement additives for fuel and lubricants, may release nanowaste into air, soil, and water systems or through various waste streams such as surface runoffs, spills, and leaks from automotive or sewage/wastewater drainage systems. In the context of the health and fitness sector, which accounts for almost 50% of the global ENM-containing products (Vance et al. 2015), there is a dramatic increase for use of NMs in sunscreens, supplements, and other personal care. The NMs like nano-silver (for antibacterial properties), titanium oxide, zinc oxide (UV absorbance and reflection), gold nanoparticles (anti-aging skin agent), and fullerene (antioxidant) have a great demand.

It is reasonable to assume, based on the demand and the constituents, that the health and personal care sector may be the most significant contributor, in terms of releasing and dispersing of NMs into aquatic and terrestrial environments through various activities such as washing, showering, and swimming. Moreover, Boxall and others have shown that cosmetic products have a higher concentration of ENMs than the other NM-containing products (Boxall et al. 2007). Most NMs of cosmetic products (95%) may likely end up their life cycle in wastewater treatment plants according to Mueller and Nowack (2008). In addition to that, mechanical release is a possible pathway for ENM release at the product disposal stage. A study, which was based on a life-cycle analysis approach and waste management data from eight different world regions (Asia, Australia and Oceania, Central America and Caribbean, Europe, Middle East and North Africa and Greater Arabia, North America, South America, Sub-Saharan Africa) for ten major ENMs, estimated the release of ENM at global levels annually at 189,200 mt into landfills, 51,600 tons into soil, 8100 tons into air, and 69,200 tons into water sinks (Keller et al. 2014). Furthermore, Younis and others described a general flow model of nanowaste into the environment throughout their life cycle (i.e., manufactures, uses, and disposal or recycling processes) (Younis et al. 2018). They have shown that the engineered NMs in waste streams can be considered a contaminant and may be accumulated without the waste being identifiable as nanowaste. This identification was suggested due to the inconsistency of data available for the same type of NMs. For example, some studies conducted on NMs show biocompatibility whereas there is evidence to prove their potential toxic nature, if the disposal level exceeds the threshold level (Sánchez et al. 2011; Boldrin et al. 2014; Mrowiec 2016; Resent 2016). Therefore, the best estimation of current and future quantities of uncontrollable nanowaste can be made based on data gathered from production volumes of NM-related development and manufacturing processes at each geographic area of production (Younus et al. 2018).

5.7 COMMON MECHANISMS OF CYTOTOXICITY OF NPS

The systematic findings on the cellular effect of nanoparticles can be used to model nanotoxicity in the human body. However, this area is still developing and the reports on findings are mostly based on experiments with cultured cells.

It has been noted that inorganic NPs play an important role in toxicology study and also inorganic nanoparticles have been applied as biomedical applications from cancer treatments to imaging techniques on complex cell environments (Soenena et al. 2011). Furthermore, the confinement of NPs in endosomes like subcellular structures can cause very high local concentrations, which cannot be achieved by free ions and which can locally exceed the LD50 values (Soenena et al. 2011).

The common mechanism of cellular level cytotoxicity of widely used inorganic nanoparticles has been reported. According to Sonesen, the mechanisms can affect the cell homeostasis. The mechanisms are generation of reactive oxygen species, cell morphology and cytoskeleton defects, intracellular signaling pathways and genotoxicity, intracellular NP degradability, and interaction with biological molecules (Soenena et al. 2011). The study showed that the minute size plays the key role in causing adverse effects regardless of the intrinsic differences between the various inorganic NPs. Being of small size (the size of natural proteins), the NPs easily can reach places where larger particles cannot enter, for instance, the nucleus or, in case of in vivo settings, NPs can transfer across the placental barrier from pregnant mice to pups (Gu et al. 2009; Chu et al. 2010).

5.8 METHODS OF NANOWASTE MANAGEMENT

Nanowaste management is a new area in waste management and there is a lack of information regarding the quantity of nanowaste generation during nanomaterial and other new product development. As a result, these NMs enter the biological systems and the environment in an uncontrollable manner, giving rise to severe environmental concerns. Further, it is notable that although the conventional waste management methods may not be effective, modification can open new ways to handle the nanowaste. Governments of developing countries should take initiatives with researchers to make systems for nanowaste detection, elimination, recycling, and prevention strategies and technologies before preparing legislative policies.

5.9   DIFFICULTIES AND CONCERNS ABOUT NANOWASTE MANAGEMENT

The behavior of nanoparticles is completely different from their bulk materials because NPs are more chemically reactive and may be more toxic than ordinary regular sized particles. The dynamic transformation taking place during particles' lifetime influences the fate and behavior of NEMs in different environments owing to the intrinsic properties of nanostructures. This problem is further enhanced due to such transformation, influences the fate and behavior of these materials in different environments owing to nanostructures intrinsic properties (e.g., surface chemistry, aggregation, agglomeration, adsorption, or absorption properties, etc.), and environmental factors (pH, presence or absence of oxidants, complexed ions, zeta potential, effects of macromolecules, presence of other chemicals, etc.). Due to this, predicting the behavior of particles under different conditions is a herculean task and this unpredictability poses serious troubles in material handling.

It has been identified that some characteristics such as high surface area, high reactivity, and ability to cross cell membranes have been a serious concern in toxicological research due to their potential toxicity (Marquis et al. 2009; Love et al. 2012).

The key challenge in nanowaste management is that the existing methods and techniques are not fully applicable in removing NWs. Development of new approaches with adequate capability is very important in this regard.

In the process of developing waste removal methods, there are two factors to be taken into consideration:

1. Amount of waste matter generated in the production stage
2. The nature of the wastewater

In some cases, the only available data source is modeling and estimated studies for nanotechnology-related industrial processes. However, the great extent of nanoproduct usage, which is possible in almost every continent, makes it harder to find the amount of nanowaste that has been added to the environment. In this case, new scientific research plans dealing with the toxicology effects of existing and newly developed forms of NMs or nanotechnologies are timely needed.

5.10   CLASSIFICATION OF NANOWASTE

Nanowaste, as a bulk material, or any waste matter can be classified as benign to extremely hazardous. However, nanowaste cannot always be treated in conjunction with bulk matter related in classification. The nature of constituents of the materials and the production life cycle need to be considered. Therefore, nanowaste classification is a challenge in toxicological science, as there is no internationally agreed classification system of nanowaste. However, the first qualitative classification approach proposed by Musee (2011) is presented in Table 5.3. For the classification

TABLE 5.3

Nanowaste Classification as a Function of Constituent Nanomaterials' (NMs) Toxicity and Exposure Potency as a Function of NM Loci in the Nanoproducts

Nanowaste Classes	Description and Disposable Methods	Examples Waste Stream
Class I	Nontoxic and low to high exposure risk is expected Concerns on waste management may only arise if the bulk parent materials (Trojan horse effects) can cause toxicity to humans and the environment through accumulation beyond a certain threshold concentration limit. Otherwise, nanowaste can be handled as benign/safe. No special disposal requirements. Risk profile: none to very low	E-solid waste such as display backplanes of television screens, solar panels, memory chips, and polishing agents
Class II	NM hazard level is harmful/toxic and exposure risk level is low to medium The toxicity of NMs may potentially cause acute or chronic effects therefore, the most suitable and optimal management approach is necessary during handling, transportation, or disposal processes. Risk profile: low to medium	E-solid waste: display backplane, memory chips, polishing agents, solar panels, paints, and coatings
Class III	NM hazard level is toxic to very toxic Exposure: low to medium The recommended appropriate protocol for the waste management chain is the managing hazardous waste streams. Need research to determine if current waste management infrastructure is adequate to deal with hazardousness of waste streams due to nanoscale. materials. Risk profile: medium to high.	Food packaging, food additives, wastewater containing personal care products, polishing agents, pesticides
Class IV	NM hazard level: toxic to very toxic Exposure: medium to high Waste streams should be disposed only in specialized hazardous wastes designated sites. Inadequate WM could lead to serious threats to humans and environmental systems. Risk profile: high.	Paints and coatings, personal care products, Pesticides.
Class V	NM hazard: very toxic to extremely toxic Exposure: medium to high Dispose only in specialized hazardous waste streams designated sites. Poor waste management can cause extensive nanopollution to diverse ecological and water systems, which may prove to be costly, laborious, and time-consuming to remediate. Immobilization and neutralization techniques are among the most effective treatment techniques. Risk profile: high to very high.	Pesticides, sunscreen lotions, and food and beverages containing fullerenes in colloidal suspensions

of NMs, it is a requirement to consider the eight factors that influence the potential exposure of nanomaterials contained in nanoproducts into the receptor organism in the environment.

1. Unknown exposure scenarios of a given product
2. The presence or absence of the coating of the nanomaterials in each product
3. The multiplicity of possible disposal pathways for a given nanowaste stream
4. Effect of the disposal media
5. Bioavailability and persistence of the nanomaterials in different media
6. The potential ease of nanomaterial release from a given product as a function of loci in the nanoproduct
7. Quantities of the nanowaste released into the environment
8. The presence or absence of other environmental contaminants that may lead to antagonistic or synergistic effects to a given set of nanomaterials

Here, the main concern is how stronger the binding of nanomaterial is to the final product. If it is a firmly bound solid nanoproduct, the potential degree of exposure is very low compared to the loosely or freely bound nanomaterials in liquid suspension.

According to the proposed classification methods, class I is potentially less toxic and class V is the most dangerous. There is another complementary classification system based on application of the nanoproducts. Class I shows the lowest risk profile whereas class IV and V have high risk profiles (Table 5.3).

According to the complementary classification suggested by Musee, five application categories are listed as personal care products: food/beverages, sunscreen lotions, automobile parts, and polishing agents. Fullerenes have been reported to have the highest risk NM out of the commonly used NMs for direct human consumption applications such as sunscreen lotions and food/beverages (Musee 2011).

5.11   RISK ASSESSMENT TECHNIQUES AND APPROACHES

In view of the environmental risk scenarios, that is, ecological and human health, involving ENMs, it is necessary to understand the final destination of nanomaterials for introducing new approaches for nanowaste management and this area has been a topic of serious attention among researchers, corporate and government organizations, and policymakers (Owen and Handy 2007). Nanowaste management in developing countries is difficult due to the absence of specific regulation on the matter. Therefore, there is a need for environmental and health risks tools to evaluate and assess these risks associated with nanowaste. In this section, the available risk assessment frameworks for nanowaste management is introduced.

According to Griger and others (2010), ecotoxicological and exposure data for NM have been a priority, but only little effort has been expended to develop tools/approaches for short-term decision-making. The chemical-based environmental risk assessment is a difficult, expensive, and time-consuming process (Owen and Handy 2007; Grieger et al. 2010; 2012; Choi et al. 2009). Also, the unknown chemical and physical properties of ENMs are another major challenge.

According to Maynard (2006), quantitative risk assessment may be a convenient and straightforward approach as feed data for the risk assessment models become available; however, this process may require a few decades to materialize (Maynard 2006). In fact, the risk assessment approaches become more complicated and difficult to manage due to the increase of different varieties of ENMs available in consumer products (Linkov et al. 2009).

There are several different frameworks and approaches proposed for ENM risk analysis strategies depending upon the objectives and scope. The proposed frameworks are categorized into four classes as screening-level frameworks, risk governance frameworks, risk assessment and management frameworks, and adaptable risk assessment tools. An analysis based on a framework, for their suitability and applicability to ENM risk analysis, showed that the frameworks may not be applicable for a specific ENM environmental risk. Therefore, careful selection should be made when determining the most appropriate strategy for a certain risk evaluation.

The principal components of risk assessment methodology for ENMs include hazard/risk identification, exposure assessment, hazard assessment, and exposure risk characterization (Savolainen et al. 2010; Aschberger et al. 2011).

The risk assessment (RA) using the aforementioned four stages helps to determine the possibility of a certain hazardous event and the risk (i.e., hazardous consequence) associated with it. Therefore, it is essential that the presence of both a hazardous element and an exposure to it are necessary for the risk to take place. Therefore, it is clear that the absence of any of the two factors results in no risk associated with the event (Musee 2011; Savolainen et al. 2010). Using an appropriate evidence-based risk management approach, it would be possible to predict the associated risk quantitatively (Saleem et al. 2018). Some of the available risk assessment frameworks are mentioned below and first two of them are described briefly. It is seen that all the frameworks consist of strengths as well as areas to be improved (Hristozov et al. 2016).

1. Nanotechnology risk governance
2. Nano risk framework
3. Scientific Committee on Emerging and Newly Identified Health Risks (SCENIHR)
4. Nano-life cycle risk analysis (nano-LCRA)
5. NMs under the Registration, Evaluation, Authorisation and Restriction of Chemicals (REACH)
6. Comprehensive environmental assessment (CEA)
7. Managing Risks of Nanoparticles (MARINA)
8. European Chemicals Agency/National Institute for Public Health and the Environment (ECHA/RIVM/C) read-across and grouping framework for NMs
9. European Centre for Ecotoxicology and Toxicology of Chemicals (ECETOC) decision analytical framework to facilitate the grouping of NMs (DF4nanoGrouping)
10. ECHA Guidance on the risk assessment of NMs under REACH

11. European Food Safety Authority (EFSA) Guidance on the risk assessment of NMs in the food and feed chain
12. Scientific Committee on Consumer Safety (SCCS) Guidance on the risk assessment of NMs in cosmetic products

5.11.1 NANOTECHNOLOGY RISK GOVERNANCE

The structure of the framework has five phases: (1) pre-assessment; (2) risk appraisal (risk and concern assessments); (3) tolerability and acceptability judgment; (4) risk management, and (5) risk communication. Further, data sharing, societal concerns, and issues are taken into account. Risk communication is considered as an integral part of all stages of the risk governance process and crucial for effectively linking the different components among the strengths of the framework (Hristozov et al. 2016).

5.11.2 NANO RISK FRAMEWORK

The framework consists of six steps: (1) Describe material and application; (2) Profile life cycle(s); (3) Evaluate risks; (4) Assess risk management; (5) Decide, document, and act; and (6) Review and adapt. It considers the potential exposure throughout the whole life cycle of the NMs. The framework is iterative and adaptive. The RA is expected to be updated when new information becomes available and reviews have to be planned when performing the first RA.

5.12 ENVIRONMENTAL EXPOSURE ASSESSMENT FRAMEWORK OF NANOPARTICLES IN SOLID NANOWASTE

This section describes the first environmental assessment framework of solid nanowaste. In 2014, Boldrin and others proposed a framework for environmental exposure assessment of nanoparticles in solid nanowaste. This can be considered as the first such attempt to develop a framework of this nature. However, later on, there were some new models that demonstrate a new approach to develop a framework for nanowaste risk assessment and those models are mentioned in the previous section (Hristozov et al. 2016).

The framework includes five steps: (1) quantification of nanowaste amount, (2) evaluation of matrix properties and nanowaste treatment processes, (3) evaluation of the nanostructures' physicochemical properties, (4) evaluation of transformation processes and release of ENMs into the environment, and (5) assessment of potential exposure. A description of each step is mentioned below.

5.12.1 STEP 1: QUANTIFICATION OF NANOWASTE AMOUNT

Nanowaste generation is quantified through market product analysis according to the life cycle stage of the nanoproduct. Further, Keller and others have shown that quantification can be classified into two categories as shown below (Keller et al. 2013). Since data related to nanowaste is not well documented, there is a need for appropriate analytical techniques for quantification of nanomaterials in waste materials (Boldrin et al. 2014).

1. Nanowaste generated as by-products from ENM manufacturing
2. Nanowaste generated by EOL (end-of-life) nanoproducts

The by-products generated from ENM manufacturing:
Nanowaste may occur as a by-product of the manufacturing process as (1) rejected material from the ENM size selection stage (Ko'hler et al. 2008), (2) residual ancillary materials used for the manufacturing and/or purification of ENMs, and (3) leftover surplus of raw material. Moreover, the top-down approach of nanomaterial production causes more waste than the bottom-up approach (Dahl et al. 2007). The nanowaste manufacturing rate can be measured/estimated by the producer or estimated assuming generation rates, which provide values for certain amounts of wastes per certain amount of ENM. The amount of nanowaste generated can be estimated by combining the generation rates with the amount of nanoproducts (or ENMs) produced (Boldrin et al. 2014).

End-of-life (EOL) nanoproducts
As stated by Campos and López (2018), there are three critical factors that determine the amount of generation of nanowaste from EOL:

1. The amount of nanoproducts produced and traded in the market (the data availability is poor).
2. The life span of ENMs or products, depending on the consumer behavior and material properties present in them.
3. The fraction of the virgin product reaching the EOL stage.

As the fraction of the virgin product reaching the EOL stage depends on the basis of the nature of the NMs and/or the hosting matrix, the amount of nanowaste generated $X_{t,p}$ (kg/year) in year t for p-type nano-product can be calculated using the relationship: $[X_{t,p} = X_{t-rt,p} \times F_{pen,p} \times F_{eol,p}]$, where rt is the duration time in the market phase, $F_{pen,p}$, and $F_{eol,p}$ [0 1] are market penetration parameters, and EOL factors correspond to the fraction of the virgin nano-product reaching the EOL phase and becoming nanowaste. In addition to that, the amount of $NMs_{t,p}$ (kg/year) in nanowaste originating from a given nano-product can be calculated, based on the amount of nanowaste $X_{t,p}$ and knowledge regarding their NM content, from $[NMs_{t,p} = X_{t,p} \times C_{NM,p} \times F_{NM,p}]$ where $C_{NM,p}$ (mg/mg) and $F_{NM,p}$ are concentration of NMs and EOL factor of NMs in their product, respectively (Boldrin et al. 2014).

5.12.2   STEP 2: EVALUATION OF MATRIX PROPERTIES AND NANOWASTE TREATMENT PROCESSES

EOL nanoproducts, which are present in regular waste flows, are not easily controllable whereas nanowaste generated as a by-product is considered to be a controllable category as it is identifiable in nature. Therefore, it is necessary to consider the properties of matrix material and properties of nanoproducts in the management process of EOL nanoproducts.

Moreover, both the physical and chemical stability of the matrix material, becoming important and decisive for managing ENM waste, and the condition, which is relevant for the type of waste treatment under consideration, may further determine the release of ENMs into the environment.

5.12.3   STEP 3: EVALUATION OF THE NANOSTRUCTURES' PHYSICOCHEMICAL PROPERTIES

In order to describe the potential release mechanism of ENM to the environment, the physical and chemical properties of the matrix materials, properties of the ENMs, and the localization in the matrix material are very important parameters to be considered. Hansen et al. (2007) provided a framework for categorizing nanoproducts according to the location of the nanoscale structure in the matrix material:

1. In the "bulk" of the material
2. On the surface of the material
3. As "free" or aggregated nanoparticles

According to Hansen and others (2007) and Boldrin and others (2014), the following ENMs are the most prone to release:

- Surface nanofilms
- ENMs bound to the surface of another solid structure
- ENMs suspended in a liquid
- ENMs suspended in solids
- Airborne ENMs (in enclosed containers)

The release amount of ENM depends on the properties of ENMs, application in the nanoproduct, and the general properties thereof.

5.12.4   STEP 4: EVALUATION OF TRANSFORMATION PROCESSES AND RELEASE OF ENMs INTO THE ENVIRONMENT

The specific conditions in the treatment technology, the potential transformations of the matrix materials, and the ENMs heavily determine the releasing process of ENMs into the environment. Nowack and others (2012) identified some of the potential processes affecting ENMs, as mentioned below (Nowack et al. (2012)).

- Photochemical transformations
- Oxidations
- Reductions
- Dissolution and precipitation
- Adsorption and desorption

- Combustion
- Biotransformation and biodegradation
- Abrasion or mechanical erosion

The processes, as aforementioned, should be evaluated based on the applicability of a variety of nanomaterials and the usefulness of specific properties of a nanomaterial. One of the fundamental principles in modern waste management systems is that waste is collected and treated as quickly as possible to minimize the impact to the environment. In that case, the photochemical method is the least relevant of the processes mentioned above. However, the potential for photochemical transformation cannot be ignored, as the exposure of nanowaste to sunlight, may still be possible (e.g., during collection, storage prior to treatment, and open-dump landfilling). In a landfill, ENMs released into infiltrating water may be determined by processes such as reduction, dissolution/precipitation, and adsorption/desorption. The waste incineration by combustion emits flue gas or the solid residues as the end product of ENMs from the incinerator. Biodegradation and biotransformation are possible processes in biological waste treatment (e.g., anaerobic digestion and composting). The necessary precautions are important during waste collection as ENMs may be emitted into the environment because of abrasion during mechanical compaction (Roes et al. 2012). Further, during recycling processes, material shredding and sorting may cause mechanical erosion (Boldrin et al. 2014).

5.12.5 Step 5: Assessment of Potential Exposure

The final step in the exposure assessment framework is to determine the potential magnitude, frequency, and duration of exposure. Even though a quantitative assessment is difficult due to the lack of relevant information, a qualitative assessment approach could be adopted to identify the level of potential exposure as low, medium, and high. After the qualitative evaluation, having increased the quality of data, the assessment can be replaced by quantitative evaluations.

The application of the assessment framework has been considered only for a few nanoproducts. Nevertheless, it has made a significant contribution to develop methods for nanowaste management.

According to the assessment framework, ENMs in individual waste materials may undergo alternative disposal routes, resulting in different exposure pathways. Moreover, an assessment of ENM exposure routes in relation to the waste management phase is complex and should include evaluations of all critical aspects in relation to the ENMs. Because even small amounts of nanomaterials may potentially have adverse environmental effects (Baun et al. 2009; Stone et al. 2010). In order to avoid such a risk, both the matrix materials and the potential transformation processes in the waste system should be taken into consideration in the evaluation processes.

The environmental exposure to ENMs as a consequence of potential effects of ENMs stemming from waste is significantly related to local conditions such as geographical, cultural context, and the local waste management system. For instance, the sunscreen bottle may undergo incineration in some regions (Boldrin et al. 2014). In order to design an effective nanowaste management system, it is a clear requirement to consider all aforementioned factors.

5.13 NANOWASTE DETECTION AND MONITORING

In the nanowaste management process, one key requirement is the precise characterization and monitoring of EOL life cycle of NMs. It is necessary to develop regulatory strategy in relation to effective nanotoxicity removal protocols. In some cases, the identification of potential hazards associated with nanowaste treatment processing is difficult (Younus et al. 2018).

The quantitative monitoring of NMs, based on their characteristic factors, is very important to develop full steps of risk-associated nanowaste. Analysis of nanomaterials, which exist in various forms, either dispersed in the form of colloids in solutions, particles (dry powders) or thin films, can be performed using various characterization techniques. Although the techniques to be used would depend upon the type of material and information required, usually there are common characteristics such as concentration related to particle number or mass, particle size, distribution, agglomeration/aggregation state, shape, and structure/crystallinity (Part et al. 2015b; Johnson 2016; Oomen et al. 2018). A list of various commonly used techniques can be found in Table 5.4.

The key characteristic parameters used in the detection of NMs are summarized in Table 5.5 with the corresponding characterization tools required for the detection of NMs (Boldrin et al. 2014; Part et al. 2015b).

The characterization of each parameter can be performed with analytical techniques as shown in Table 5.5. The analytical techniques shown in the table can be used for the characterization of nanomaterials but not as tools to treat the type of nanowaste. However, it should be mentioned that there are still limitations and obstacles in nanowaste monitoring in different environmental systems. The accurate quantification of NMs from natural NMs (e.g., Al_2O_3, CeO_2, CuO, SiO_2, TiO_2 or ZnO), which are likely present in higher concentration in different unknown matrices, is almost impossible.

The separation techniques such as AF4, CE, HDC, or SEC can be used to determine the particle size distribution of single NMs in the liquid streams. However, according to Younis and others (2018), there are limitations of these methods regarding particle size of NMs in liquid streams due to clogging of the nano-membrane by unwanted conventional particles that are larger than the pore size. Therefore, it is an obvious fact that current analytical tools are not fully capable of providing evidence for comprehensive nanowaste characterization and nanowaste monitoring. Furthermore, it is a requirement to develop and standardize reliable analytical protocols for nanowaste sampling, characterization, and nano-risk quantification (Younus et al. 2018). In order to assess nano-risk quantification, in-depth analysis of NMs regarding elemental composition or surface properties, at the point of producing NMs, is a main requirement in nanowaste management. The advanced approaches such as use of nanotraces or NMs with properties like using a unique element or isotopic ratios as fingerprints (Part et al. 2015b) will facilitate the improvement of the quality of nano risk assessment and monitoring of nanoproducts.

TABLE 5.4
Analytical Tools Used for Monitoring of NMs in the Environment

Abbreviation	Analytical Tools
AAS	Atomic absorption spectrometry
AES	Atomic emission spectrometry
AFM	Atomic force microscopy
AF4	Asymmetric flow field-flow fractionation, aerodynamic particle sizer
APS	Brunauer–Emmett–Teller theory,
BET	Capillary electrophoresis
CE	Confocal laser scanning microscopy
CLSM	Condensation particle counter
CPC	Dynamic light scattering
DLS	Electronic diffusion battery
EDB	Energy-dispersive X-ray spectroscopy,
EDX	Electron energy loss spectroscopy
EELS	Electrical low-pressure impactor
ELPI	Field-flow fractionation
FFF,	Fast mobility particle sizer
FMPS	Fluorescence spectroscopy
FS	Gas chromatography mass spectrometry
GC-MS,	Scanning transmission electron microscopy
HAADF-STEM	with a high-angle annular dark field detector
HDC	Hydrodynamic chromatography
HPLC	High-performance liquid chromatography
HR-TEM	High-resolution transmission electron microscopy
ICP-MS	Inductively coupled plasma mass spectrometry, inductively coupled plasma
ICP-OES	atomic emission spectroscopy
MALS (Nano)	Multi-angle light scattering (Nano)
DMA (nano)	Differential mobility analyzer
NTA	Nano tracking analysis
SAED	Selected area electron diffraction
SAXS	Small-angle x-ray Scattering
SEC	Size exclusion chromatography
SEM	Scanning electron microscopy
SIMS	Secondary ion mass Spectrometry
SMPS	Scanning mobility particle sizer
SP-ICP-MS	Single-particle inductively coupled plasma mass spectrometry
STM	Scanning tunneling microscope
TD-GC-MS	Thermal-desorption GC-MS
UV/vis	Ultraviolet–visible spectrophotometry
XPS	X-ray photoelectron spectroscopy
XRD	X-ray diffraction

TABLE 5.5
Analytical Tools and Key Parameters for Monitoring of NMs in the Environment

Characteristic Parameters of NM	Analytical Tools	Properties Characterized	Complexities Associated with NM	Reference Source
Concentration related to particle number or mass	FS, GC-MS, MALS, ICP-MS, SP-ICP-MS, UV/vis	Identifying quantities of NMs and their individual level and potential exposure level upon disposal and in nanowastes	Typical high doses may elicit a chronic toxicological effect in nature than doses at lower, more realistic exposure concentrations, even in long time	Part et al. 2015b, Johnson 2016, Oomen et al. 2018, Younis et al. 2018
Composition	EDX, GC-MS, ICP-MS, SP-ICP-MS, XPS	Identifying the NM concentrations and their fate behavior in different leachate streams.	Toxicity study is a factor on the chemical elements that are used for particle's core, shell, or coating (in case of coated nanocomposites) due to their effect on the fate and behavior in environment. In contrast, the results from this sort of analysis is difficult to interpret in the laboratory, much less in the real world	
Particle size	AFM, APS, CE, CLSM, CPC, DMA, DLS, ELPI, FFF, FMPS, FS, HDC, NTA, MALS, SAXS, SEC, SEM, SMPS, SP-ICP-MS, TEM, XRD	Particles size of NM	Particles size can clearly show different toxicities by wide mechanisms of bioavailability and bioaccumulation actions	
Particle size distribution	AFM, APS, CE, CLSM, CPC, DMA, DLS, ELPI, FFF, FMPS, FS, HDC, NTA, MALS, SAXS,SEC, SEM, SMPS, SP-ICP-MS, TEM, XRD	Particle size distribution	Particle size distribution is a critical toxicity key as it influences uptake and toxicity mechanisms in biological systems. It is difficult to characterize nano-size deposition of NMs in biological systems.	

Property	Techniques	Measurement	Description
Agglomeration/ aggregation state	AFM, DLS, CLSM, FFF, FS, NTA, SEM, SP-ICP-MS, TEM	Agglomeration and aggregation of particles	Both agglomeration and aggregation of particles can result in larger particles that may elicit different exposure routes and ecotoxicological effect
Shape	AFM, CLSM, FFF, SEM, TEM	Shape of NM	Understanding the synergistic impacts of NMs and their transportation in environment. The shape and crystallinity of NMs can influence stability, toxicity, and behavior of NMs in their environmental and biological systems
Structure/ crystallinity	HR-TEM, SAED, SAXS, XRD	The shape and crystallinity of NMs	Understanding the process of NM transformation in environment and the impact of released NMs from nano-products and nanowastes. Surface properties such as surface area, charge, functionality, and speciation influence bioavailability, toxicity, and aggregation kinetics of NMs in both environmental and biological systems.
Surface area	AFM, BET, SEM, TEM, XPS	Surface area	However, the natural and anthropogenic activities make it difficult to distinguish the specific source of nano-sized type (natural or engineered), and therefore specific surface entities can enable adequate identification and monitoring the nanotoxicity related to engineered NM

Once ENMs enter the environment or waste streams, their characterisation and detection can be divided into three steps (Bandyopadhyay et al. 2013):

First step: sampling including sample preparation and preservation.
Second step: sample pre-fractionation and/or digestion, followed by separation.
Third step: characterization and/or quantification of the nano-analytes of interest.

All these steps are very important and should be considered in detail, during method development and adaptation, because each stage of the characterization process can cause measurement artifacts, loss of material, or alteration of the analytes (Part et al. 2015).

The sampling of representative waste samples, which are very heterogeneous with respect to element composition and size fractions, is particularly challenging in this regard and usually the most error-prone step in waste analysis (Ferrari et al. 2006). Moreover, it is extremely difficult to differentiate ENMs that are expected in low concentrations from NNMs and the natural background during analysis (Hassellov et al. 2008).

5.14   NANOWASTE TREATMENT PROCESSING

In this section, some of the processing treatments for specific materials are presented. There are many methods for nanowaste treatment as there is a diversity of nanomaterials. These methods broadly can be emphasized in recycling nanomaterials and in designing methods to eliminate nanowaste.

Engineered NMs (ENMs) do not resemble any other known pollutants due to their unique characteristics and their potential toxicity. There is a continuous environmental input of specific NMs containing leachates by intentional or unintentional release. The release can be due to mainly three disposal methods: (1) atmospheric emission during incineration or calcination (inhalation), (2) soil (when the nanowaste is discarded in the landfill), (3) liquid waste streams (from production facilities and purification processes and from nanotechnology clothes that are washed) (Klaine et al. 2008; Musee 2011; Resent 2016).

These disposal routes of NMs and the amount that they carry to the environments vary and depend on the following factors.

1. Operations related to the NM production such as synthetic routes, purification, and cleaning of production chambers
2. Spills from production, storage, transport, and disposal of NMs or nano-products
3. The use and disposal of nano-products including incomplete waste incineration, sewage (washing sunscreen and nano-clothes), and landfill unable to hold NMs back or degrade them
4. Degradation/accumulation affinity of nanotechnology containing NMs in the environmental systems.

A deactivation of NMs containing waste is necessary to minimize the effect of toxicity toward macroscale pollutants in the environment due to the fact that combined NMs with conventional leachate can cause severe problematic conditions. Therefore, we can identify four specific processes that are used to deactivate NM-containing waste streams, depending on the type and risk class of the nanowaste. The four categories of nanowaste treatment are briefly explained below.

Recycling
The problem of recycling NMs is the identification and differentiation of natural and engineered NMs. Further, the dust can occur in the NM residual handling and thus would require specific safety conditions, both to prevent human contact and contact with the environment. Therefore, the outcome and behavior of NMs in the recycling process are unclear due to the challenges in the exposure of nano-products to the real working environment (Mrowiec 2016; Resent 2016; Engelmann et al. 2017).

Incineration (thermal)
In incineration plants, the nanowaste is mixed and thermally treated to destroy the flammable parts, while the thermally undestroyed nanoparticles are left in the combustion chamber. Therefore, specific cleaning procedures and sophisticated nanofilters are required to reduce the amount of nano-hazardous waste. In the worst case, undissolved NMs would remain in the environment as an influence of cleaning on the treatment of persisting NMs in the chimneys of plants. As a result, the removal of the NMs remaining and their efficiency was reported in various ways in several studies. Still, even with this treatment, 20% of the total material would go through the method, which would require additional preventive mechanisms (Mrowiec 2016; Resent 2016; Engelmann et al. 2017).

Landfill deposition (physical-chemical)
Landfilling of waste with untreated (biodegradable) NMs is the main and the most commonly used management technique in the developing countries. The main disadvantage is that the nanoparticles, depending on geography, can leave the landfill by being emitted to the atmosphere, seeped into water, and left on the ground. Consequently, the capture would occur before the aggregation or agglomeration with organic matter and bacteria, which is similar to water treatment. However, the effectiveness of landfill linings in maintaining NMs for the environment and the extent to which landfill or release gas surfaces have not been studied in depth (Mrowiec 2016; Resent 2016; Engelmann et al. 2017).

Sewage treatment (physical-chemical)
As a result of particle release during use and as well as in contact with water, NMs can be found in wastewater treatment, including sludge incinerated and used as fertilizer for agriculture. Therefore, the absence of knowledge exists regarding the environmental impacts resulting from the use of this sludge as fertilizer. As a result, the nanoparticle residues would remain in the surface waters and bulk NMs in solid sludge (Mrowiec 2016; Resent 2016; Engelmann et al. 2017).

5.15   DRAWBACKS OF NANOWASTE TREATMENT METHODS

Since the diversity and complexity of nanowaste is so high, handling the nanowaste is a more difficult task than normal waste. As a result, the applicability of universal methods for nanowaste management is impossible and it is obvious that any waste treatment procedure may have certain drawbacks, some of which are shown below as examples.

1. The combustion processes, involving carbon-based NMs (CNT, graphene, C60 fullerene, etc.); incineration releases subsequent bottom carbon-rich NM ash from incineration due to the effects of process conditions. The heterogeneity of these leaching nanowaste residues has a potential to fabricate or degrade other contaminants and may reduce the efficiency of emission reduction technologies (Boldrin et al. 2014; Mrowiec 2016).
2. It has been shown that the physicochemical and hydraulic conditions of the landfill (i.e., the composition, type, and thickness of the soil liners and pH control of the landfill body as defined by the European Landfill Directive (1999/ 31/EC)) can affect the deactivation processes of the NMs. The landfill condition may also affect the drainage and treatment of nano-leachate and capture of gases in the landfill design (Boldrin et al. 2014; Part et al. 2015b; Mrowiec 2016). As an example, Mrowiec showed that organic acids are able to reduce emissions of CNTs in the landfill leachate through decrease of the agglomeration and immobilization of CNTs and thus their diffusion through a HDPE membrane did not take place (Mrowiec 2016).
3. In case of NMs containing liquid stream treatments, according to some scientific sources, a certain percentage of NM passes untreated and potentially can adversely impact on the life forms as well as can aid in the transfer of nanowaste pollutants from the liquid (influent) to the solid phase (adsorption on sludge matrix), depending on the surface charge, surface functionality, dissolution, nano-size, and presence of stabilizing surfactants (Limbach et al. 2008; Bystrzejewska-Piotrowska et al. 2009; Musee 2011).

The synthetic nanowaste does not disappear even when exposed to harsh environmental conditions due to the unique properties of nanoparticles. For instance, cerium oxide (CeO_2) nanoparticles do not burn or change even when exposed to heat in a waste incineration plant; instead particles remain intact on combustion residues or in the incineration system. The incineration treatment (Mitrano et al. 2017) of cerium oxide was tested using two experiments (Vejerano et al. 2015). In the first experiment, 10 kg of cerium oxide particles of 80 nm in diameter onto communal refuse was incinerated in a waste incineration plant equipped with modern filters and fly-ash separation systems based on electrostatic filters and a wet scrubber. It is important to mention that up to 8 tons of waste were incinerated at the plant per hour. In the second experiment, the particles were sprayed directly into the combustion chamber, thereby simulating a future worst-case scenario with massive nanoparticle release during incineration. The results reported that unfortunately, in both cases, treatment did not remove the nanoparticles.

Olapiriyakul and Caudill reported a case where a high-temperature metal recovery process was used for battery recycling, which may be inadequate for nano-enabled lithium-ion batteries. The nanomaterials in those batteries may require smelting temperatures that are significantly higher than current operating conditions, which results in higher energy consumptions and overall emissions (Olapiriyakul and Caudill 2009).

In addition to the heat treatment methods mentioned above, new treatments have been developed especially for metal recovery. For example, nano-SnO_2 was recovered from industrial electroplating waste sludge using the selective crystallization and growth of acid-soluble amorphous SnO_2 into acid-insoluble SnO_2 nanowires. In another study, absorption-induced crystallization of uranium-rich nanocrystals was used for uranyl enrichment. Also, thermo-reversible liquid–liquid phase transition and cloud point extraction hold promise for the successful separation and recovery of critical, high-value, and resource-limited materials from nanowaste (Zhuang et al. 2012).

5.16 POTENTIAL OPPORTUNITIES FOR THE RECOVERY AND REUSE OF NANOWASTE TECHNOLOGIES

In terms of technology and health, the recycling of nanomaterials is challenging, but there is a keen interest and motivation among manufacturers and users for reuse via EOL nanoproduct cycle and even from the origin of nanowaste streams (NanoTrust Dossier Nr. 009/2009 and Nr. 040/2014 (Greßler et al. 2014); Directive 2008/98/EG).

The limited understanding of the behavior of NMs during the release into the environment has recorded only recovery procedures of certain NMs (OECD 2004/ 2007; OECD and de Tilly 2007; Yamaguchi 2015; BAuA 2013). The consideration of many NMs as hazardous waste may cause the loss of valuable NMs. Consequently, it should be highlighted that the attentiveness to manufacture NMs is very important and they might remain individually or form agglomerates during the recycling process. Some NMs may not be toxic because they may have different physicochemical properties from the "parent" NMs.

Three effects could be identified about recycling of nano-processes: (1) occupational health effects in relation to the recycling processes themselves, (2) environmental impacts related to the residues generated or treated during the recycling processes (i.e., incineration, landfill, or sewage treatment), and (3) introduction of recovered NMs into products containing recycled NMs (Nowack 2009; Mrowiec 2016; Younis et al. 2018). Additionally, there are other factors that affect the effectiveness of NM recycling/recovery processes such as the types and class of NMs and matrices (i.e., pure NMs, nano-by-products, liquid suspension containing NMs, other contaminants with NMs, and solid matrices with integrated NMs) and the technology (Younius et al. 2018).

In the past, conventional techniques such as separation by centrifugation and solvent evaporation were basically tested for nanowaste recycling processes. However, researchers recently have recommended alternative methods that are mainly based on target NM properties such as magnetic fields, pH, and thermo sensitivity of

NMs or by using selective biological extraction procedures, cloud point extraction (CPE), molecular anti-solvents, or nanostructured colloidal solvents. These separation processes are more favorable for the separation process of recycling NMs in terms of effectiveness, cost reduction, time consumption, and energy demand (2011/ 696/EU; OECD 2004/2007; OECD and de Tilly 2007; Kakhki 2015; BAuA 2013; Yamaguchi 2015). Table 5.6 provides examples of selected nanowaste streams with possible recycling procedures (Yamaguchi 2015).

TABLE 5.6
Selected Nanowaste Streams and Current Recycle Procedures for a Given NMs

Existing NMs	Industrial Waste	Recycling Procedure
Metal oxides, CNTs, SiO_2	Metal waste (scrap)	Shredding, smelting
Carbon black (from the ink), TiO_2 (except for special papers, TiO2 is not in the nanoform)	Paper and cardboard	Pulping, de-inking (wet processes)
CNT, SiO_2, TiO_2	Plastic	Collect and sorting or separate collection (e.g., for PET bottles) Mechanical recycling, shredding, washing, and re-granulation Feedstock recycling: depolymerization and cracking (for basic chemicals)
CNTs, Ag nanoparticles	Textiles	Collect, reuse, sort, prepare for reuse, shredding to get fiber NMs
Carbon black (in plastic and in toners), CNTs (in electronic devices and in plastic housings), nano-iron oxide, ZnO, SiO2, Ag (in coatings)	Waste of electronic and electrical equipment (WEEE)	Collecting, dismantling, sorting by hand, shredding, and separating of the fractions, processing of fractions (non-magnetic metals, iron, glass, plastics etc.), further processing of the components (metal melting, material recovery of iron and non-iron metals, extraction of metals from circuit boards
Electrodes with CNTs or nanophosphate® (nLiFePO4)	Batteries	Collect, sorting. Mechanical/chemical and/ or thermal treatment (various procedures, e.g., BATREC (Switzerland) for alkali and mercury batteries; Battery Solutions (USA), Toxco (USA), for lithium batteries; or INMETCO (USA) for Ni-Cd batteries
CNTs, SiO_2, TiO_2, Fe_2O_3, Cu, Ag nanoparticles	Construction and demolition wastes	Reuse of components, sorting of fractions (wood, concrete, brick, metal etc.), metal recycling, secondary building materials, incineration and landfill

TABLE 5.6 (Continued)
Selected Nanowaste Streams and Current Recycle Procedures for a Given NMs

Existing NMs	Industrial Waste	Recycling Procedure
CNTs, SiO_2, TiO_2, (in plastics, coatings, and paints)	End-of-life vehicles (ELV)	Dismantling for reusable parts (including tires), Removal of hazardous components (e.g., batteries), shredding and separation of fractions, Metals go to smelting and refining, Glass is recycled or landfilled Nonmetallic shredder residues for incineration or landfill
Carbon black, silica; there are indications that future developments will include others, e.g., CNTs, nanoclay (SiO_2), or organic nano-polymers	Tires	Collect, storage (danger of ignition), refurbish and reuse, shredding of metal, reuse of rubber for downcycled products or for energy recovery
NMs from nanowaste in the municipal waste that are not destroyed or evaporated may stay in the bottom ash	Recycling of residues from waste incineration	Separation of metal bottom ash from municipal solid waste (MSW), (ca. 220 kg of bottom ash is produced when incinerating tons of MSW, these contain metal residues (iron, aluminum, copper, even gold) from MSW.

5.17 ENVIRONMENTAL IMPLICATIONS OF NANOTECHNOLOGY IN SOME DEVELOPING COUNTRIES

There is a big gap between developed countries and developing countries in terms of handling nanowaste. However, nanotechnology research areas, industrial activities, and medical applications are growing rapidly even in developing countries. It has been reported that different international organizations such as Organisation for Economic Co-operation and Development (OECD), International Union for Conservation of Nature (IUCN), and developed countries have taken necessary actions to manage nanowaste. Developing countries also should seriously consider the importance of nano contamination and nanowaste management, especially because developed countries do not provide efficient and effective protection against this type contamination and adverse effects worldwide (Erdogan et al. 2019). Here, a few suggestions like workshops, education, and research activities, can be made based on a study done from Turkey (Erdogan et al. 2019).

5.17.1 India

India is the largest country in the South Asian region and has a population of over 1.3 billion people. The Indian government is the main contributor to nanotechnology development in the country. However, industry participation is also increasing, especially as a public–private partnership mainly focusing on pharmaceutical/nanomedicine, textiles, nanocomposites, and agriculture opportunities (see http://nanomission.gov.in). There are at least 170 institutions and universities involved in nanotechnology research, international nanotechnology conferences are being organized on a regular basis, and there are two Indian scientific periodicals dedicated exclusively to nanotechnology.

As a well-established nanotechnological member of the region, in order to regulate the environmental impact, there are several institutions involved in nanotoxicology and other aspects of nano risk assessment in India. Table 5.7 shows the institutions and their research areas in the context of nanowaste . In addition to the institutions mentioned in the table, there are few other institutions involved in

TABLE 5.7
The Institutions and Their Research Areas in the Context of Nanowaste in India

Institute	Focus
Indian Institute of Toxicology Research (IITR)	The development and validation of methods for nanomaterial toxicology in vitro and in vivo toxicity of nanoparticles used in consumer products and therapeutics
	Eco-toxicity of certain nanostructured polymers, metals, and metal oxides
	Efficacy and safety of nano-based herbal products
	Guidelines for the safe handling of nanomaterials
National Institute of Pharmaceutical Education and Research (NIPER)	Test the toxicity of newly developed drugs based on nanotechnologies and hosts the Center for Pharmaceutical Nanotechnology regulatory guidelines for approving nanotechnology-based drugs
	Protocols and guidelines for toxicity
	Testing software to analyze links between nanomaterial properties and physiological responses
Indian Institute of Chemical Technology (IICT)	Nano science research protocols
	In vivo and in vitro toxicology studies
Indian Council of Medical Research (ICMR)	Formulates, coordinates, and promotes nanomaterial safety-related biomedical research
	Nanomaterial safety studies
	Home to the National Institute of Occupational Health (NIOH)
Central Drug Research Institute (CDRI)	Nanotoxicological studies
Bureau of Indian Standard Committee MTD33	Fund toxicology studies on various nanomaterials

risk assessment and risk management including the Center for Occupational and Environmental Health (COEH), the Bureau of Indian Standards (BIS) sectional committee on nanotechnology (MTD 33), the Central Food Technology Research Institute (CFTRI), the National Environmental Engineering Research Institute (NEERI), the National Chemical Laboratory (NCL), the National Institute of Oceanography (NIO), the Technology Information, Forecasting and Assessment Council (TIFAC),and the Indian Council of Agricultural Research (ICAR) (Azoulay et al. 2013).

Even though India has established a network of institutes, the country needs to develop a framework of legislation that empowers the legal provisions by which nanotechnological risks and the potentially detrimental effects of nanotechnologies on health/environment could be addressed. Furthermore, the lack of coordination among a multitude of government departments is the main issue in dealing with risk identification and management, which is the key in nanowaste management. David Azoulay and others mentioned, in their report on social and environmental implications of nanotechnology development in the Asia-Pacific region, that none of the acts or legislation of Ministry of Environment and Forest (MOEF) have explicitly identified nanoparticles as a major category of risk-related areas under the responsibility of MoEF (Azoulay et al. 2013).

5.17.2 PAKISTAN

Pakistan is a semi-industrial country with a considerable potential to grow in terms of nanotechnology. The following research institutes are involved in nanotechnology research: the Pakistan Institute of Nuclear Science and Technology (PINSTECH), COMSATS Institute of Information Technology (CIIT), National Institute of Biotechnology and Genetic Engineering (NIBGE), Quaid-i-Azam University, Pakistan Institute of Engineering and Applied Sciences (PIEAS), GIK Institute of Engineering Sciences and Technology (GIKI), and Pakistan Council of Scientific and Industrial Research (PCSIR).

According to www.nanostar.com, Pakistan has been recorded in the 20th place in terms of number of nanotechnology-related articles indexed in Web of Science (Azoulay et al. 2013).

5.17.3 BRAZIL

The first nanotoxicology network in Brazil was launched in 2011, with the collaboration of the National Council for Scientific and Technological Development (CNPq), together with the Ministry of Science and Technology and Innovation (MCTI), Brazil. One of the significant works presented was the project titled "Occupational and environmental nanotoxicology: scientific subsidies to establish regulatory frameworks and risk assessment" (MCTI / CNPq process 552131/2011-3), pointing to toxic effects and confirmation of evidence that the carbon nanotubes are potentially hazardous in aquatic environments. It shows possible neurotoxicity effects on zebrafish (Engelmann et al. 2017).

5.18   GREEN SYNTHESIS OF NANOMATERIALS: ZERO NANOWASTE

The manufacturing of nano-products by processes that are energy efficient with minimal waste is one of the highlighted areas in environmental science as it is a well-known fact that nanowaste treatment to eliminate all health hazards associated with NMs can be achieved only to a certain extent. Therefore, there is a need to seek alternative approaches to minimize nanowaste streams even at production stages. This section presents an alternative, but a potential approach to synthesis of NMs, identified as a novelty to nanotechnology and indirectly as an alternative solution to overcome health risks due to nanowaste.

Green synthesis, which is mainly based upon the principles of green chemistry, is a new platform to design novel non-hazardous nanomaterials to human and environmental health and has the extensive potential to revolutionize large-scale nanomanufacturing processes (Duan et al. 2015). Furthermore, it is believed that the application based on these greener nanomaterials in the field of nanomedicine as drug carriers may appear in the near future (Nath and Banerjee 2013). According to Gawanda, the greener synthesis of nanomaterials sets the benchmark for the development of cleaner, safer, and sustainable nanoproducts and nanomaterials (Gawande et al. 2013).

Several research reports published have shown the possibility of applying green synthetic processes in developing green NMs from natural resources using eco-friendly and sustainable methodologies such as biological processes, bio-extracts, aerosol technologies, ultrasonic radiations (i.e., UV and gamma radiations, laser ablation, microwave) and photochemical reduction techniques, and other synthetic methods (Eckelman et al. 2008; Dhingra et al. 2010; Smith and Granqvist 2010; Kharissova et al. 2013; Virkutyte and Varma 2013; Rickerby 2014; Basiuk and Basiuk 2015; Kanagaraj et al. 2015; Al-ruqeishi et al. 2016).

The most studied natural substance categories, which serve as stabilization, capping, and reducing agents in green synthesis of metal nanoparticles for electronics and medical applications, are vitamins, proteins, plant extracts, carbohydrates (e.g., glucose), biopolymers (e.g., lignin, cellulose, chitosan), peptides (e.g., glutathione), algae, bacteria, and fungi (Iravani 2011; Lu and Ozcan 2015).

Most green nanotechnology methods are limited to the laboratory scale without the feasibility of expanding to industrial scale due to their novelty and high cost.

It is important to regulate the exciting policies associated with "greener" nano-technology production. In this context, Michael Berger of Nanowerk LLC (2017) has suggested a potential solution for "greener" nano-synthesis methodology called "Green Alternatives Methodology," by applying the 12 principles of green chemistry. The US Environmental Protection Agency (EPA) has also outlined ten steps of combined life cycle and risk assessment and Comprehensive Environmental Assessment to "green" nanomanufacturing processes. The development of benign nano-innovation design using green nanotechnologies would apply to a wide range of economic sectors to provide sustainable solutions for global issues such as food packaging, biomedicine, energy shortages, and clean water scarcity, electronics, and many other health and environmental areas of concern (Younis et al. 2018).

5.19 SUMMARY

In the future, the nanotechnology field will expand continuously providing solutions for many problems related to energy, medicine, cost reduction, crop production, and pollution control. Nevertheless, nanowaste is a consequence of the development of nanotechnology and also new approaches are necessary to find long-lasting solutions to that problem. The current situation in developing countries should be improved as NM products are spreading worldwide without a limit to certain regions or countries. Therefore, developing countries should be able to develop their own nanowaste management and risk assessment framework in line with international policies that have been created by international legislation for nanowaste treatment, especially in industries, where a large amount of nanowaste is released.

REFERENCES

Adams FC and Barbante C (2013) Nanoscience, nanotechnology and spectrometry. *Spectrochim Acta Part B* 86:3–13. https://doi.org/10.1016/j.sab.2013.04.008

Ålander T, Antikainen E, Raunemaa T, Elonen E, Rautiola A, and Torkkell K (2005) Particle emissions from a small two-stroke engine: effects of fuel, lubricating oil, and exhaust after treatment on particle characteristics. *Aerosol Sci Technol* 39:151–161. https://doi.org/10.1080/ 027868290910224

Al-Mubaddel FS, Haider S, Al-Masry WA, Al-Zeghayer Y, Imran M, Adnan H, and Zahoor U (2017) Engineered nanostructures: a review of their synthesis, characterization and toxic hazard considerations. *Arab J Chem* 10:S376–S388. https://doi.org/10.1016/j.ara bjc.2012.09.010

Al-ruqeishi MS, Mohiuddin T, and Al-saadi LK (2016) Green synthesis of iron oxide nanorods from deciduous Omani mango tree leaves for heavy oil viscosity treatment. *Arab J Chem* 12(8):4080–4090. https://doi.org/10.1016/j.arabjc.2016.04.003

Anderson PJ, Wilson JD, and Hiller FC (1989) Particle size distribution of mainstream tobacco and marijuana smoke: analysis using the electrical aerosol analyzer. *Am Rev Respir Dis* 140:202–205. https://doi.org/10.1164/ajrccm/140.1.202

Aschberger K, Micheletti C, Sokull-Klüttgen B, Christensen FM (2011) Analysis of currently available data for characterising the risk of engineered nanomaterials to the environment and human health—lessons learned from four case studies. *Environ Int* 37:1143–1156. https://doi.org/10.1016/j.envint.2011.02.005

Azoulay D, Senjen R, and Foladori G (2013) *Social and Environmental Implications of Nanotechnology Development in Asia-Pacific*. IPEN, Gothenburg, Sweden.

Bandyopadhyay S, Peralta-Videa JR, and Gardea-Torresdey JL (2013) Advanced analytical techniques for the measurement of nanomaterials in food and agricultural samples: a review. *Environ Eng Sci* 30:118–125.

Basiuk VA and Basiuk EV (2015) *Green Processes for Nanotechnology: From Inorganic to Bioinspired Nanomaterials*. Springer, Cham.

BAuA (2013) German Federal Institute for Occupational Safety and Health (BAuA), Committee on Hazardous Substances (AGS), Announcement on Hazardous Substances, Manufactured Nano-materials, Announcement 527, Berlin. www.baua.de/en/Topics-from-A-to-Z/Hazardous-Substances/TRGS/Announcement-527.html

Baun A, Hartmann NIB, Grieger KD, and Hansen SF (2009) Setting the limits for engineered nanoparticles in European surface waters—are current approaches appropriate. *J Environ Monit* 11:1774–1781.

Berger M (2008) *Nanotechnology—Not That Green?* Nanowerk LLC, Berlin. Available online: www.nanowerk.com/spotlight/spotid=7853.php. Accessed June 16, 2010.

Bhatia S (2016) Nanoparticles types, classification, characterization, fabrication methods and drug delivery applications. In *Natural Polymer Drug Delivery Systems.* Springer International Publishing, Switzerland, pp. 33–93. https://doi.org/10.1007/978-3-319-41129-3_2

Biswas P and Wu C-Y (2005) Nanoparticles and the environment. *J Air Waste Manag Assoc* 55:708–746. https://doi.org/10.1080/10473289.2005.10464656

Boldrin A, Hansen SF, Baun A, et al. (2014) Environmental exposure assessment framework for nanoparticles in solid waste. *J Nanopart Res* 16:2394. https://doi.org/10.1007/s11051-014-2394-2

Boxall AB, Chaudhry Q, Sinclair C, Jones A, Aitken R, Jefferson B, and Watts C (2007) *Current and Future Predicted Environmental Exposure to Engineered Nanoparticles.* Central Science Laboratory, London.

Bystrzejewska-Piotrowska G, Golimowski J, and Urban PL (2009) Nanoparticles: their potential toxicity, waste and environmental management. *Waste Manag* 29(9): 2587–2595. https://doi.org/10.1016/j.wasman.2009.04.001

Cabada JC, Rees S, Takahama S, Khlystov A, Pandis SN, Davidson CI, and Robinson AL (2004) Mass size distributions and size resolved chemical composition of fine particulate matter at the Pittsburgh supersite. *Atmos Environ* 38:3127–3141. https://doi.org/10.1016/j. atmosenv.2004.03.004

Campos A, and López I (2018) *Handbook of Environmental Materials Management.* Springer, Cham. https://doi.org/10.1007/978-3-319-58538-3_161-1

Chang M-CO, Chow JC, Watson JG, Hopke PK, Yi S-M, and England GC (2004) Measurement of ultrafine particle size distributions from coal-, oil-, and gas-fired stationary combustion sources. *J Air Waste Manag Assoc* 54:1494–1505. https://doi.org/10.1080/10473289.2004.10471010

Choi J-Y, Ramachandran G, and Kandlikar M (2009) The impact of toxicity testing costs on nanomaterial regulation. *Environ Sci Technol* 43:3030–3034. https://doi.org/10.1021/es802388s

Chu M, Wu Q, Yang H, Yuan R, Hou S, and Yang Y (2010) Transfer of quantum dots from pregnant mice to pups across the placental barrier. *Small* 6(5):670–678.

Chung A, Herner JD, and Kleeman MJ (2001) Detection of alkaline ultrafine atmospheric particles at Bakersfield, California. *Environ Sci Technol* 35:2184–2190. https://doi.org/10.1021/es0018791

Clarke AD (1993) Atmospheric nuclei in the Pacific midtroposphere: their nature, concentration, and evolution. *J Geophys Res* 98:20633. https://doi.org/10.1029/93JD00797

Covert DS, Wiedensohler A, Aalto P, Heintzenberg J, Mcmurry PH, and Leck C (1996b) Aerosol number size distributions from 3 to 500 nm diameter in the arctic marine boundary layer during summer and autumn. *Tellus Ser B Chem Phys Meteorol* 48:197–212. https://doi.org/10.3402/tellusb.v48i2.15886

Dahl JA, Maddux BLS, and Hutchison JE (2007) Toward greener nanosynthesis. *Chem Rev* 107(6):2228–2269

Dhingra R, Naidu S, Upreti G, and Sawhney R (2010) Sustainable nanotechnology: through green methods and life-cycle thinking. *Sustainability* 2:3323–3338.

Dowling A, Clift R, Grobert N, et al. (2004) Nanoscience and nanotechnologies: opportunities and uncertainties. *Lond R Soc R Acad Eng Rep* 46:618–618. https://doi.org/10.1007/s00234-004-1255-6

Duan H, Wang D, and Li Y (2015) Green chemistry for nanoparticle synthesis. *Chem Soc Rev* 44:5778–5792. https://doi.org/10.1039/C4CS00363B

Eckelman MJ, Zimmerman JB, and Anastas PT (2008) Toward green nano. *J Ind Ecol* 12:316–328.

Engelmann W, Leal DWS, and Von Hohendorff R (2017) The nanotechnological revolution and the complexity of waste: the possibility of using the OECD protocol as an alternative to nanowaste risk management. *Adv Recycl Waste Manag* 2:136. https://doi.org/10.4172/2475-7675.1000136

Erdogan O and Kara M (2019) Analytical approach to the waste management of nanomaterials in developing countries. *Front Drug Chem Clin Res* 2:1–5. doi: 10.15761/FDCCR.1000117

Faiz A, Weaver CS, and Walsh MP (1996) *Air Pollution from Motor Vehicles*. World Bank. https://doi.org/10.1596/0-8213-3444-1

Ferrari BJ, Masfaraud JF, Maul A, and Férard JF (2006) Predicting uncertainty in the eco-toxicological assessment of solid waste leachates. *Environ Sci Technol* 40(22):7012–7017. doi:10.1021/es052491z

Fine PM, Chakrabarti B, Krudysz M, Schauer JJ, and Sioutas C (2004) Diurnal variations of individual organic compound constituents of ultrafine and accumulation mode particulate matter in the Los Angeles Basin. *Environ Sci Technol* 38:1296–1304. https://doi.org/10.1021/es0348389

Future Markets Inc. (2012) *The Global Market for Nanomaterials 2002–2016: Production Volumes, Revenues and End User Market Demand*. Future Markets, Edinburgh.

Gawande MB, Branco PS, and Varma RS (2013) Nano-magnetite (Fe_3O_4) as a support for recyclable catalysts in the development of sustainable methodologies. *Chem Soc Rev* 42:3371–3393.

Gottschalk F, Scholz RW, and Nowack B (2010) Probabilistic material flow modeling for assessing the environmental exposure to compounds: methodology and an application to engineered nano TiO_2 particles. *Environ Model Softw* 25:320–332

Greßler S, Part F, and Gazsó A (2014) "Nanowaste": nanomaterial-containing products (NanoTrust-Dossier Nr. 040–August 2014). https://doi.org/10.1553/ita-nt-040

Grieger KD, Baun A, and Owen R (2010) Redefining risk research priorities for nanomaterials. *J Nanopart Res* 12:383–392. https://doi.org/10.1007/s11051-009-9829-1

Grieger KD, Linkov I, Hansen SF, and Baun A (2012) Environmental risk analysis for nanomaterials: review and evaluation of frameworks. *Nanotoxicology* 6:196–212. https://doi.org/10.3109/17435390.2011.569095

Gu X, Zheng Y, Cheng Y, Zhong S, and Xi T (2009) In vitro corrosion and biocompatibility of binary magnesium alloys. *Biomaterials* 30(4):484–498. https://doi.org/10.1016/j.biomaterials.2008.10.021

Hansen SF, Larsen BH, Olsen SI, and Baun A (2007) Categorization framework to aid hazard identification of nanomaterials. *Nanotoxicology* 1:243–250.

Harris SJ, and Maricq MM (2001) Signature size distributions for diesel and gasoline engine exhaust particulate matter. *J Aerosol Sci* 32:749–764. https://doi.org/10.1016/S0021-8502(00)00111-7

Hassellov M, Readman JW, Ranville JF, and Tiede K (2008) Nanoparticle analysis and characterization methodologies in environmental risk assessment of engineered nanoparticles. *Ecotoxicology* 17:344–361.

Hristozov D, Gottardo S, Semenzin E, et al. (2016) Frameworks and tools for risk assessment of manufactured nanomaterials. *Environ Int* 95:36–53. https://doi.org/10.1016/j.Envint.2016.07.016

Hughes LS, Cass GR, Gone J, Ames M, and Olmez I (1998) Physical and chemical characterization of atmospheric ultrafine particles in the Los Angeles area. *Environ Sci Technol* 32:1153–1161. https://doi.org/10.1021/es970280r

Iravani S (2011) Green synthesis of metal nanoparticles using plants. *Green Chem* 13:2638–2650.

Jimenez JL, Bahreini R., David RC III, Zhuang H, Varutbangkul V, Flagan, RC, Seinfeld JH, O'Dowd CD, and Hoffmann T (2003) New particle formation from photooxidation of diiodomethane (CH_2I_2). *J Geophys Res* 108:4318. https://doi.org/10.1029/2002J D002452

Johnson DR (2016) Nanometer-sized emissions from municipal waste incinerators: a qualitative risk assessment. *J Hazard Mater* 320:67–79. https://doi.org/10.1016/j.jhaz mat.2016.08.016

Kakhki RM (2015) Recent advances in removing nanowastes by the cloud point. *Jordan J Chem* 10:149–160.

Kanagaraj J, Senthilvelan T, Panda RC, and Kavitha S (2015) Eco-friendly waste management strategies for greener environment towards sustainable development in leather industry: a comprehensive review. *J Clean Prod* 89:1–17. https://doi.org/10.1016/j.jcle pro.2014.11.013

Keller A, McFerran S, Lazareva A, and Suh S (2013) Global life cycle releases of engineered nanomaterials. *J Nanopart Res* 15(6):1692.

Keller AA and Lazareva A (2014) Predicted releases of engineered nanomaterials: from global to regional to local. *Environ Sci Technol Lett* 1:65–70. https://doi.org/10.1021/ez400106t

Khan I, Saeed K, and Khan I (2017) Nanoparticles: properties, applications and toxicities. *Arab J Chem* 12(7):908–931. https://doi.org/10.1016/j.arabjc.2017.05.011

Khan IA, Berge ND, Sabo-Attwood T, et al. (2013) Single-walled carbon nanotube transport in representative municipal solid waste landfill conditions. *Environ Sci Technol* 47: 8425–8433.

Kharissova OV, Dias HVR, Kharisov BI, et al. (2013) The greener synthesis of nanoparticles. *Trends Biotechnol* 31:240–248. https://doi.org/10.1016/j.tibtech.2013.01.003

Klaine SJ, Alvarez PJJ, Batley GE, Fernandes TF, Handy RD, Lyon DY, Mahendra S, McLaughlin MJ, and Lead JR (2008) Nanomaterials in the environment: behavior, fate, bioavailability, and effects. *Environ Toxicol Chem* 27(9):1825–1851.

Kohler AR, Som C, Helland A, and Gottschalk F (2008) Studying the potential release of carbon nanotubes throughout the application life cycle. *J Clean Prod* 16:927–937.

Kulmala M, Vehkamäki H, Petäjä T, Dal Maso M, Lauri A, Kerminen V-M, Birmili W, and McMurry PH (2004) Formation and growth rates of ultrafine atmospheric particles: a review of observations. *J Aerosol Sci* 35:143–176. https://doi.org/10.1016/ j.jaerosci.2003.10.003

Kwak JI and An YJ (2016) The current state of the art in research on engineered nanomaterials and terrestrial environments: different-scale approaches. *Environ Res* 151:368–382. https://doi.org/10.1016/j.envres.2016.08.005

Lecoanet HF and Wiesner MR (2004) Velocity effects on fullerene and oxide nanoparticle deposition in porous media. *Environ Sci Technol* 38:4377–4382. https://doi.org/10.1021/ es035354f

Li C-S, Jenq F-T, and Lin W-H (1992) Field characterization of submicron aerosols from indoor combustion sources. *J Aerosol Sci* 23:547–550. https://doi.org/10.1016/ 0021-8502(92)90470-G

Limbach LK, Bereiter R, Müller E, Krebs R, Gälli R, and Stark WJ (2008) Removal of oxide nanoparticles in a model wastewater treatment plant: influence of agglomeration and surfactants on clearing efficiency. *Environ Sci Technol* 42(15):5828–5833.

Linkov I, Steevens J, Adlakha-Hutcheon G, Bennett E, Chappell M, Colvin V, Davis JM, Davis T, Elder A, Foss Hansen S, Hakkinen PB, Hussain SM, Karkan D, Korenstein R, Lynch I, Metcalfe C, Ramadan AB, and Satterstrom FK (2009) Emerging methods and tools for environmental risk assessment, decision-making, and policy for nanomaterials: summary of NATO Advanced Research Workshop. *J Nanopart Res* 11:513–527. https://doi.org/10.1007/s11051-008-9514-9

Love SA, Maurer-Jones MA, and Thompson JW (2012) Assessing nanoparticle toxicity. *Annu Rev Anal Chem* 5:181–205.

Lu Y and Ozcan S (2015) Green nanomaterials: on track for a sustainable future. *Nano Today* 10:417–420. https://doi.org/10.1016/j.nantod.2015.04.010

Luther W (2004) Industrial application of nanomaterials: chances and risks technological analysis. *Futur Technol* 54:1–112.

Maguhn J, Karg E, Kettrup A, and Zimmermann R (2003) On-line analysis of the size distribution of fine and ultrafine aerosol particles in flue and stack gas of a municipal waste incineration plant: effects of dynamic process control measures and emission reduction devices. *Environ Sci Technol* 37:4761–4770. https://doi.org/10.1021/es020227p

Marquis BJ, Love SA, Braun KA, and Haynes CL (2009) Analytical methods to assess nanoparticle toxicity. *Analyst* 134:425–439.

Maynard AD (2006) Nanotechnology: assessing the risks. *Nano Today* 1:22–33. https://doi.org/ 10.1016/S1748-0132(06)70045-7

Mitrano D, Mehrabi K, Arroyo Y, and Nowack C (2017) Mobility of metallic (nano) particles in leachates from landfills containing waste incineration residues. *Environ Sci Nano* 4:480.

Mrowiec B (2016) Directions and possibilities of the safe nanowaste management. *CHEMIK* 70:593–596.

Mueller NC and Nowack B (2008) Exposure modeling of engineered nanoparticles in the environment. *Environ Sci Technol* 42:4447–4453. https://doi.org/10.1021/es7029637

Musee N (2011) Nanowastes and the environment: potential new waste management paradigm. *Environ Int* 37:112–128.

Nanowerk (2017) Nanomaterial DatabaseTM. Nanowerk LLC ("NANOWERK"). www.nanowerk.com/phpscripts/n_dbsearch.php. Accessed 10 September 2017.

Nath D and Banerjee P (2013) Green nanotechnology—a new hope for medical biology. *Environ Toxicol Pharmacol* 36(3):997–1014.

Nowack B (2009) Is anything out there? What life cycle perspectives of nano products can tell us about nanoparticles in the environment. *Nano Today* 4:11–12. https://doi.org/10.1016/j.nantod.2008.10.001

Nowack B, et al. (2012) Potential scenarios for nanomaterial release and subsequent alteration in the environment. *Environ Toxicol Chem* 31(1):50–59.

O'Dowd CD, Jimenez JL, Bahreini R, Flagan RC, Seinfeld JH, Hameri K, Pirjola L, Kulmala M, Jennings SG, and Hoffmann T (2002) Marine aerosol formation from biogenic iodine emissions. *Nature* 417:632–636.

OECD, de Tilly S, et al. (2007) *Guidance Manual for the Implementation of the OECD Recommendation C(2004)100 on Environmentally Sound Management.* OECD, Paris. www.oecd.org/env/waste/39559085.pdf

Olapiriyakul S and Caudill RJ (2009) Thermodynamic analysis to assess the environmental impact of end-of-life recovery processing for nanotechnology products. *Environ Sci Technol* 43(21):8140–8146.

Oomen AG, Steinhäuser KG, Bleeker EAJ, van Broekhuizen F, Sips A, Dekkers S, Wijnhoven SWP, and Sayre PG (2018) Risk assessment frameworks for nanomaterials: scope, link to regulations, applicability, and outline for future directions in view of needed increase in efficiency. *NanoImpact* 9:1–13. https://doi.org/10.1016/j.impact.2017.09.001

Owen R and Handy R (2007) Viewpoint: formulating the problems for environmental risk assessment of nanomaterials. *Environ Sci Technol* 41:5582–5588. https://doi.org/10.1021/es072598h

Part F, Zecha G, Causon T, Sinner E-K, and Huber-Humer M (2015) Current limitations and challenges in nanowaste detection, characterisation and monitoring. *Waste Manag* 43:407–420.

Pokropivny V, Lohmus R, Hussainova I, Pokropivny A, and Vlassov S (2007) *Introduction to Nanomaterials and Nanotechnology*. Tartu University Press, Ukraine, pp. 1–225.

Purohit R, Mittal A, Dalela S, Warudkar V, Purohit K, and Purohitc S (2017) Social, environmental and ethical impacts of nanotechnology. *MaterToday* Proc 4(4):5461–5467. https://doi.org/10.1016/j.matpr.2017.05.058

Resent I (2016) Risk of nanowastes. *Inżynieria i Ochr Środowiska* 19:469–478. https://doi.org/10.17512/ios.2016.4.3

Rickeard DJ, Bateman JR, Kwon YK, McAughey JJ, and Dickens CJ (1996) Exhaust particulate size distribution: vehicle and fuel influences in light duty vehicles. SAE Technical Paper 961980. https://doi.org/10.4271/ 961980

Rickerby D (ed) (2014) *Nanotechnology for Sustainable Manufacturing*. CRC Press, Boca Raton, FL.

Ristovski ZD, Morawska L, Bofinger ND, and Hitchins J (1998) Submicrometer and supermicrometer particulate emission from spark ignition vehicles. *Environ Sci Technol* 32:3845–3852. https://doi.org/10.1021/es980102d

Roes L, Patel MK, Worrell E, and Ludwig C (2012) Preliminary evaluation of risks related to waste incineration of polymer nanocomposites. *Sci Total Environ* 417–418:76–86.

Rühle T, Schneider H, Find J, Herein D, Pfänder N, Wild U, Schlögl R, Nachtigall D, Artelt S, and Heinrich U (1997) Preparation and characterisation of Pt/Al$_2$O$_3$ aerosol precursors as model Pt-emissions from catalytic converters. *Appl Catal B Environ* 14:69–84. https://doi.org/10.1016/S0926-3373(97)00013-1

Saleem J, Shahid UB, and McKay G (2018) Environmental nanotechnology. In C Hussain (ed), *Handbook of Environmental Materials Management*. Springer, Cham, pp. 1–32. https://doi.org/10.1007/978-3-319-58538-3_94-1

Sánchez A, Recillas S, Font X, Casals E, González E, and Puntes V (2011) Ecotoxicity of, and remediation with, engineered inorganic nanoparticles in the environment. *TrAC – Trends Anal Chem* 30:507–516. https://doi.org/10.1016/j.trac.2010.11.011

Savolainen K, Alenius H, Norppa H, Pylkkänen L, Tuomi T, and Kasper G (2010) Risk assessment of engineered nanomaterials and nanotechnologies—a review. *Toxicology* 269:92–104. https:// doi.org/10.1016/j.tox.2010.01.013

Schulte PA, McKernan LT, Heidel DS, Okun AH, Dotson GS, Lentz TJ, Geraci CL, Heckel PE, and Branche CM (2013) Occupational safety and health, green chemistry, and sustainability: a review of areas of convergence. *Environ Health* 12(1):1–9. https://doi.org/10.1186/1476-069X-12-31

Seinfeld JH and Pandis SN (1998) *Atmospheric Chemistry and Physics: From Air Pollution to Climate Change*, 2nd edn. Wiley, New York.

Shi JP, Evans DE, Khan A, and Harrison RM (2001) Sources and concentration of nanoparticles (<10nm diameter) in the urban atmosphere. *Atmos Environ* 35:1193–1202. https://doi.org/10.1016/S1352-2310(00)00418-0

Smith GB and Granqvist C-GS (2010) *Green Nanotechnology: Solutions for Sustainability and Energy in the Built Environment*. CRC Press, Boca Raton, FL.

Soenena J, Rivera-Gil P, Montenegro J, Wolfgang JP, Stefaan CS, and Kevin B (2011) Cellular toxicity of inorganic nanoparticles: common aspects and guidelines for improved nanotoxicity evaluation. *Nano Today* 6:446–465. https://doi.org/10.1016/j.nan tod.2011.08.001

Stone V, Nowack B, Baun A , van den Brink N, von der Kammer F, Dusinska M, Handy R, Hankin S, Hassello ̈vM, Joner E, and Fernandes TF (2010) Nanomaterials for environmental studies: classification, reference material issues, and strategies for physicochemical characterisation. *Sci Total Environ* 408:1745–1754.

Tobias HJ, Beving DE, Ziemann PJ, Sakurai H, Zuk M, McMurry PH, Zarling D, Waytulonis R, and Kittelson DB (2001) Chemical analysis of diesel engine nanoparticles using a nano-DMA/thermal desorption particle beam mass spectrometer. *Environ Sci Technol* 35:2233–2243. https://doi.org/10.1021/es0016654

Vejerano EP, Yanjun M, Holder AL, Pruden A, Elankumaran S, and Marr LC (2015) Toxicity of particulate matter from incineration of nanowaste. *Environ Sci: Nano* 2, 143–154.

Venkataraman C (2005) Residential biofuels in South Asia: carbonaceous aerosol emissions and climate impacts. *Science* 307(5714):1454–1456. https://doi.org/10.1126/scie nce.1104359

Vincent JH and Clement CF (2000) Ultrafine particles in workplace atmospheres. *Philos Trans R Soc A Math Phys Eng Sci* 358:2673–2682. https://doi.org/10.1098/rsta.2000.0676

Virkutyte J and Varma RS (2013) Green synthesis of nanomaterials: environmental aspects. In N Shamim and VK Sharma (eds), *Sustainable Nanotechnology and the Environment: Advances and Achievements*. ACS Publications, pp. 11–39.

Wallace LA, Emmerich SJ, and Howard-Reed C (2004) Source strengths of ultrafine and fine particles due to cooking with a gas stove. *Environ Sci Technol* 38:2304–2311. https://doi.org/10.1021/es0306260

Weber RJ, McMurry PH, Mauldin L, Tanner DJ, Eisele FL, Brechtel FJ, Kreidenweis SM, Kok GL, Schillawski RD, and Baumgardner D (1998) A study of new particle formation and growth involving biogenic and trace gas species measured during ACE 1. *J Geophys Res Atmos* 103:16385–16396. https://doi.org/10.1029/97JD02465

Wiedensohler A, Covert DS, Swietlicki E, Aalto P, Heintzenberg J, and Leck C, (1996) Occurrence of an ultrafine particle mode less than 20 nm in diameter in the marine boundary layer during Arctic summer and autumn. *Tellus B* 48:213–222. https://doi.org/10.1034/j.1600-0889.1996.t01-1-00006.x

Yamaguchi S (2015) Unclassified: recycling of waste containing nanomaterials (WCNM). ENV/EPOC/WPRPW(2013)2/FINAL.

Younis SA, El-Fawal EM, and Serp P (2018) Nano-wastes and the environment: potential challenges and opportunities of nano-waste management paradigm for greener nanotechnologies. In CM Hussain (ed), *Handbook of Environmental Materials Management*. Springer, Cham. https://doi.org/10.1007/978-3-319-58538-3_53-1

Zhuang Z, Xu X, Wang Y, Wang Y, Huang F, and Lin Z (2012) Treatment of nanowaste via fast crystal growth: with recycling of nano-SnO_2 from electroplating sludge as a study case. *J Hazard Mater* 211–212:414–419.

6 Microbiology of Wastewater Management

Challenges, Opportunities, and Innovations

Oshani Ratnayake[1], Dinuka Lakmali Jayasuriya[1], Dilani K. Hettiarachchi[1], and Neelamanie Yapa[1]*
[1]Department of Biological Sciences, Faculty of Applied Sciences, Rajarata University Sri Lanka, Mihintale, 50300, Sri Lanka
*Corresponding author: Neelamanie Yapa; neelamanie@as.rjt.ac.lk

CONTENTS

6.1 INTRODUCTION

Water undoubtedly plays a significant role in the lives of human beings, plants, and animals. According to the United Nations World Development Report in 2018, the world is experiencing a 1% increase in water demand annually with the increasing population, demands for clean water for drinking, sanitation, irrigation, and inevitable other uses (UNESCO 2018). Reduction of water resources and increasing water pollution are the major consequences arising from dynamic population increase and economic development. It is evident that the ever -increasing depletion and contamination of groundwater resources undoubtedly put water security at risk (Hanjra and Qureshi 2010; Biswas and Tortajada 2019). Reports have suggested the possibility of water scarcity arising in different parts of the world, mainly due to factors discussed above and due to natural reasons like the variability of rainfall and climatic changes (Ciais et al. 2005). Water scarcity imposes negative consequences in terms of social integrity and economic development of the world (Dalezios et al. 2018). Water scarcity directly affects agriculture as 80% of the total water usage is mainly for farming (Paranychianakis et al. 2015; Vörösmarty et al. 2018). Human involvement is identified as the main cause for water scarcity as anthropogenic activities lead to water wastage and pollution and inappropriate management (Ranade and Bhandari 2014). Furthermore, agricultural intensification has increased in past years, both in conventional and organic farming, resulting in loads of water pollutants, which affect ecosystems and human health (Bagdi et al. 2015). At the same time, industries are also expanding and contributing to water pollution (Zhu et al. 2008).

Water is supporting the growth of many types of microbes, which are beneficial as well as pathogenic (Tsai et al. 1998; Chattopadhyay and Taft 2018). For instance, the metabolic activities of many beneficial microorganisms in water can aid in contaminated water remediation. Nevertheless, the presence of pathogenic microorganisms can spread the diseases as well. Many microorganisms are found naturally in fresh and saltwater including bacteria, cyanobacteria, protozoa, algae, and virus (Chapelle 2000; Chattopadhyay and Taft 2018).

In general, one of the greatest microbial risks is associated with the ingestion of water that is contaminated with human or animal feces. Wastewater discharges in to fresh and marine water are a major cause of these contaminations (WHO 2008; Payment and Locas 2011). The wastewater treatment plants receive pathogens through human gastrointestinal tract releases, contaminated wastewater, release of non-treated hospital wastewater, and surface water runoff (Chattopadhyay and Taft 2018). Researchers have found that some pathogenic bacterial strains have developed in wastewater treatment plants (Osuolale and Okoh 2017). These hazardous effluents contain bacteria such as *Escherichia coli* O157:H7, *Shigella* spp., *Salmonella* spp., *Proteus vulgaris*, *Mycobacteria* spp., *Campylobacter jejuni*, *Helicobacter pylori*, *Legionella* spp., and *Vibrio* spp. (Tsai et al. 1998; Chapelle 2000; Caicedo et al. 2016; Chattopadhyay and Taft 2018), yeasts, and fungal species such as *Penicillium* spp. and *Aspergillus* spp. (Arvanitidou et al. 1999; Grabińska-Łoniewska et al. 2007). While *Ascaris lumbricoides*, *Ancylostoma duodenale*, and *Taenia* are some examples of the helminths present in the contaminated wastewater. Protozoa are another group of microbes that causes prolonged diarrhea in humans. *Giardia* and *Cryptosporidium* are two most concerned protozoa in water, which cause diseases (Akpor et al. 2010). Apart from that, the intestinal tract of warm-blooded animals also contains viruses that can contaminate water and cause diseases such as rotavirus, enteroviruses, coxsackievirus, adenovirus, astrovirus, calicivirus, coronavirus, enteroviruses, poliovirus, and hepatitis A and E (Chapelle 2000). Hence, efficient wastewater management techniques and strategies are needed to remove such microbes to enhance the quality of wastewater treatment, which can reduce the risk of the growth of pathogenic and drug-resistant microorganisms in water.

The alteration of chemical, biological, and physical factors of a water body rendering the water unsuitable for consumption is identified as water pollution (Schweitzer and Noblet 2018). Emerging trends of chemical and pharmaceutical contaminants in the aquatic environments raise the concern for wastewater treatment globally (Larsson et al. 2007). Due to the severe scarcity of the water, there is a vital need of effective method for water reuse. Apart from that, the quality of treated wastewater is responsible for the degradation of receiving water bodies, which can cause severe health risk to environment as well as the public health (Larsson et al. 2007). According to Batstone et al. (2015) approximately 20% of manufactured nitrogen (N) and phosphorous (P) is included in wastewater. The removal of pollutants and toxicants from wastewater include two main processes of chemical and biological treatments. Yet, biological treatments are employed due to the high risk and drawbacks of the chemical treatment (Akpor et al. 2010).

Microorganisms also play a vital role in wastewater treatment, which includes elimination of toxic materials such as ammonia, nitrite, and hydrogen sulfide (Nathanson and Ambulkar 2021). The removal of P and N are mostly done by bacteria while protozoa effectively purify by feeding on bacteria, thereby reducing their number (El Fantroussi and Agathos 2005; Demnerova et al. 2005; Saeed and Sun 2012). Fungi has also been reported to aid in the biological treatment of wastewater by increasing the degradability of wastewater sludge and contributing to the sludge management strategy (Haritash and Kaushik 2009).

6.2   WASTEWATER TREATMENT

Designing and operating a wastewater treatment plant needs a basic understanding of the composition and characteristics of wastewater. This is important because the water treatment process and special facilities needed for treatment are dependent on the proportions and types of pollutants present in wastewater. If further explained, the types and amounts of solids in wastewater determine the sludge handling and transportation. The biological oxygen demand (BOD) is a measure of organic material in water and removal of this depends on the form and nature of organic material present. pH of water is an indication of metal speciation, toxicity, and gives an idea of the need for neutralization. Special operational mechanisms and facilities would be required if the treated water contains greater proportions of oils and grease. When all these are considered, it is important to know the characteristics and composition of wastewater treated (Englande et al. 2015). It is of utmost importance to recycle and reuse wastewater in order to maintain ambient environmental health while meeting with the water demand of the populations (Larsson et al. 2007; Yadav et al. 2020).

A paradigm shift toward the wastewater treatment and effective management is necessary not only to protect sensitive ecosystems but also to emphasize the wastewater is a resource for many agricultural and industrial purposes.

6.2.1 THE CONVENTIONAL PROCESS OF WASTEWATER TREATMENT

The conventional wastewater treatment process is organized in three operational levels as the primary treatment, secondary treatment, and tertiary treatment depending on two factors, the nature of the separation processes, and the outcome of the processes (Ranade and Bhandari 2014). The primary treatment step is a size-based separation, which generally comes under physical methods (filtration, sedimentation). The secondary treatment phase is important to remove most of the total suspended solids (TSS) while reducing the BOD/COD in wastewater. This phase employs many physicochemical processes as well as biological processes, or a combination of both. The tertiary phase of wastewater treatment is responsible for the removal of last traces pollutants and/or toxic contaminants from treated water. By the end of the tertiary treatment, treated water would be containing contaminants in acceptable levels to be released back to the environment (Rizzo et al. 2013; Ranade and Bhandari 2014).

6.2.2 BIOLOGICAL TREATMENT PROCESSES

The biological mode of wastewater treatment is one of the most popular modes of treatment, which is similar to or mimics some of the natural water purification processes. The secondary treatment phase of wastewater treatment basically employs microorganisms to fulfill the tasks of purification such as degradation of organic contaminants as well as conversion of nitrogenous contaminants. At this stage, most of the biodegradable organic contaminants are converted to more stable and simpler biomass with the aid of various microorganisms (Lee et al. 2001). This biological treatment can be of two types as follows:

1. Fixed film process—trickling filter system
2. Suspended growth processes—activated sludge system

6.2.2.1 Fixed Film Processes

The classical fixed film process utilizes microorganism attached tightly to a rigid surface where biofilm formation is facilitated. Some examples for fixed film processes are trickling filter system and rotating biological contactors (RBC). A trickling filter system is a three-phase system containing several important components like a biofilm career, a distribution system through which wastewater is supplied to the bioreactor, ventilation system, and an underdrain. In addition to these, a secondary clarifier is also incorporated into the trickling filter system in order to separate the total suspended solids produced during the treatment process (Daigger and Boltz 2011).

The biofilms play a critical role in trickling filter systems. The total biology of the biofilm does not only contain aerobic, facultative bacteria and fungi but also algae, protozoa, and other higher animals. The fungi in the biofilm are important to stabilize the waste and are usually critically important in industrial wastes having low pH values. However, under high growth conditions of fungi, the filter could be blocked, reducing the efficiency of the process (Wik 2003). Many of the bacteria found in the trickling filter biofilms are facultative and they decompose organic material in the water along with other aerobic and anaerobic bacteria inhabiting the oxygen-poor inner layers of the biofilm. Some examples for biofilm inhabiting bacteria are *Pseudomonas, Flavobacterium, Acromobacter*, and *Alcaligenes* (Wik 2003; Akunna et al. 2017).

The process of contaminant removal within a biofilm has been studied to be a complex one involving several steps. If the mechanism is simply described, it includes the transport of wastewater and oxygen to the surface of the biofilm followed by internal transport of contaminants in wastewater to biofilms via diffusion. The diffused contaminants are oxidized within the biofilm finally releasing the diffusion by-products back to the wastewater (Lessard and Bihan 2003).

6.2.2.2 Suspended Growth Processes

Suspended growth processes basically involve microorganisms in suspension and examples are the conventional activated sludge process and membrane bioreactors (Abu Bakar et al. 2018). The activated sludge (AS) process is the most popular biological mode of wastewater treatment in the world and the process includes mixing of beneficial microorganisms and wastewater, facilitating aerobic digestion of organic waste in water. In addition to AS process, constructed treatment wetlands are also rapidly evolving as a biological mode of wastewater treatment due to its low cost and suitability to warm and developing countries (Scholz 2016).

The AS process contains two separate phases: aeration and sludge settlement. Water settled from primary treatment is sent to the aeration tank with a consortium of microorganisms and the tank is aerated either by mechanical agitation or by directly pumping oxygen. Aeration ensures maximum contact between microorganisms and surfaces of flocks while providing enough oxygen to aerobic microorganisms in the reactor (Scholz 2016). Part of the sludge produced during the process is returned to the aeration tank as an inoculant to maintain the microbial population to ensure the oxidation of a maximum proportion of wastewater.

6.3 STRATEGIC WASTEWATER MANAGEMENT

The current trend is more focused toward wastewater management rather than treatment. Wastewater management implies the concept of 3Rs—reduce, reuse, and recycle—to achieve a sustainable level in water usage. For proper utilization and conservation of the limited available water resources, several questions must be addressed. The successful management of wastewater is dependent upon finding answers to questions like the possibility of avoiding the discharged liquids, the possibility of reusing the treated water, management of cost and space for wastewater treatment, the possibility of recovering important chemicals or energy from wastewater, and the possible conservation aspects of water (Daigger 2009; Rizzo et al. 2013; Ranade and Bhandari 2014; Batstone et al. 2015).

A hierarchical approach could be taken to address the problem of wastewater management where at each level solutions are provided to minimize and eliminate the generation and disposal of wastes to the environment (Ranade and Bhandari 2014). In hierarchical approach in wastewater management, waste is taken care of or minimized at each operational level finally focusing on minimal disposal of waste to the environment and enable long-term societal sustainability (Mccarty et al. 2011; Batstone et al. 2015) (Figure 6.1).

When following a hierarchical approach, care must be taken to select the most efficient technology suitable for the process while paying attention to economic and social feasibilities (Corcoran et al. 2010), Moreover, there should be enough opportunities to introduce novel technologies or to develop/change the prevailing technologies in order to accomplish maximum efficiency in the process with reduced waste

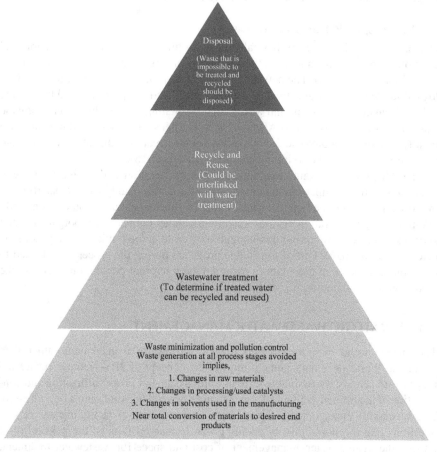

FIGURE 6.1   Hierarchical approach of pollution control in wastewater.

generation (Daigger 2009; Ranade and Bhandari 2014). Novel methodologies are constantly being introduced for wastewater treatment to achieve the goal of strategic water management. This is aimed at reducing the substantial energy consumption in existing wastewater treatment methodologies and enabling the recovery of inherent resources in wastewater (Daigger 2009). Among them, high-footprint passive systems (wetlands, lagoons) and low-energy anaerobic systems together with nitrogen removal methods are introduced (Wett et al. 2013).

6.3.1 Integrated Fixed Film-Activated Sludge Process

Integrated fixed film activated sludge (IFAS) process is identified as one of the leading-edge technologies in wastewater treatment for providing a sustainable solution for the prevailing treatment systems. This IFAS process is identified as a potential and effective process of removal of dissolved organic carbon with the ability to remove significant level of nitrification and denitrification (Arias et al. 2018). IFAS employs a combined process of fixed film and suspended growth systems to enhance the wastewater treatment (Jabari and Oleszkiewicz 2014). IFAS process is also known for its improved biological nitrogen removal, which is accomplished by nitrifiers attached onto the career media improving system's robustness and compactness (Bai et al. 2015).

One major characteristic of IFAS is the presence of both biofilm and flocks, in the process representing two unique microbial communities with distinct properties. The presence of different microbial communities pave the way for the coexistence of both autotrophic and heterotrophic bacteria, which promotes the removal of phosphorous and biological nitrogen in the wastewater (Boltz et al. 2009). Many advantages like reduced footprint, enhanced nutrient removals, longer solid retention times, and stronger removal of anthropogenic contaminants are offered by IFAS process (Moretti et al. 2015; Wang et al. 2006)

6.3.2 Recombinant Bacteria in Biofilms

The production of recombinant bacteria with the potential of degrading chemical substances in wastewater is a promising technology for sustainable wastewater treatment. Many studies have been conducted on the feasibility of using degrading bacteria in removing toxic pollutants in wastewater. However, one disadvantage of using most of the biodegrading bacteria is the fact that they don't exhibit strong biofilm-forming capabilities on carrier surfaces. As a result, they would be washed out from the system during the repeated wastewater replacement process (Li et al. 2018). In some instances, wastewater treatment efficiency can decrease due to several reasons such as the following:

1. Failure to form a complete biofilm due to competition between biofilm-forming bacteria and degrading bacteria mixed together
2. Inability of biofilm-forming bacteria to tolerate high concentrations of pollutants in wastewater
3. Degrading bacteria not being fully immobilized in biofilms

Taking all these factors into account, it is possible to deduce that production of recombinant bacteria having capabilities of both biofilm formation and contaminant degradation could increase the efficiency of wastewater treatment (Li et al. 2016). However, only a limited number of studies have been conducted to explore the potential of using recombinant bacteria in wastewater treatment.

A study conducted by Fujita et al. (1991) has demonstrated the ability of using genetically engineered microorganisms in degradation of xenobiotics in wastewater. The study has investigated the degradation potential and plasmid stability in some bacterial strains (*Escherichia coli* and *Pseudomonas putida*) harboring a recombinant plasmid (containing salicylate oxidase gene or catechol 2,3-oxygenase gene). Comparatively higher degradation rates have been observed in recombinant strains in comparison to wild relatives. However, the recombinant plasmids have displayed an instability in the absence of selective pressure. In another experiment of the same study, *Pseudomonas lemoignei*, a flock-forming bacteria has been transformed with NAH plasmid to generate a flock-forming and salicylate-degrading bacterium. This recombinant bacterium has shown to be growing flocculently maintain stability in activated sludge and in the wastewater system (Fujita et al. 1991).

6.4 CHALLENGES OF WASTEWATER TREATMENT METHODOLOGY

6.4.1 BIOAEROSOLS

Bioaerosols are identified as minute particles suspended in the atmosphere containing matter of microbial, plant, or animal origin. These bioaerosols are prone to contain pathogenic, nonpathogenic, live or dead bacteria and fungi, viruses, bacterial endotoxins, mycotoxins, and allergens (Han et al. 2019). Studies focused on exposure to bioaerosols have rapidly increased, mainly because it is now clearly understood that this exposure can lead to adverse health conditions including contagious infections, cancers, and acute toxic effects. Exposure to bioaerosols can happen in both residential and occupational setting and waste recycling/ treatment process is identified as one prominent setting, leading to this exposure (Han et al. 2020). According to a study by Kowalski et al. (2017), bioaerosols are found to be frequently associated with bacteria and fungi in high concentrations (8.00×10^1–6.90×10^3 and 5.10×10^2–3.90×10^3 CFU/m^3 respectively) in wastewater treatment facilities.

Wastewater treatment plants can serve as good reservoirs of hazardous components and pathogenic microorganisms. Ingestion, inhalation, and dermal contact of bioaerosols are considered to be a potential health risk for employees of wastewater treatment plants and residents living in surrounding areas (Sánchez-Monedero et al. 2008; Stellacci et al. 2010). Many epidemiological studies have exhibited the tendency of wastewater treatment plant workers to be affected by diseases and complications like gastrointestinal symptoms, respiratory tract infections, and allergies and hypersensitivities due to exposure to bioaerosols (Gangamma et al. 2011; Masclaux et al. 2014). Many studies also have reported the presence of numerous pathogenic bacterial genera *Pseudomonas, Serratia,*

Enterobacter, Alcaligenes, Acinetobacter, and *Bacteroides* as well as fungi *Aspergillus, Cladosporium,* and *Penicillium* in bioaerosols (Korzeniewska et al. 2009; Korzeniewska et al. 2011; Han et al. 2019; Yang et al. 2019). The factors such as the type of the wastewater treated, treatment process, and meteorological parameters determine the concentrations, size distribution, and microbial population of bioaerosols escaped from each wastewater treatment process (Kowalski et al. 2017; Szyłak-Szydłowski et al. 2016).

6.4.2 Antibiotic Resistance Genes

The development of antibiotic resistance and spreading of genes responsible for antibiotic resistance have become a global concern. Most of the antibiotics prescribed to humans and animals eventually end up in sewage, making the wastewater treatment plants the leading sources of releasing antibiotic-resistant bacteria (ARB) and antibiotic-resistant genes (ARGs) to the environment (Bouki et al. 2013; Rizzo et al. 2013). Wastewater treatment plants have been recognized as hotspots for horizontal gene transfer broadening the dissemination of ARGs in the environment (Berendonk et al. 2015; Martinez 2009). An optimum environment for horizontal gene transfer (HGT) among environmental bacteria and human pathogens is established within the wastewater treatment plant as it contains a high density of bacteria (Watkinson et al. 2007).

A well-established fact about ARGs and their selection in wastewater is that even very low concentrations of antibiotics promote the selection of ARGs, rendering the defining a safe concentration of antibiotics in wastewater difficult (Andersson and Hughes 2014). In several studies regarding wastewater treatment plants, ARBs have been detected in the treatment plant as well as in their effluent depicting the inability of the treatment plant to fully eliminate these bacteria (Bergeron et al. 2015; Łuczkiewicz et al. 2010; da Silva et al. 2006; Yang et al. 2014). However, since the bacterial loads are reduced during the treatment process, the overall resistance levels and ARGs are also reduced but not completely eliminated (Yang et al. 2014; Bengtsson-Palme et al. 2016).

Advanced treatment technologies are employed to get rid of emerging contaminants from wastewater treated by active sludge processes. Personal care products, endocrine-disrupting compounds, surfactants, pesticides, and pharmaceuticals are considered as some of the emerging contaminants. Wastewater treatment plants employ a range of advanced techniques such as photo-catalysis, activated carbon adsorption, membrane filtration, and advanced oxidation processes to get rid of these emerging contaminants (Ahmed et al. 2017). Although these technologies are efficient at removal of emerging contaminants, they fail when in the process of removing ARGs. Along with that, some of these technologies give rise to conditions that induce SOS responses in bacteria, further increasing the mutations in bacteria and facilitating HGT of ARGs (Beaber et al. 2004; Qin et al. 2015). Therefore it is essential to evaluate the possible outcomes of advanced treatment technologies before implementing them at a large scale (Karkman et al. 2018).

6.5 CONTROL OF PATHOGENIC MICROORGANISMS IN WASTEWATER RECYCLING

When treated wastewater is reused in agriculture and industry, the major obstacle faced in developing countries is the presence of pathogenic microorganisms and health risks associate with them (Toscano et al. 2013). Therefore, it is essential to understand the health risks involved and develop proper treatment processes to minimize the microbial loads of the treated wastewater (Harwood et al. 2005; Quiñónez-Díaz et al. 2001; Sánchez-Monedero et al. 2008).

Pathogenic microorganism removal of wastewater systems can be accomplished by following physical, chemical, and biological mechanisms. Among them physical mechanisms include sedimentation and filtration of microbes in treated wastewater, while adsorption to the surface or particles, UV irradiation by sunlight, and exposure to root exudates are considered as the chemical processes. Predation, natural die-off, and retention in the biofilms are some of the methods of biological removal (Jasper et al. 2013; Shingare et al, 2019).

6.6 RESOURCE RECOVERY FROM WASTEWATER

Considering the growing demand of water globally, wastewater is gaining momentum as an alternative source of water. However, finding the methods for wastewater management is more challenging. Therefore, instead of wastewater treatment and disposal, it should be more appropriate to find methods for wastewater reuse, recycle, and resource recovery from wastewater. Then undoubtedly wastewater is no longer a problem, rather it is a challenge that the society can gain the maximum benefits out of it.

Wastewater can be considered as a sustainable resource of energy, plant nutrients, and organic matter. The potential benefits of obtaining such resources from wastewater will secure the food and energy sustainably. Although 50–100% of the waste resources are contained in wastewater, until recently there was no technological advancements developed for recovering useful substances and energy from wastewater (House 2012). Therefore, it is essential to design next-generation wastewater treatment facilities that produce energy, organic compounds, and recovery of nutrients (Batstone et al. 2015; Bai et al. 2015). However, method development is necessary to improve the recovery of substances from wastewater using microbial transformations. The different microbial populations can be used in such resource partitioning in modern wastewater treatment processes to recover products from wastewater. These include heterotrophic bacteria where required energy and substances for their growth are taken from the chemicals recovered from the wastewater (Jimenez et al. 2015) and phototrophic anaerobic bacteria where energy sourced from light and nutrients from the wastewater (Hülsen et al. 2014). Moreover, aerobic photosynthetic bacteria and algae, which produce CO_2, have shown a wider potential in resource recovery from wastewater (Cai et al. 2004).

6.7 BIOFUEL PRODUCTION

Certain new methodologies were developed recently by scientists to recover biofuels such as biogas, bioethanol, biohydrogen, biodiesel, and microbial fuel cells (Mccarty et al. 2011). The development of anaerobic membrane bioreactors facilitate the advancement of biofuel production from wastewater and sludge (Dereli et al. 2012). Methane-rich biogas can be produced from the anaerobic digestion of wastewater by methanogenic archaea (Batstone and Viridis 2014). Anaerobic membrane bioreactors (AnMBR) can further expand the application of anaerobic digestion systems to be used in wastewater contained different substrates (Dereli et al. 2012).

6.7.1 MICROBIAL FUEL CELLS

In microbial fuel cells (MFCs), electric energy is produced by converting the chemical energy stored in organic pollutants in wastewater through a series of electrochemical reactions catalyzed by microorganisms (Li et al. 2013). When considering sustainable energy production and wastewater treatment, extracting energy from wastewater is considered to be a promising technology (Angenent et al. 2004). However, many challenges, both technological and economical, must be addressed before venturing into a commercialized application of microbial fuel cells. Various issues like low power density (PD), energy efficiency, high construction cost, problems related to scaling up, and fouling problems arising during long-term operation are to be addressed before commercially establishing MFCs as a technology in wastewater treatment (Li et al. 2013).

Several significant advantages are offered by MFCs, which make the technology a promising tool for sustainable wastewater management. MFCs ensure an appreciable effluent quality by enabling more efficient and complete removal of pollutants (Li et al. 2013). Usually the effluent from anaerobically treated/fermentation processes contain significant levels of volatile fatty acids and the chemical oxygen demand (COD) tends to be high (Yu et al. 2002). This creates the need for a subsequent aerobic treatment to satisfy the discharge limits, which is adding extra cost on aeration and sludge removal (Cusick et al. 2010). Another critical issue associated with traditional wastewater treatment is the production of hazardous by-products that can have adverse impacts on treating facilities (Hamelin et al. 2011). In comparison to traditional treatment methods, MFCs have a higher potential of COD removal even at low influent concentrations (Kim et al. 2010). If toxic substances like sulfides are formed during the process, they can be used as electron donors to generate electricity (Rabaey et al. 2006). Due to many reasons, MFCs are highly suitable for energy generation from organic wastewaters; for example, from the perspective of energy recovery and utilization, electricity is a better option than the production of methane or hydrogen and a minimum amount of waste is generated in MFCs.

6.8 CHAPTER SUMMARY

Water is precious for all the living organisms and with the increasing population and developments in the economy, water is slowly becoming a scares resource for the inhabitants of the earth. The major reasons for diminishing the quality of water are pollution occurring due to industrial development and effluent release from health care facilities and domestic effluents. Conventional water treatment processes are practiced in treating water before releasing it back to the environment. However, the latest focus is on managing wastewater, which is a sustainable solution for wastewater treatment. A hierarchical approach is taken toward minimizing wastewater generation and minimizing disposable effluent. Novel technologies like recombinant bacteria in biofilms, MFCs, and IFAS are employed in wastewater treatment, leading to a sustainable wastewater management system. Although microorganisms play an important role in wastewater treatment, they could also generate problems in the process. The major problems associated with the involvement of microorganisms in wastewater treatment are bioaerosols and ARGs and ARB. When implementing wastewater treatment plants, careful evaluation and selection of proper technologies and processes can increase the efficiency of the process while reducing operational cost and release of effluent to the environment.

REFERENCES

Abu Bakar, S.N.H., A.H. Hassimi, A.W. Mohammad, S.R.S. Abdullah, T.Y. Haan, N. Rahmat, et al. 2018. A review of moving-bed biofilm reactor technology for palm oil mill effluent treatment. *Journal of Cleaner Production* 171: 1532–45. https://doi.org/https://doi.org/ 10.1016/j.jclepro.2017.10.100

Ahmed, M.B., J. L. Zhou, H.H. Ngo, W.G. Nikolaos, S. Thomaidis, and J. Xu. 2017. Progress in the biological and chemical treatment technologies for emerging contaminant removal from wastewater: a critical review. *Journal of Hazardous Materials* 323: 274–98. https:// doi.org/https://doi.org/10.1016/j.jhazmat.2016.04.045

Akpor, O.B., and M. Muchie. 2010. Bioremediation of polluted wastewater influent: phosphorus and nitrogen removal. *Scientific Research and Essays* 5(21): 3222–30. https:// doi.org/10.5897/SRE.9000241

Akunna, J.C., J.M. O'Keeffe, and R. Allan. 2017. Reviewing factors affecting the effectiveness of decentralised domestic wastewater treatment systems for phosphorus and pathogen removal. *Desalination and Water Treatment* 91:40–47. doi: 10.5004/dwt.2017.20750

Andersson, D.I., and D. Hughes. 2014. Microbiological effects of sublethal levels of antibiotics. *Nature Reviews Microbiology* 12(7): 465–78. https://doi.org/10.1038/nrmicro3270

Angenent, L.T., K. Karim, M.H. Al-Dahhan, B.A. Wrenn, and R. Domíguez-Espinosa. 2004. Production of bioenergy and biochemicals from industrial and agricultural wastewater. *Trends in Biotechnology* 22(9): 477–85. https://doi.org/10.1016/j.tibtech.2004.07.001

Arias, A., T. Alvarino, T. Allegue, S. Suárez, J.M. Garrido, and F. Omil. 2018. An innovative wastewater treatment technology based on UASB and IFAS for cost-efficient macro and micropollutant removal. *Journal of Hazardous Materials* 359: 113–20. https://doi.org/ https://doi.org/10.1016/j.jhazmat.2018.07.042

Arvanitidou, M., K. Kanellou, T.C. Constantinides, and V. Katsouyannopoulos. 1999. The occurrence of fungi in hospital and community potable waters. *Letters in Applied Microbiology* 29 (2): 81–84. https://doi.org/10.1046/j.1365-2672.1999.00583.x

Bagdi, G.L., P.K. Mishra, R.S. Kurothe, S.L. Arya, S.L. Patil, A.K. Singh, et al. 2015. Post-adaptation behaviour of farmers towards soil and water conservation technologies of watershed management in India. *International Journal of Soil & Water Conservation Research* 3: 161–69. https://doi.org/10.1016/j.iswcr.2015.08.003

Bai, Y., Y. Zhang, X. Quan, and S. Chen. 2015. Enhancing nitrogen removal efficiency and reducing nitrate liquor recirculation ratio by improving simultaneous nitrification and denitrification in integrated fixed-film activated sludge (IFAS) process. *Water Science and Technology* 73 (4): 827–34. https://doi.org/10.2166/wst.2015.558

Batstone, D.J., T. Hülsen, C.M. Mehta, and J. Keller. 2015. Platforms for energy and nutrient recovery from domestic wastewater: a review. *Chemosphere* 140: 2–11. doi: 10.1016/j.chemosphere.2014.10.021

Batstone, D.J., and B. Virdis, 2014. The role of anaerobic digestion in the emerging energy economy. *Current opinion in biotechnology* 27: 142–49. https://doi.org/10.1016/j.copbio.2014.01.013

Beaber, J.W, B. Hochhut, and M.K. Waldor. 2004. SOS response promotes horizontal dissemination of antibiotic resistance genes. *Nature* 427: 72–74. https://doi.org/10.1038/nature02241

Bengtsson-Palme J., R. Hammarén, C. Pal, M. Östman, B. Björlenius, C.F., Flach, et al. 2016. Elucidating selection processes for antibiotic resistance in sewage treatment plants using metagenomics. *Science of the Total Environment* 572: 697–712. doi: 10.1016/j.scitotenv.2016.06.228

Berendonk T.U., C.M. Manaia, C. Merlin, D. Fatta-Kassinos, E. Cytryn, F. Walsh, et al. 2015. Tackling antibiotic resistance: the environmental framework. *Nature Reviews Microbiology* 13 (5): 310–17. https://doi.org/10.1038/nrmicro3439

Bergeron, S., R. Boopathy, R. Nathaniel, A. Corbin, and G. LaFleur. 2015. Presence of antibiotic resistant bacteria and antibiotic resistance genes in raw source water and treated drinking water. *International Biodeterioration & Biodegradation* 102: 370–74. https://doi.org/https://doi.org/10.1016/j.ibiod.2015.04.017

Biswas, A.K., and C. Tortajada. 2019. Water quality management: a globally neglected issue. *International Journal of Water Resources and Development* 35: 913–16.

Boltz, J.B., B.R. Johnson, G.T. Daigger, and J. Sandino. 2009. Modeling integrated fixed-film activated sludge and moving-bed biofilm reactor systems i: mathematical treatment and model development. *Water Environment Research: A Research Publication of the Water Environment Federation* 81 (July): 555–75. https://doi.org/10.2175/106143008X357066

Bouki, C., D. Venieri, and E. Diamadopoulos. 2013. Detection and fate of antibiotic resistant bacteria in wastewater treatment plants: a review. *Ecotoxicology and Environmental Safety* 91: 1–9. https://doi.org/https://doi.org/10.1016/j.ecoenv.2013.01.016

Cai, M., J. Liu, and Y. Wei. 2004. Enhanced biohydrogen production from sewage sludge with alkaline pretreatment. *Environment Science and Technology* 38: 3195–202. doi: 10.1021/es0349204

Caicedo, C., S. Beutel, T. Scheper, K.H. Rosenwinkel, and R. Nogueira. 2016. Occurrence of *Legionella* in wastewater treatment plants linked to wastewater characteristics. *Environment Science and Pollution Research International* 23: 16873–81. doi: 10.1016/j.jiph.2016.11.012

Chapelle, F.H. 2000. *Ground Water Microbiology and Geochemistry.* New York: Wiley. ISBN: 978-0-471-34852-8

Chattopadhyay, S., and S Taft. 2018, August. *Exposure Pathways to High-Consequence Pathogens in the Wastewater Collection and Treatment Systems.* Washington, DC: U.S. Environmental Protection Agency. https://cfpub.epa.gov/si/si_public_record_report.cfm?dirEntryId=341856

Ciais, P., M. Reichstein, N. Viovy, A. Granier, J. Ogée, V. Allard, et al. 2005. Europe-wide reduction in primary productivity caused by the heat and drought in 2003. *Nature* 437: 529–33. www.ncbi.nlm.nih.gov/pubmed/16177786

Corcoran, E., C. Nellemann, E. Baker, R. Bos, D. Osborn, and H. Savelli (eds). 2010. *Sick Water? The central role of waste-water management in sustainable development. A Rapid Response Assessment.* United Nations Environment Programme, UN-HABITAT, GRID-Arendal. Norway: Birkeland Trykkeri AS, Norway. www.grida.no. ISBN: 978-82-7701-075-5.

Cusick, R.D., P.D. Kiely, and B.E. Logan. 2010. A monetary comparison of energy recovered from microbial fuel cells and microbial electrolysis cells fed winery or domestic wastewaters. *International Journal of Hydrogen Energy* 35 (17): 8855–61. https://doi.org/https://doi.org/10.1016/j.ijhydene.2010.06.077

Daigger, G.T. 2009. Evolving urban water and residuals management paradigms: water reclamation and reuse, decentralization, and resource recovery. *Water and Environment Research* 81: 809–23. doi: 10.2175/106143009X425898

Daigger, G.T., and J. Boltz. 2011. Trickling filter and trickling filter-suspended growth process design and operation: a state-of-the-art review. *Water Environment Research: A Research Publication of the Water Environment Federation* 83: 388–404. https://doi.org/10.2175/106143010X12681059117211

Dalezios, N.R., A.N. Angelakis, and S. Eslamian. 2018. Water scarcity management: part 1: methodological framework. *International Journal of Global Environmental Issues* 17 (1): 1–40.

da Silva, M.F., I. Tiago, A. Veríssimo, R.A.R. Boaventura, O.C. Nunes, and C.M. Manaia. 2006. Antibiotic resistance of Enterococci and related bacteria in an urban wastewater treatment plant. *FEMS Microbiology Ecology* 55 (2): 322–29. https://doi.org/10.1111/j.1574-6941.2005.00032.x

Demnerova K., M. Mackova, V. Spevakova, K. Beranova, L. Kochankova, P. Lovecka, et al. 2005. Two approaches to biological decontamination of groundwater and soil polluted by aromatics characterization of microbial populations. *International Microbiology* 8: 205–11. PMID: 16200499.

Dereli, R.K., M.E. Ersahin, H. Ozgun, I. Ozturk, D. Jeison, F. Van Der Zee, et al. 2012. Potentials of anaerobic membrane bioreactors to overcome treatment limitations induced by industrial wastewaters. *Bioresource Technology* 122: 160–70. doi: 10.1016/j.biortech.2012.05.139

El Fantroussi, S., and S.N. Agathos. 2005. Is bioaugmentation a feasible strategy for pollutant removal and site remediation? *Current Opinion in Microbiology* 8: 268–75. doi: 10.1016/j.mib.2005.04.011

Englande, J., A.J. Krenkel, and J. Shamas. 2015. Wastewater Treatment &Water Reclamation. *Reference Module in Earth Systems and Environmental Sciences* 1–32. https://doi.org/10.1016/B978-0-12-409548-9.09508-7

Fujita, M, M. Ike, and S. Hashimoto. 1991. Feasibility of wastewater treatment using genetically engineered microorganisms. *Water Research* 25 (8): 979–84. https://doi.org/https://doi.org/10.1016/0043-1354(91)90147-I

Gangamma, S., R.S. Patil, and S. Mukherji. 2011. Characterization and proinflammatory response of airborne biological particles from wastewater treatment plants. *Environmental Science & Technology* 45 (8): 3282–87. https://doi.org/10.1021/es103652z

Grabińska-Łoniewska, A., T. Koniłłowicz-Kowalska, G. Wardzyn´ska, and K. Boryn. 2007. Occurrence of fungi in water distribution system. *Polish Journal of Environmental Studies* 16 (4): 539–47.

Hamelin, L., M. Wesnæs, H. Wenzel, and B.M. Petersen. 2011. Environmental consequences of future biogas technologies based on separated slurry. *Environmental Science & Technology* 45 (13): 5869–77. https://doi.org/10.1021/es200273j

Han, Y., K. Yang, T. Yang, M. Zhang, and L. Li. 2019. Bioaerosols emission and exposure risk of a wastewater treatment plant with A2O treatment process. *Ecotoxicology and Environmental Safety* 169: 161–68. https://doi.org/https://doi.org/10.1016/j.ecoenv.2018.11.018

Han, Y., T. Yang, G. Xu, L. Li, and J. Liu. 2020. Characteristics and interactions of bioaerosol microorganisms from wastewater treatment plants. *Journal of Hazardous Materials* 391: 122256. https://doi.org/https://doi.org/10.1016/j.jhazmat.2020.122256

Hanjra, M.A., and M.E. Qureshi. 2010. Global water crisis and future food security in an era of climate change. *Food Policy* 35 (5): 365–77. http://doi.org/10.1016/j.food pol.2010.05.006

Haritash, A.K., and C.P. Kaushik. 2009. Biodegradation aspects of polycyclic aromatic hydrocarbons (PAHs): a review. *Journal of Hazardous Materials* 69 (1–3): 1–15. doi: 10.1016/j.jhazmat.2009.03.137

Harwood, V.J., A.D. Levine, T.M. Scott, V. Chivukula, J. Lukasik, S.R. Farrah, and J.B. Rose. 2005. Validity of the indicator organism paradigm for pathogen reduction in reclaimed water and public health protection. *Applied and Environmental Microbiology* 71 (6): 3163–70. https://doi.org/10.1128/AEM.71.6.3163-3170.2005

House, T.W. 2012. National bioeconomy blueprint, April 2012. *Industrial Biotechnology* 8: 97–102. doi: 10.1089/ind.2012.1524

Hülsen, T., D.J. Batstone, and J. Keller. 2014. Phototrophic bacteria for nutrient recovery from domestic wastewater. *Water Research* 50: 18–26. doi: 10.1016/j.watres. 2013.10.05

Jabari, P., G. Munz, and J.A. Oleszkiewicz. 2014. Selection of denitrifying phosphorous accumulating organisms in IFAS systems: comparison of nitrite with nitrate as an electron acceptor. *Chemosphere* 109: 20–27. https://doi.org/https://doi.org/10.1016/j.chemosph ere.2014.03.002

Jasper, J.T., M.T. Nguyen, Z.L. Jones, N.S. Ismail, D.L. Sedlak, J.O. Sharp, et al. 2013. Unit process wetlands for removal of trace organic contaminants and pathogens from municipal wastewater effluents. *Environment Engineering Science* 30: 421–36. doi:10.1089/ ees.2012.0239

Jimenez, J., M. Miller, C. Bott, S. Murthy, H. De Clippeleir, and Wett, B. (2015). High-rate activated sludge system for carbon management—evaluation of crucial process mechanisms and design parameters. *Water Research* 87: 476–82. doi: 10.1016/ j.watres

Karkman, A.V., T.T. Do, F. Walsh, and M.P.J. Virta. 2018. Antibiotic-resistance genes in waste water. *Trends in Microbiology* 26 (3): 220–28. https://doi.org/10.1016/j.tim.2017.09.005

Kim, J.R., G.C. Premier, F.R. Hawkes, J. Rodríguez, R.M. Dinsdale, and A.J. Guwy. 2010. Modular tubular microbial fuel cells for energy recovery during sucrose wastewater treatment at low organic loading rate. *Bioresource Technology* 101 (4): 1190–98. https:// doi.org/10.1016/j.biortech

Korzeniewska, E. 2011. Emission of bacteria and fungi in the air from wastewater treatment plants—a review. *Frontiers in Bioscience* (Scholar Edition) 3 (January): 393–407. https://doi.org/10.2741/s159

Korzeniewska, E., Z. Filipkowska, A. Gotkowska-Płachta, W. Janczukowicz, B. Dixon, and M. Czułowska. 2009. Determination of emitted airborne microorganisms from a bio-pak wastewater treatment plant. *Water Research* 43 (11): 2841–51. https://doi.org/https:// doi.org/10.1016/j.watres

Kowalski, M., J. Wolany, J.S. Pastuszka, G. Płaza, A. Wlazło, K. Ulfig, and A. Malina. 2017. Characteristics of airborne bacteria and fungi in some polish wastewater treatment plants. *International Journal of Environmental Science and Technology* 14 (10): 2181–92. https://doi.org/10.1007/s13762-017-1314-2

Larsson, D.G.J., C. Pedro, and N. Paxeus. 2007. Effluent from drug manufactures contains extremely high levels of pharmaceuticals. *Journal of Hazardous Materials* 148 (3): 751–55. https://doi.org/https://doi.org/10.1016/j.jhazmat.2007.07.008

Lee, J., W.Y. Ahn, and C.H. Lee. 2001. Comparison of the filtration characteristics between attached and suspended growth microorganisms in submerged membrane bio-reactor. *Water Research* 35 (4): 2435–45. https://doi.org/https://doi.org/10.1016/ S0043-1354(00)00524-8

Lessard, P., and Y.L. Bihan. 2003. Fixed Film Processes. In *Handbook of Water and Wastewater Microbiology*, edited by Duncan Mara and Nigel Horan, 317–18. United Kingdom: Academic Press.

Li, W.W., G.P. Sheng, and H.Q. Yu. 2013. Chapter 14—Electricity Generation from Food Industry Wastewater Using Microbial Fuel Cell Technology. In *Food Industry Wastes Webb*, edited by Maria R. Kosseva and B.T. Colin, 249–61. San Diego: Academic Press. https://doi.org/https://doi.org/10.1016/B978-0-12-391921-2.00014-7

Li, C., Y. Sun, Z. Yue, M. Huang, J. Wang, X. Chen, et al. 2018. Combination of a recombinant bacterium with organonitrile-degrading and biofilm-forming capability and a positively charged carrier for organonitriles removal. *Journal of Hazardous Materials* 353: 372–80. https://doi.org/10.1016/j.jhazmat.2018.03.058

Li, C., Z. Yue, F. Feng, C. Xi, H. Zang, X. An, and K. Liu. 2016. A novel strategy for acetonitrile wastewater treatment by using a recombinant bacterium with biofilm-forming and nitrile-degrading capability. *Chemosphere* 161: 224–32. https://doi.org/https://doi.org/10.1016/j.chemosphere.2016.07.019

Łuczkiewicz, A., S. Fudala-Książek, K. Jankowska, B. Quant, and K. Olańczuk-Neyman. 2010. Diversity of fecal coliforms and their antimicrobial resistance patterns in wastewater treatment model plant. *Water Science and Technology* 61 (6): 1383–92. https://doi.org/10.2166/wst.2010.015

Martinez, J.L. 2009. Environmental pollution by antibiotics and by antibiotic resistance determinants. *Environmental Pollution* 157 (11): 2893–902. https://doi.org/https://doi.org/10.1016/j.envpol.2009.05.051

Masclaux, F.G., P. Hotz, D. Gashi, D. Savova-Bianchi, and A. Oppliger. 2014. Assessment of airborne virus contamination in wastewater treatment plants. *Environmental Research* 133: 260–65. https://doi.org/https://doi.org/10.1016/j.envres.2014.06.002

Mccarty, P.L., J. Bae, and J. Kim. 2011. Domestic wastewater treatment as a net energy producer—can this be achieved? *Environment Science and Technology* 45: 7100–06. doi: 10.1021/es2014264

Moretti, P., J.M. Choubert, J.P. Canler, O. Petrimaux, P. Buffiere, and P. Lessard. 2015. Understanding the contribution of biofilm in an integrated fixed-film-activated sludge system (IFAS) designed for nitrogen removal. *Water Science and Technology* 71 (10): 1500–06. https://doi.org/10.2166/wst.2015.127

Nathanson, J.A., and A. Ambulkar. 2021. *Wastewater Treatment*. Encyclopaedia Britannica. www.britannica.com/technology/wastewater-treatment

Osuolale, O., and A. Okoh. 2017. Human enteric bacteria and viruses in five wastewater treatment plants in the Eastern Cape, South Africa. *Journal of Infections and Public Health* 10: 541–47. doi: 10.1016/j.jiph.2016.11.012

Paranychianakis, N.V., M. Salgot, S.A. Snyder, and A.N. Angelakis. 2015. Water Reuse in EU States: necessity for uniform criteria to mitigate human and environmental risks. *Critical Review of Environment Science and Technology* 45: 1409–68.

Payment, P., and A. Locas. 2011. Pathogens in water: value and limits of correlation with microbial indicators. *Ground Water* 49: 4–11. doi: 10.1111/j.1745-6584.2010.00710.x

Qin, T.T., H.-Q. Kang, P. Ma, P.-P. Li, L.-Y. Huang, and B. Gu. 2015. SOS response and its regulation on the fluoroquinolone resistance. *Annals of Translational Medicine* 3 (22): 358. https://doi.org/10.3978/j.issn.2305-5839

Quiñónez-Díaz, M.D.J., M.M. Karpiscak, E.D. Ellman, and C.P. Gerba. 2001. Removal of pathogenic and indicator microorganisms by a constructed wetland receiving untreated domestic wastewater. *Journal of Environmental Science and Health Part A* 36: 1311–20. doi: 10.1081/ese-100104880

Rabaey, K., K. Van de Sompel, L. Maignien, N. Boon, P. Aelterman, P. Clauwaert, et al. 2006. Microbial fuel cells for sulfide removal. *Environmental Science & Technology* 40 (17): 5218–24. https://doi.org/10.1021/es060382u

Ranade, V.V., and V.M. Bhandari. 2014. Chapter 1—Industrial wastewater treatment, recycling, and reuse: an overview. In *Industrial Wastewater Treatment, Recycling and Reuse*, edited by Vivek V. Ranade and Vinay M. Bhandari, 1–80. Oxford: Butterworth-Heinemann. https://doi.org/https://doi.org/10.1016/B978-0-08-099968-5.00001-5

Rizzo, L., C. Manaia, C. Merlin, T. Schwartz, C. Dagot, M.C. Ploy, I. Michael, and D. Fatta-Kassinos. 2013. Urban wastewater treatment plants as hotspots for antibiotic resistant bacteria and genes spread into the environment: a review. *Science of the Total Environment* 447: 345–60. https://doi.org/https://doi.org/10.1016/j.scitotenv.2013.01.032

Saeed, T., and G.A. Sun. 2012. A review on nitrogen and organics removal mechanisms in subsurface flow constructed wetlands: dependency on environmental parameters, operating conditions and supporting media. *Journal of Environment Management* 112: 429–48. doi: 10.1016/j.jenvman.2012.08.011

Sánchez-Monedero, M.A., M.I. Aguilar, R. Fenoll, and A. Roig. 2008. Effect of the aeration system on the levels of airborne microorganisms generated at wastewater treatment plants. *Water Research* 42 (14): 3739–44. https://doi.org/https://doi.org/10.1016/j.watres.2008.06.028

Scholz, M. 2016. Chapter 15—Activated sludge processes. In *Wetlands for Water Pollution Control* (Second Edition), edited by Miklas B. T. Scholz, 91–105. Amsterdam: Elsevier. https://doi.org/https://doi.org/10.1016/B978-0-444-63607-2.00015-0

Schweitzer, L., and Noblet, J. 2018. Water contamination and pollution. *Green Chemistry: An Inclusive Approach*, 261–90. https://doi.org/10.1016/B978-0-12-809270-5.00011-x

Shingare, R.P., P.R. Thawale, K. Raghunathan, A. Mishra, and S. Kumar. 2019. Constructed wetland for wastewater reuse: role and efficiency in removing enteric pathogens. *Journal of Environment Management* 246: 444–61. doi: 10.1016/j.jenvman.2019.05.157

Stellacci, P., L. Lorenzo, M. Notarnicola, and C.N. Haas. 2010. Hygienic sustainability of site location of wastewater treatment plants: a case study. i. Estimating odour emission impact. *Desalination* 253 (1): 51–56. https://doi.org/https://doi.org/10.1016/j.desal.2009.11.034

Szyłak-Szydłowski, M., A. Kulig, and E. Miaśkiewicz-Pęska. 2016. Seasonal changes in the concentrations of airborne bacteria emitted from a large wastewater treatment plant. *International Journal of Biodeterioration & Biodegradation* 115: 11–16. https://doi.org/https://doi.org/10.1016/j.ibiod.2016.07.008

Toscano, A., C. Hellio, A. Marzo, M. Milani, K. Lebret, G.L. Cirelli, and G. Langergraber. 2013. Removal efficiency of a constructed wetland combined with ultrasound and UV devices for wastewater reuse in agriculture. *Environmental Technology* 34: 2327–36. doi: 10.1080/09593330.2013.767284

Tsai, C.T, J.S. Lai, and S.T. Lin. 1998. Quantification of pathogenic microorganisms in the sludge from treated hospital wastewater. *Journal of Applied Microbiology* 85 (1): 171–76. https://doi.org/10.1046/j.1365-2672.1998.00491.x

UNESCO. 2018. *The United Nations World Water Development Report 2018: Nature-Based Solutions for Water*. UNESCO.

Vörösmarty, C.J., V.R. Osuna, A. Cak, A. Bhaduri, S.E. Bunn, F. Corsi, et al. 2018. Ecosystem-based water security and the sustainable development goals. *Ecohydrology and Hydrobiology* 18 (4): 317–33. http://doi 10.1016/jecohyd.2018.07.004

Wang, X.J., S.Q. Xia, L. Chen, J.F. Zhao, N.J. Renault, and J.M. Chovelon. 2006. Nutrients removal from municipal wastewater by chemical precipitation in a moving bed biofilm reactor. *Process Biochemistry* 41 (4): 824–28. https://doi.org/https://doi.org/10.1016/j.procbio.2005.10.015

Watkinson, A.J., G.B. Micalizzi, G.M. Graham, J.B. Bates, and S.D. Costanzo. 2007. Antibiotic-resistant *Escherichia coli* in wastewaters, surface waters, and oysters from an urban riverine system. *Applied and Environmental Microbiology* 73 (17): 5667–70. https://doi.org/10.1128/AEM.00763-07

Wett, B., A. Omari, S.M. Podmirseg, M. Han, O. Akintayo, M. Gómez Brandón, et al. 2013. Going for mainstream deammonification from bench to full scale for maximized resource efficiency. *Water Science and Technology* 68: 283–89. doi: 10. 2166/wst.2013.150

WHO (World Health Organization). 2008. *Guidelines for Drinking-water Quality, Incorporating 1st and 2nd Addenda, Volume 1, Recommendations.* Geneva, Switzerland: WHO.

Wik, T. 2003. Trickling filters and biofilm reactor modelling. *Reviews in Environmental Science and Biotechnology* 2 (2): 193–212. https://doi.org/10.1023/B:RESB.0000040470.48460.bb

Yadav, B., A.K. Pandey, L.R. Kumar, R. Kaur, S.K. Yellapu, B. Sellamuthu, et al. 2020. 1— Introduction to wastewater microbiology: special emphasis on hospital wastewater. In *Current Developments in Biotechnology and Bioengineering Pandey*, edited by R.D. Tyagi, Balasubramanian Sellamuthu, Bhagyashree Tiwari, Song Yan, Patrick Drogui, Xiaolei Zhang, and Ashok B T, 1–41. Amsterdam: Elsevier. https://doi.org/https://doi.org/10.1016/B978-0-12-819722-6.00001-8

Yang, T., Y. Han, J. Liu, and L. Li. 2019. Aerosols from a wastewater treatment plant using oxidation ditch process: characteristics, source apportionment, and exposure risks. *Environmental Pollution* 250: 627–38. https://doi.org/https://doi.org/10.1016/j.envpol.2019.04.071

Yang, Y., B. Li, S. Zou, H.P. Fang, and T. Zhang. 2014. Fate of antibiotic resistance genes in sewage treatment plant revealed by metagenomic approach. *Water Research* 62: 97–106. https://doi.org/10.1016/j.watres.2014.05.019

Yu, H., Z. Zhu, W. Hu, and H.R. Zhang. 2002. Hydrogen production from rice winery wastewater in an upflow anaerobic reactor by using mixed anaerobic cultures. *International Journal of Hydrogen Energy* 27: 1359–65.

Zhu, W., J. Graney, and K. Salvage. 2008. Land-use impact on water pollution: Elevated pollutant input and reduced pollutant retention. *Journal of Contemporary Water Research & Education* 138: 15–21. doi: 10.1111/j.1936-704X.2008.00004.x

7 Phytoremediation
A Green Tool to Manage Waste

N. D. A. D. Wijegunawardana[1],
E. G. Perera[1], and M. S. Ekanayake[2]
[1]Department of Bioprocess Technology, Faculty of
Technology, Rajarata University Sri Lanka, Mihintale,
Sri Lanka
[2]Department of Aquaculture & Fisheries, Faculty of
Livestock Fisheries & Nutrition, Wayamba University
of Sri Lanka

CONTENTS

DOI: 10.1201/9781003132349-8

7.1 INTRODUCTION

Due to its applicability, cost-effectiveness, and eco-friendly nature, phytoremediation has increasingly become popular in developed countries to treat chemical-contaminated soil and water. Hyperaccumulator plants, which are known for their capability of extracting and accumulating high amounts of contaminants, have shown considerable potential in phytoremediation. The most common phytoremediation techniques include phytostabilization, phytodegradation, rhizofiltration, phytoextraction, and phytovolatilization (Table 7.1). The factors that affect the employed techniques depend on the nature of the chemical contaminant, soil condition, and plant species used. Phytoremediation techniques mentioned above have limitations under certain situations. These include the slow-growing and adaptions of plants to the various environments and soil conditions (e.g., poor nutrients in soil). Hence, some measures have been taken to improve the efficiency of traditional phytoremediation techniques. These approaches include the application of genetically engineered plants and

TABLE 7.1
Different Methods of Phytoremediation

Method of Phytoremediation	Details	Example/s	References
Phytoextraction	Techniques by which chemical pollutants in soil/water are concentrated in harvestable plant parts by introducing plants into the contaminated site. This is a common technique to treat heavy metal contaminated soil and water.	*Parthenium hysterophorous, Cannabis sativa, Solanum nigrum, Ricinus communis*	Chandra, Dubey & Kumar, 2017; Blaylock & Huang, 2000; Pivetz, 2001; Selvi et al., 2019; Mao et al., 2016.
Rhizofiltration	A technique that is employed to treat chemical contaminants, especially in water and liquid waste. Chemical contaminants adsorb and precipitate in plant roots or are absorbed by plant roots. Plants with the highest spread fibrous root system are advantageous when employing this technique.	*Helianthus annuus, Ulex europaeus*	Dhanwal et al., 2017; Eapen et al., 2003; Selvi et al., 2019.

TABLE 7.1 (Continued)
Different Methods of Phytoremediation

Method of Phytoremediation	Details	Example/s	References
Phytostabilization	A strategy that is used to immobilize heavy metals by storing them in plant roots or precipitating them in the rhizosphere by root exudates. This decreases the movement of contaminants and thereby avoids entering them into the groundwater. Plants having extensively spread deep fibrous root systems are highly desirable to employ this technique.	*Brassica juncea, Festuca rubra, Zea mays*	Banuelos et al., 1993; Ginn et al., 2008; Gerhardt et al., 2017; Kumpiene et al., 2012; Mahar et al., 2016; Marques et al., 2009; Wong, 2003.
Phytodegradation	Technique in which either metabolic process inside the plant degrades the absorbed pollutants into non-hazardous compounds or chemical contaminants in soil is breakdown by root secreted enzymes.	*Populus tremuloides, Populus deltoids, Populus alba, Populus nigra*	Assunção et al., 2001; Bizily et al., 2000; Burges et al., 2018.
Rhizodegradation	Use of roots for degradation or breakdown of organic pollutants. Explain thes as others the process	*Mentha spicata, Morus rubra, Medicagosativa* L., *Typha latifolia*	Adiloğlu, 2017; Assunção et al., 2001;DalCorso et al., 2019.
Phytovolatilization	The process in which absorbed chemical contaminants from roots is converted into less harmful volatile compounds and released to the air through stomata in leaves.	*Brassica juncea, Arabidopsis thaliana, Populus (poplars)* and *Salix (willows)*	Mahar et al., 2016; Banuelos and Meek, 1990; Terry et al., 1992; Banuelos et al., 1993; de Souza et al., 2000; Terry et al., 2000.

phytoremediation assisted with phytohormones, plant growth-promoting bacteria, and Arbuscular Mycorrhizal Fungi (AMF) inoculation.

During phytoremediation, the absorption of the contaminants is completed in three stages:

(1) The transportation of the nutrients to the root circle and root surface;
(2) The absorption of the nutrient ions into the roots; and
(3) The transportation of the nutrient ions, which entered into the root to the necessary parts by the transmission branches.

The transportation of the nutrients to the root surface follows two basic theories either in "intersection and contact change" and/or "carbonic acid theory" (Karaman et al., 2012). Phytoremediation has been used to clean up metals, pesticides, solvents, explosives, crude oil, polyaromatic hydrocarbons, landfill leachates, agricultural runoff, and acid mines.

The majority of heavy metals accumulate in plants through a series of processes that include heavy metal mobilization, deposition in the root system, dispersion in the shoots, storage compartments within the stems, and compartmentation in the roots. For the vast majority of its existence, heavy metals are contained in the soil in a form that is inaccessible to plants. A plant may generate a variety of roots and rhizosphere exudates, which change the pH of the rhizosphere and increase the solubility of heavy metals (Dalvi & Bhalerao, 2013). When metal sorbs at the root surface, it enters the root cells. Heavy metals mostly pass into the apoplastic or symplaxation pathways (active transport against electrochemical potential gradients and concentration across the plasma membrane) (Figure 7.1a). It depends on whether the heavy metals are taken up by symplastic means and which complexing agents are used (Peer et al., 2005). When roots are present, heavy metal ions may form different chelates (in addition to the chelates already in the root environment, which are organic acids). In addition, these heavy metal ion–chelate complexes; carbonate, sulfate, and phosphate precipitates are trapped in the extra- or intracellular spaces (sympotic) (Ali et al., 2013). Within the vacuoles, they can join the root symplasm and be translocated to the shoots, after which they can be transported into the xylem through xylem vessels. Thus, they are moved around the leaves and sequestered in the cell vacuole, avoiding free metal ions accumulating in cytosol (Tong et al., 2004) (Figure 7.1b).

7.2 WASTE ACCUMULATION AND GENERATION IN DEVELOPING COUNTRIES

The most recent research study from BCC, "Bioremediation: Global Markets and Technologies in 2023", predicts that the industry will reach $186.3 billion by 2023, up against $91 billion in 2018. The largest rise in the market is expected before it can be treated compared with *ex-situ* bioremediation, pumping of soil water or excavation, in situ bioremediation—treatment of polluted earth or groundwater at the locations in which it occurs. The projected growth for bioremediation in situ and former local compound annual growth rates (CAGRs) between 2018 and 2023 are

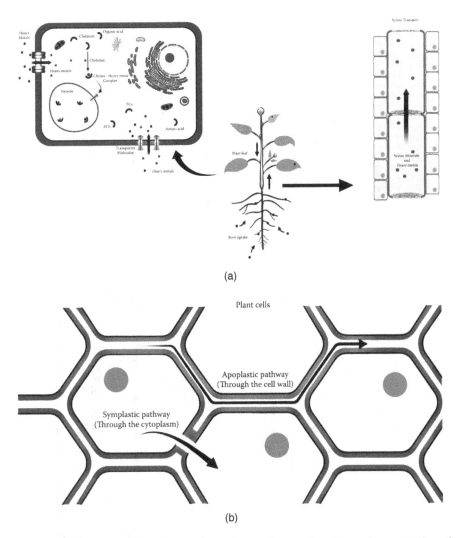

(a)

(b)

FIGURE 7.1 (a) Schematic diagram shows the uptake, translocation, and sequestration of heavy metals in plants, (b) demonstrate how metals can enter a plant, pass through the plant and be removed from the body (absorbed and excreted), and (c) shows the critical steps involved in hyperaccumulation.

15.0% and 16.7%, respectively. Furthermore, it is projected that the Asia-Pacific region will become the fastest-growing bioremediation market and would expand at a CAGR of 16.7% during 2018–2023. This will include phytoremediation as major tool of bioremediation market. The rapid increase of phytoremediation technology uses for the detoxification of man-made contaminants, increased military activities, and increase of automotive stations in India, China, Japan, and South Korea located in the Asia-Pacific region is considered as the main reason for the fastest growth in the bioremediation market in this region. Therefore, a visible clear demand for

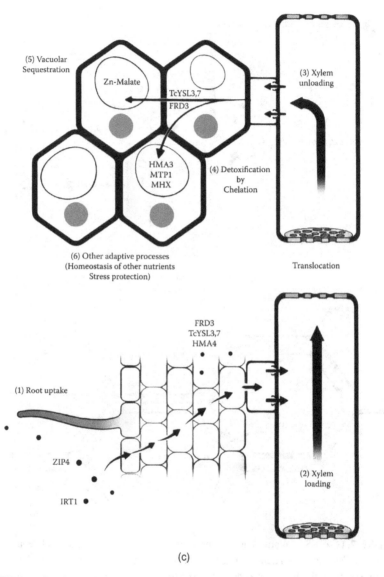

(c)

FIGURE 7.1   (Continued)

the bioremediation market including phytoremediation tools in the future is another advantage for adopting phytoremediation practices and findings of new ways to improve its efficacy and acceptability.

7.3   AVAILABLE PHYTOREMEDIATION METHODS USED TO SOLVE WASTE ACCUMULATION AND GENERATION

Phytoremediation uses green plants to clean up contaminated soil, groundwater, and wastewater. To minimize the hazardous effects of pollutants, this strategy employs plant interactions (physical, biochemical, biological, chemical, and micro-biological) in contaminated areas. Depending on the pollutant type (elemental or organic), phytoremediation involves a number of methods (accumulation or extraction, degradation, filtering, stabilization, and volatilization). The removal of elemental contaminants (toxic heavy metals and radionuclides) is mostly accomplished by extraction, transformation, and sequestration. Some important factors to consider when selecting a plant as a phytoremediator include root system, which may be fibrous or tap depending on pollutant depth, above ground biomass, which should not be available for animal consumption, pollutant toxicity to plant, plant survival, and adaptability to prevailing environmental conditions, plant growth rate, site monitoring, and, most importantly, time required to achieve the desired level of cleanliness. Furthermore, the plant should be disease- and insect-resistant (Lee, 2013). In some polluted settings, the process of contaminant removal by plant includes the following steps: absorption, which is mostly passive; translocation from roots to shoots, which is carried out by xylem flow; and accumulation in shoots. Translocation and accumulation are also affected by transpiration and the partitioning of xylem sap and neighboring tissues, respectively. Nonetheless, the procedure is likely to vary depending on other parameters such as the nature of the contamination and the species of plant. Most plants growing in contaminated areas are likely to be strong phytoremediators. As a result, the success of any phytoremediation strategy is essentially dependent on improving the remediation potentials of native plants growing in contaminated areas, either by bioaugmentation using indigenous or exogenous plant rhizobacteria or by biostimulation. It has been suggested that the use of plant growth-promoting rhizobacteria (PGPR) may play an essential role in phytoremediation since PGPR increases biomass output and plant tolerance to heavy metals and other unfavorable soil (edaphic) conditions (Yancheshmeh et al., 2011; de-Bashan et al., 2012).

The depth of anthropogenic pollution is extensively established in the developing world. Environmental pollution has been highlighted as one of the main concerns that need to be addressed urgently by most of the countries in the developing world.

Contaminants of most concern are metals, polyaromatic hydrocarbons (PAHs), and mineral oils (Lawal, 2017). The existing technique of heavy metal pollution remediation elsewhere depends primarily on digging-and-dumping or encapsulation, which fails to deal with soil decontamination (Wuana & Okieimen, 2011). Physicochemical mobilization or extraction can only be costly and appropriate in small regions when full decontamination is needed quickly and efficiently (Cisneros, 2011). For most underdeveloped nations, this technique may not be viable.

Natural processes in phytotechnology can be found in wetlands in developing nations. These wetlands are a transitional system between ground and aquatic systems where the water table is generally at or close to the ground level. It is a generic term that refers to aquatic vegetation and provides the biofilters for water contaminants to be removed by these wetlands (Omondi & Navalia, 2020). They can be natural marsh and swamp settings, but they can also be intentionally created storage basins or ponds. They have also been dubbed "the landscape kidneys" (Omondi & Navalia, 2020). In certain operating circumstances, natural humidity reduces diffuse pollution. Hydrological or pollutant loads are rapidly strained and degenerated when they surpass their assimilative capability. In the majority of developing countries, artificial-built wetlands (storage basins or ponds) produce "typical" wetland ecosystems and serve a role in controlling flooding and pollution. Artificial wetlands, created, planned, and designed for the treatment of wastewater comparable to the natural wetland is termed constructed wetlands. It is a feasible alternative for developing nations to the functional part of traditional wastewater treatment plants (Omondi & Navalia, 2020). Due to their "strong," "low-tech" character with no or few moving components such as pumps, they may be employed as part of decentralized wastewater treatment systems and function on relatively minimal operating needs. Phyto-technologies for household and municipal wastewater, gray waters, and selected industrial and commercial wastewaters can be employed in these built wetlands. For this aim, several famous metal-accumulating plant types might be planted in such constructed wetlands, such as *Sedum alfredii*, *Thlaspi caerulescens*, *Helianthus annuus*, *Brassica juncea*, and *Salix* (Hooda, 2007; Yan et al., 2020). These hyperaccumulators can provide higher performance but are less expensive compared to standard metal removal methods. Since photoreduction (phytomining) is cheaper and more efficient than chemical processes, the remediation of heavy metal-contaminated sites provides comparatively inexpensive sustainable solutions for the treatment of wastewater in these developing countries (Omondi & Navalia, 2020).

Besides food products, biofortification, and biofuel production as a new energy resource, the acquisition of reclaimed soil for agricultural and commercial purposes, and climate change biochar, offer fresh insight into heavy-metal phytoremediation through the use of constructed wetland environments (Omondi & Navalia, 2020). Furthermore, owing to its great potential, ease, efficiency, and economic benefits than other methods, plant restoration attracts the attention of scientists, remediation specialists, and environmental professionals of many sectors of industry and government in these developing countries (Singh, 2021).

7.4 WASTE CATEGORIES MANAGED BY PHYTOREMEDIATION

7.4.1 Soil Polluted by Heavy Metals

As a result of industrialization and urbanization, heavy metal pollution has greatly increased over the years. Heavy metals and metalloids include cadmium (Cd), mercury (Hg), lead (Pb), zinc (Zn), copper (Cu) (noble), and chromium (Cr).

Biological systems classify metals into two categories: "essential" and "nonessential". Essential heavy metals such as Cu, Mn, Ni, and Zn are needed for life cycle processes. However, when they are in excessive quantities, they can poison the organism. It is possible for these toxins to reach the food chain via crop processing and bioaccumulate, thereby posing a serious health risk to humans. This means that heavy metals in the atmosphere needs to be kept in check at the source, removed from the air and water, and not allowed to contaminate land.

There are many remediation methods used to return soil polluted by heavy metals. Any of these processes are mechanical or physiochemical, including soil incineration, cleaning, solidification, and use of an electric fields. In a similar way, phytotherapy, a plant-based method, uses plants to strip the elements or alter their bioavailability in the soil. Plants can absorb ionic compounds from their root system, even at a low concentration. When plants expand their root system into the rhizosphere to pull up and recycle unnecessary contaminants, they become part of the soil biogeochemical processes, preserving the stability of the soil's rhizosphere (DalCorso et al., 2019).

7.4.2 MUNICIPAL DISCHARGES AND INDUSTRIAL DISCHARGES

Municipal wastewater is considered as point source pollution collected in the sewage system. The main environmental issues relate to urban wastewater, even when those treated, brings water containing used compounds, including pollutants, into the body of water. Urban wastewater is never treated to restore its original quality (as it was a water source), as natural self-cleaning and diluting capacity is used to get the job done.

Industrial wastewater varies greatly in quality and quantity, depending on the type of industry in which it is produced. It may or may not be highly biodegradable and may or may not contain compounds that resist treatment. These include synthetic organics and heavy metals, and the content (quantity and quality) of wastewater from developing countries can be significantly different from that of developed countries. The main problem with industrial wastewater is the increase in the amount and type of synthetic compounds that are present and released into the environment.

7.4.3   LEACHATE CONTAMINATION FROM LANDFILL SITES

Leachate is a liquid that forms when rainwater filters through waste in a landfill. When this liquid comes into touch with buried waste, it leaches or pulls chemicals or elements from the trash. Leachate is a source of environmental concern because the pollutant combination it contains can harm ecosystems and public health when it reaches the soil, surface, and groundwater near the landfill (Baderna et al., 2019). Leachate is toxic, and elements including ammonia, organic compounds, and heavy metals contribute to its toxicity and carcinogenic potential even at low levels (Benfenati et al., 2007). According to medical literature, some general health concerns induced by drinking polluted leachate water might include sweating, bleeding stomach illnesses, blood problems, congenital defects, and even cancer. However, the impacts vary depending on the varied harmful components of these leachates. Accordingly, the leachate pollution index (LPI) assesses a landfill's total polluting potential. It is stated that the LPI value may be used as a tool to assess the potential for leachate contamination from landfill sites, particularly in areas where there is a significant risk of leachate migration and groundwater pollution. Phytoremediation, a low-cost and ecologically beneficial way of dealing with leachate, is one option (Ribé et al., 2012).

7.5   SUCCESSFULNESS AND FAILURES OF THESE METHODS

Phytoremediation can be used in a variety of ways in the field. The efficacy of this approach is dependent on the plant's capacity to withstand stress induced by leachate's harmful substances (Ignatius et al., 2014). Pollutants can collect inside the bodies of some plants, and there is a possibility of pollutants entering the food chain (Vaverková & Adamcová, 2015). Ram Chandra et al., (2017) explained the performance of heavy metal accumulation of 15 potential native plants growing on sludge. Further, transmission electron microscopic observations of the root of *Parthenium hysterophorous*, *Cannabis sativa*, *Solanum nigrum*, and *Ricinus communis* plants showing the formation of multi-nucleolus, multi-vacuoles, and deposition of metal granules in cellular component of roots as a plant adaptation mechanism for phytoextraction of heavy metal-rich polluted site were explained by them. Hence, they have suggested these native plants may be used as a tool for in situ phytoremediation and eco-restoration of industrial waste-contaminated sites in India (Chandra et al., 2017).

The Cu absorption in corn (*Zea mays* L.) was studied by inoculating various pot dosages of *Acaulospora mellea*, an arbuscular mycorrhizal fungus (Wang et al., 2007). They came to the conclusion that poor plant absorption in the high Cu pots was due to the acidic soil. The study found that the organic acids such as malic acid, citric acid, and oxalic acid tended to be concentrated in the soil structure due to fungal effects. It has been observed that *A. mellea* is not efficient for phytoextraction; however, mycorrhizal plants have a greater copper concentration in their roots (Wang et al., 2007; Lingua et al., 2012). However, in addition to soil movement, wind, water, and degradation of pollutants may be blocked or mitigated. The plant's root atmosphere and plant chemistry are strongly linked, which allows it to modulate the shape of contaminants from insoluble to non-transported.

7.6 PHYTOREMEDIATION TECHNIQUES

To remove heavy metals from the environment, many separate interventions including physical, chemical, and biological applications are proposed. Integrated processes are gaining popularity since they are claimed to successfully fulfill the aim of phytoremediation in many environmental matrixes and will solve a significant obstacle of large-scale application. However, integrating two distinct processes necessitates a thorough grasp of the goals of the processes. Two processes should be integrated in such a way that they are experimentally possible even in large-scale applications, economically viable, and comparably efficient against the individual processes. As a result of several findings, integrated techniques for removing contaminants like heavy metals removal from diverse environmental matrixes are gaining in popularity (Huang et al., 2012; Chen et al., 2003).

7.6.1 ELECTROKINETIC (EK) PHYTOREMEDIATION APPROACH

The recovery yield and the process rate both need to be significantly improved. However, this can be improved by combining phytoremediation with other techniques. Electrokinetic (EK) method is a novel form of remediation that has shown to be more effective in terms of metal recovery and more cost-efficient than the other integrated techniques. EK remediation and its compatibility with phytoremediation have been proven (Rosestolato et al., 2015; Selvi et al., 2019). Laboratory research on EK and phytoremediation approaches have demonstrated a promising method in heavy/trace metal remediation of Zn, Pb, Cu, Cd, and As. In this approach the organic and inorganic molecules are separated using a direct current carried between electrodes positioned vertically in soil (Cao et al., 2003; Santos et al., 2008). In such studies, the gathered samples were first remediated using the EK technique, then phytoremediation was performed by extruding the remediated soil samples from the electrokinetic cell (Bhargavi & Sudha, 2015). This EK-remediated soil used for phytoremediation demonstrated a promising buildup of Cd and Cr in a single harvest, which increased in consecutive harvests. A similar approach helped to remove Pb from contaminated soil using mustard plant (Lim et al., 2004). Furthermore, studies revealed other types of metals that can remove using the same way such as Cd, Cu, Pb, and Zn from soil (Cang et al., 2012). In this approach, voltage used play a vital role as a direct effect on soil properties, and that plant growth boosted soil enzymatic activity to achieve maximal heavy metal cleanup (Cang et al., 2012). Similarly, the use of EK-coupled phytoremediation by using the species *Lemna minor* to treat harmful As in water was also found to be effective (Kubiak et al., 2012). Another approach reported lead removal from the soil in which the EK remediation was carried out by utilizing multiple electrode configurations to increase phytoremediation by increasing the depth of soil to avoid the leaching of mobile metals on the ground surface (Hodko et al., 2000).

7.6.2 BIOAUGMENTATION WITH PHYTOREMEDIATION APPROACH

Phytobial remediation is an effective and environmentally acceptable method of removing heavy metals from soil and water using both plants and microbes. In this approach, plants to absorb heavy metals and bacteria are used to facilitate the breakdown of those metallic elements (Lynch & Moffat, 2005). Figure 7.2 outlines the different mechanisms of phytoremediation.

These methods can be improved by including an appropriate bacterium capable of secreting a variety of plant growth promoting substance (Martin & Ruby, 2004). Organic acids, ACC deaminase, siderophores, and biosurfactants are among the

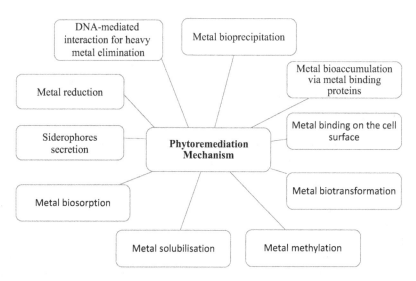

FIGURE 7.2   Different mechanisms of phytoremediation (Ahemad, 2015).

chemicals that will convert metals into bioavailable forms with this approach. (Roy et al., 2015).

7.6.3 SYNERGISTIC BIOSORPTION WITH PHYTOREMEDIATION APPROACH

Heavy metal complexes can function additively, synergistically, or antagonistically in nature (Wu et al., 1995). As a protective strategy, these poisonous heavy metals will cause the generation of reactive oxygen species (ROS) and inducers in a plant's antioxidant system (Foyer & Noctor, 2005; Foyer et al., 2009). The effects of Pb–Zn uptake and their activities on antioxidant enzymes that protect plants in different compartments of the plant cell were studied (Singh et al., 2006; Singh et al., 2017). The synchronized actions of antioxidant activities are components of the system that defends plants against ROS in distinct plant cell compartments (Singh et al., 2006; Singh et al., 2017).

7.6.4 ENHANCING PLANTS FOR METAL TRANSPORT, TRANSFORMATION, VOLATILIZATION WITHIN THE PLANT AND TOLERANCE TO TOXICITY

It is evident that substantial technology is required to remove heavy metals from the environment and reduce them to allowable levels. To boost metabolism and tolerance to heavy metals, recombinant deoxyribonucleic acid technology has been employed to modulate the expression of several plant-specific genes (Dowling & Doty, 2009). Composting polluted soil prior to planting increased total petroleum hydrocarbon (TPH) degradation, which favored rhizodegradation by *Suaeda glauca* (Wang et al., 2011). It follows that pretreatment and/or amendment of extremely contaminated sites prior to plant planting will aid the increase in phytoremediation efficiency by enhancing microbial diversity and activity while minimizing pollutant harmful effects on plants. Previous studies have suggested a competition-driven model for rhizosphere–microbiome interaction to better understand and identify elements that contribute to the assembly of beneficial (plant-growth promoting (PGP) and degrading) microbiota during phytoremdiation processes (Thijs et al., 2016). Four major strategies (plant selection based on microbiome, root exudate interference, disturbance, and supply line feeding) were identified as the strategies to be employed in polluted sites to ensure that opportunistic and pathogenic microbial populations are kept in check, allowing improved phytoremediation processes by degradative and PGP microbes. Furthermore, it was hypothesized that plant–microbiome interactions are not always optimum for phytoremediation; hence, human interventions are necessary to optimize such interactions to improve pollutant removal. Furthermore, the addition of organic waste (brewery spent grains) to waste lubricating oil-contaminated soil accelerated the growth of *Jatropha curcas* and microbial proliferation at the rhizosphere, resulting in an additional 33% contaminant removal from 2.5% used lubricating oil-contaminated soil (Agamuthu et al., 2010).

The genetic engineering of bacteria and plants has also proven to be worthwhile in terms of heavy metal removal applications. When microbes are genetically modified, they can perform better than the natural isolates, which has enormous remedial potential. Similarly, genetic engineering can be used to stimulate phytoremediation to increase heavy metal accumulation and absorption. The "ars" operon in the Arsenic Resistance Gene encodes a regulatory protein that aids in the detection of arsenic contamination. Recombinant *E. coli* with the *arsR* gene accumulated 5- and 60-fold higher levels of arsenate and arsenite than the control strain has been developed (Kostal et al., 2004). The recombinant strain with the "ars" operon is most suited for in situ remediations to undertake bioremediation in real-world settings (Ryan et al., 2007). Transgenic canola plants with *Enterobacter cloacae* CAL2 inserted acquired four times more heavy metals than the control cells. It has been claimed that the introduction of transgenic plants improves the plant's ability to remove heavy metals from soil (Eapen et al., 2003).

Plant genetic engineering has proven to be a promising method for the improvement of the plants' phytoremediation capacity against heavy metals. Besides, genetic modification may also transmit the desired genes to sexually undesirable plants from the hyperaccumulator, which cannot be achieved by conventional breeding practices such as crossing (Berken et al., 2002). Technically, it is more applicable than engineering hyperaccumulators to adjust fast-growing, high-biomass organisms to achieve a high resistance and high heavy metal accumulation potential to generate high biomass. High-biomass plants are either designed to improve the resistance of heavy metals or to improve the potential of heavy metals aggregation, which are the main properties of hyperaccumulators, in many applications. The selection for genetic engineering should also be based on information on the resistance and aggregation processes of heavy metals in plants.

The promising strategies to increase the accumulation of heavy metals promote the development of metal chelators through genetic engineering as metal-binding ligands that boost the bioavailability of metal ions, promoting heavy metal uptake, root-to-shoot translocation in organelles, and mediate intracellular sequestration. Heavy metal uptake and translocation can be improved by overexpression of the genes, which encode natural chelators (Wu et al., 2010). While the approach to genetic engineering offers enticing prospects of optimizing plant production in heavy metal phytoremediation, there are also a few setbacks.

Another approach to improving plant production for phytoremediation is the use of plant microorganisms (rhizospheric microorganisms). The rhizosphere's microbial community will directly induce root proliferation to encourage plant development, improve tolerance of heavy metals, and health for plants (Gupta et al., 2013). A strong capacity for enhancing the effective phytoremediation of plant growth-promoting rhizobacteria (PGPR) has been proven. PGPR can encourage growth and fitness in plants, protect plants from infections, enhance plant resistance to heavy metals, enhance plant nutrient absorption, and increase heavy-metal absorption and translocation (Ma et al., 2011). Different chemicals, including amino acids, siderophores, antibiotics, enzymes, and phytohormones are produced (Ma et al., 2011). A deaminase degrading the ethylene precursor ACC, PGPR can synthesize 1-aminocyclopropane-1-carboxylate (ACC). PGPR can reduce the production of ethylene and thus encourage plant growth by developing ACC deaminase (Glick, 2014). Plants inoculated with ACC deaminase PGPR demonstrated an improved production of biomass by high root and fire densities, which resulted in improved heavy metal uptake as well as increased efficiency in phytoremediation (Arshad et al., 2007).

Another essential microbial group that may aid phytoremediation plants is arbuscular mycorrhizal fungi (AMF). In rhizospheres, the presence of AMF increases the absorbent surface area of plant roots through the extensive hyphal network and improves the absorption of water and nutrients as well as the bioavailability of heavy metals (Göhre & Paszkowski, 2006). AMF also can produce plant growth and help plant remedies (Vamerali et al., 2010) by producing phytohormones. In addition to plant selection and safety, increasing the bioavailability of heavy metals is another effective strategy for increasing plant extraction quality.

Microorganisms can produce enzymes and chelate to a rhizosphere, thereby increasing the absorption and translocation of heavy metal–chelate complexes (Clemens et al., 2002). PGPR and PGPE (plant growth- promoting endophytes, for example) can boost the solubility of water-insoluble Zn, Ni, and Cu through the secretion of organic protons or anions. PGPR also secrete siderophores and biosurfactants to mobilize heavy metals in the soil. Siderophores are Fe chelators with high ferric iron (Fe^{3+}) affinity and a variable heavy metal affinity like Cd, Ni, As, and Pb (Schalk et al., 2011). Siderophores can improve bioavailability for rhizobacteria and plants through chelation with these heavy metals. It was proven successful to use rhizobacteria to supply heavy metal ions. The study by Braud et al. (2009) showed that the disposition of Cr and Pb by siderophore-producing bacteria inoculated maize, in an agricultural soil contaminated with Cr and Pb. Furthermore, the mycorrhizal fungi can also alter the physical and chemical properties of the soil and plant root exudates, which influence the bioavailability of heavy metals in the soil (Sarwar et al., 2017). It has been shown that red clover (*Trifolium pratensus* L.), inoculated with shrubby mycorrhiza, is far more effective when grown in soil with Zn than uninoculated controls (Chen et al., 2003). Another study showed that mycorrhizal hyphae could accumulate Zn directly from the ground and transfer it to the roots, thus enhancing its accumulations (Chen et al., 2003). Chelating agents are another commonly considered bioavailability technique for heavy metals. The added chelating soil modifications form a water-soluble heavy metal–chelate complex that is more mobile and quickly absorbable in plants (Wuana & Okieimen, 2011).

Chemical chelating agents such as citric acid, malic acid, acetic acid, and oxalic acid have been shown as an alternative for this to efficiently shape complexes in heavy metals and improve the bioavailability of heavy metals (Sarwar et al., 2017). These organic chelators are natural and biodegradable in soil, which may cause less environmental risk than synthetic chelating agents (Souza et al., 2013).Using organic chelating agents for phyto-assisted chelate extraction is therefore a promising approach.

7.7   ADVANTAGES OF PHYTOREMEDIATION TECHNIQUES

Phytoremediation is acknowledged as the least invasive and least expensive option, as opposed to other invasive technologies. It also has the benefit of being able to be applied to large regions of polluted groundwater, soil, and sediment. Furthermore, it's an in situ application discovered to reduce heavy metal dispersion in the soil and help in the preservation of the topsoil.

Microbes that are free to move help in phytoremediation by mobilizing, immobilizing, and volatilizing. Metals are mobilized by a variety of processes, including volatilization, redox transformation, leaching, and chelation. Arsenic is removed by bacteria such as *Sulfurospirillum barnesii*, Geobacter, and *Bacillus selenatarsenatis*. The use of anaerobic bioleaching and electrokinetics to create a hybrid approach was discussed by Lee in 2013. This study revealed that heavy metals accumulated in the harvested tissue of the plant have potential to be used for phytoremediation, which may then be disposed. Introducing mobilizing microorganisms into polluted water accelerates the buildup of heavy metals (Wang et al., 2005). During the immobilization procedure, the contaminant's mobility is restricted by modifying its physical and chemical characteristics (Leist et al., 2000). The oxidase enzymes found in bacteria oxidize metals, rendering them immobile and less poisonous. To remove heavy metals, microbes such as *Sporosarcina ginsengisoli*, *Candida glabrata*, *Bacillus cereus*, and *Aspergillus niger* were utilized in an immobilization approach.

On the other hand, the primary benefits of utilizing plants to repair contaminated sites is that some valuable metals may bioaccumulate in plants and be retrieved after remediation, a process known as phytomining. The prospective uses of selenium-enriched material recovered from phytoremediation sites includes food, feedstuff, and biofortification of agricultural products (Wu et al., 2015). Other benefits of phytoremediation include environmental friendliness, large-scale operation, inexpensive installation and maintenance costs, soil structure conservation, avoidance of erosion, and metal leaching (Van, 2009; Mench et al., 2009).

Furthermore, higher soil fertility may result from phytoremediation due to organic matter intake (Ow, 1996). The resultant biomass after phytoremidation may be utilized for the generation of heat and power in specialized facilities. In this sense, phytoremediation is also a potentially profitable technique over the other bioremediation methods. Since phytoremediation is a more preferred green tool over the other bioremediation practices, the advantage of market expansion is more for novel phytoremediation practices. On the other hand, parallel to the rapid growth of the industrial cities in developing countries and heavy urbanization, the need for expansion of phytoremediation practices and increase of the phytoremediation market size is crucial.

The other advantage of phytoremediation approach is the use of invasive plants in phytoremediation. The rhizosphere can get adsorbed to the surface or impregnated into the roots during adsorption. Shift in the rhizosphere shifts the pH, makes them more soluble, which stops them from leaching into groundwater (Javed et al., 2019). Various species of hyacinth, cattails, duckweed, and poplar are widely used to cleanse and manage wetlands because of their rapid growth and their capacity to accumulate vast amounts of biomass (Hooda, 2007). Aquatic plants have long and fine roots, such as those of the water parsnip and lotus compared to terrestrial plants such as sunflower (*H. annuus*). It has been shown that these plants have the potential to absorb toxic metals after rhizofiltration (Dhanwal et al., 2017). A list of some plants, which show a high capacity for heavy metal accumulation, is given in Table 7.2. Conversely, it is undesirable to use plants for phytase because of the possibility of heavy metals entering the food chain through ingestion by humans or animals. Thus, plant accretion of heavy metals is essential for effective and safe phytoremediation.

TABLE 7.2
List of Plant Species Tested for Heavy Metals Accumulation

Heavey Metal	Plant Species	
As	Pteris vittate	Pteris quadriaurita
	Pteris ryukyuensis	Pteris Biaurita
	Pteris cretica	Corrigiola telephiifolia
	Eleocharis acicularis	
Cd	Phytolacca Americana	Sedum alfredii
	Prosopis laevigata	Arabis gemmifera
	Salsola Kali	Deschampsia cespitosa
	Thlaspi caerulescens	Turnip landraces
	Azolla pinnata	
Co	Haumaniastrum robertii	
Cr	Pteris vittate	
Cu	Eleocharis acicularis	Haumaniastrum
	Aeolanthus biformifolius	Katangense
	Ipomoea alpine	·Pteris vittate
Hg	Achillea millefolium	Silene vulgaris
	Marrubium vulgare	Festuca rubra
	Rumex induratus	Helianthus tuberosus
Pb	Medicago sativa	Helianthus annuus
	Brassica juncea	Euphorbia cheiradenia
	Brassica nigra	

In various invasive plant species, trace elements have been tolerated to a high degree. This kind of herb is also known as a hyperaccumulator, with unusually high levels of foliar toleration. Any hyperaccumulators (species of succulent plants) accumulate extraordinary levels of trace minerals in their aerial parts without sign of toxicity. A small number of plants (2%) of them hyperaccumulate trace metals (nickel, lead, chromium, and cadmium) (Milner & Kochian, 2008).

While several hyperaccumulators have been found, most are low biomass producers and slow growers, which hinder their ability to remove heavy metals. Using non-hyperaccumulators instead to remove toxic materials from the soil, phytoextraction of heavy metals could be practiced.

Grasses are good biofilters because of their short life cycle, fast growth, and abundant biomass because of their ability to withstand harsh environmental stresses (Malik et al., 2010). One of the fastest-growing plants, *Trifolium alexandrinum* is a Cd, Pb, Cu, and Zn binder, and it has been suggested for phytoextraction (Ali et al., 2012). One of the benefits of using woody plants for phytoextraction is that they help to remove toxins from water. Woody plants can sequester a much greater amount of heavy metals above-ground than do plants and shrubs. Their deep root system keeps dirt in place, preventing it from being carried out into the surroundings. This illustrates why farm residues are superior to chemical fertilizers for plant phytoremediation due to their non-transferable characteristics (Burges et al., 2018).

7.8 DISADVANTAGES OF PHYTOREMEDIATION TECHNIQUES

Phytoremediation has its own limitations. These include the following:

- taking at least several growing seasons to clean up a site;
- plants that absorb toxic heavy metals or persistent chemicals posing a risk to wildlife and contaminating the food chain;
- intermediates formed from those organic and inorganic contaminants may be cytotoxic to plants and animals including humans;
- limited to shallow aquifers and soil due to plant root length limitations;
- requires regular monitoring (due to litter fall); and
- lack safe proper disposal method, difficult metal recovery procedures, and recycle economy.

Disposal of phytoremediation plant waste is another problem faced by the utilizers of this technique for remediating contaminated grounds. The growth of phytoremediation is therefore heavily reliant on research and technical development, which should enhance both the pre-harvest (processes preceding the removal of plant biomass from the remediated region) and post-harvest phases (destination given to the biomass after its removal from the area). There are some phytoremediation research and development tactics in the phases before the harvest of biomass created in the polluted region. They are as follows:

(a) selecting and testing different species for different contaminants;
(b) optimizing the use of fertilizers (biostimulation), irrigation, and pest control;
(c) optimizing rhizosphere processes by inoculating seeds or root systems with bioremediation bacteria or fungi (bio-increase);
(d) increasing knowledge about the succession processes of the ecosystem and plant communities that favor phytoremediation;
(e) making use of plant metabolism and genetic engineering;
(f) using mass balance; and
(e) uncovering the path and transformation of pollutants in the plant body.

These strategies aid in the creation of post-harvest methods to keep the dangers associated with the produced metabolites at tolerable levels, since waste management remains a limiting factor in the development of phytoremediation (Ghosh & Singh, 2005; Ginneken et al., 2007). Previous toxicity and bioaccumulation experiments, as well as monitoring, must be performed until the pathway is thoroughly mapped and dose–response relationships are identified for each important by-product across a wide variety of environmental circumstances. Post-harvest possibilities for safely disposing of biomass-derived from contaminated include the following:

(a) sanitary landfill disposal of contaminated plant residues
(b) composting to degrade xenobiotic waste and reduce the final volume to be disposed of
(c) biomass incineration to destroy organic compounds and reduce the volume to be disposed of (in the case of metals that are not released into the atmosphere in significant quantities)
(d) energy use in the case of fast-growing woody plants such as resin, raw materials to manufacture various products
(f) smelting, for the economical recovery of metals, when economically viable.

When biomass is processed with phystabilization, no hazardous waste is produced. However, the selection of plant species is critical for phytostabilization. Plants must be able to withstand high metal stress for phytostabilization to be successful. Plant root systems are critical for heavy metal immobilization, soil composition, and erosion prevention. If an effective vegetation cover is to be produced, a large amount of plant biomass must be grown quickly. Most significantly, it should be reasonably simple to keep running on the field. Organic and inorganic amendments can both help with phytostabilization.

The efficiency of the phytoremediation system is governed by both plant factors (e.g., root depths) and environmental factors (e.g., climatic conditions). Principally, the soil must be appropriate to the needs of the plant for the removal of the contaminants from the soil by the plant. The pH levels of the area must be between 5.8 and 6.5 for the nutrient elements to be taken (Vanli, 2007) and it is the most important soil parameter for the efficiency of the phytoremediation system.

The main drawback of the use of this technology is that it is not good for all sites. If the contamination is too deep or the concentration of the contaminant is too high, plants alone cannot effectively remediate the contaminated site. Phytoremediation techniques are very effective in sterilizing areas that are moderately contaminated and are at low risk. Besides, there are problems identified with these physicochemical methods such as unnecessary costs when dealing with low concentrations of pollutants and the shift in the chemical and biological properties of the ecosystem's structures makes them nearly irrecoverable (DalCorso et al., 2019). As a result, reclamation technology for degraded soil must be built that is cost-effective, safe, and has a lower environmental impact. Figure 7.3 details some of the selection and performance criteria for choosing the bioremediation technique.

Challenges for phytoremediation come with the types of pollutant, which may include agrochemicals, chlorinated compounds, dyes, heavy metals, hydrocarbons, or nuclear waste that determine the technique of pollutant removal. According to the place of application, and the complexity of environmental pollution, bioremediation methods are required to be chosen from either an *ex-situ* or in situ method.

Although it is the most basic bioremediation technique, land farming, like other *ex-situ* bioremediation techniques, has some limitations including a large operating space, a reduction in microbial activity due to unfavorable environmental conditions, additional costs due to excavation, and reduced efficacy in inorganic pollutant removal (Khan et al., 2004; Maila & Colete, 2004). Furthermore, due to its design and

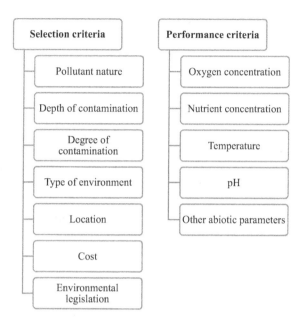

FIGURE 7.3 Selection and performance criteria for choosing bioremediation technique (Frutos et al., 2012).

technique of pollutant removal (volatilization), it is not ideal for treating soil polluted with harmful volatiles, particularly in hot (tropical) climatic zones. These and other constraints make land farming-based bioremediation time-consuming and inefficient in comparison to other methods.

7.9 POLICIES CREATED WITH PHYTOREMEDIATION IN DEVELOPING COUNTRIES

The International Environmental Technology Centre (IETC) was established in April 1994 by the United Nations Environment Programme (UNEP) with the aim of promoting the use of environmentally sound technologies (ESTs) in developing and transitional countries. Under this, several promising policies were created with phytoremediation in developing nations as per the instruction given by the World Bank and UNEP. EST is a technology that has the potential to significantly improve environmental performance compared to other technologies. EST is a more environmentally friendly way to protect the environment, reduce pollution, use resources more responsibly, recycle more waste and products, and replace all residual waste with the technology they replace. The process with IETC pays particular attention to urban environmental issues and the management of freshwater resources. IETC's work supports decision makers in government and other organizations by

* raising awareness,
* promoting information exchange,

- building local capacity to make informed decisions, and
- encouraging the development and demonstration of environmentally friendly technologies, leading to their adoption and use.

Developing countries have introduced conceptual overviews of plant engineering and guidelines to provide their importance in relation to various environmental problems and possible solutions. While these guidelines do not cover all aspects of plant engineering, they provide the basis for a conceptual framework for understanding the importance of ecosystem approaches to achieving the Sustainable Development Goals. It also considers the role of plant technology in relation to some key issues such as integrated water resources management, urban environmental management, biodiversity, and climate change.

However, developing technologies in this field opens up several options to mitigate the drawbacks present in phytoremediation. For example, its restriction to sites with shallow contamination within the rooting zone of remediation plants could be minimized by modifying the ground surface at the site to prevent flooding or erosion. Restriction to sites with low contaminant concentrations could be minimized by adding soil amendments. Conversely, climatic conditions are a limiting factor for any phytoremediation technique. Climatic or hydrologic conditions may restrict the rate of growth of plants that can be utilized for phytoremediation. The introduction of non-native species may affect biodiversity. Although climatic conditions have been identified as a major factor for the implementation of phytoremediation, research owing to its influence on plant growth. As most of the developing countries have ideal climatic conditions for plant growth all year round, which makes it an ideal environment for phytoremediation research.

7.10 CHAPTER SUMMARY

Owing to the poisonous and accelerated deposition in the environment, heavy metal contamination is a critical problem for agricultural production and food safety. Various methods have been used to help limit or clean up heavy metal pollution. Applied phytotherapy has shown positive results for the restoration of contaminated soil and has several benefits over other biochemical and physicochemical treatments. More than a hundred heavy metal hyperaccumulators have been discovered for phytoremediation so far. However, treatment of these hyperaccumulators may also have its drawbacks, since it takes time to remove metal contamination, as is the case with heavily polluted soil. Often, sluggish growth and low biomass production are partly responsible for this. Plant performance enhancement is essential in achieving plant phytoremediation. Genetic engineering has proven to be very effective in improving the capabilities of plants such as rapid growth, heavy metal resistance, and suitability to different weather and geological conditions. For phytoremediation, a thorough understanding of the processes of heavy metal absorption, and characterization of various molecules and signaling pathways, would be essential. Certain genes involved in heavy metal ingestion and sequestration, particularly those which are related to metal translocation, may be used to increase metal accumulation in plants. Besides, chelating agents and microorganisms may also be used to handle the heavy metal status of the soil or aid metal uptake in the plants. Effective and efficient cleaning of soil that is highly contaminated with metals cannot be achieved by any single process. The low-cost, in-situ method of phytoremediation is desirable for developing countries because it offers site restoration, partial decontamination, biological activity maintenance, physical soil structure, cost-effectiveness, and bio-recovery of metals. Phytoremediation can help preserve the natural resources of developing nations since it has proven to be very successful for the on-site and local cleaning of huge volumes of polluted soil and/or waterways, including groundwater without excavation or other precursor trains.

REFERENCES

Adiloğlu, S. (2017). "Heavy metal removal with phytoremediation," in Advances in Bioremediation and Phytoremediation, ed. Naofumi Shiomi (London: IntechOpen). doi: 10.5772/intechopen.70330

Agamuthu, P., Abioye, O. P., and Abdul Aziz, A. (2010). "Phytoremediation of soil contaminated with used lubricating oil using *Jatropha curcas*." *J. Hazard. Mater.* 179 (1–3), 891–894. https://doi.org/10.1016/j.jhazmat.2010.03.088.

Ahemad, M. (2015). Phosphate-solubilizing bacteria-assisted phytoremediation of metalliferous soils: a review. *3 Biotech* 5, 111–121. doi: 10.1007/s13205-014-0206-0

Ali, H., Khan, E., and Sajad, M. A. (2013). Phytoremediation of heavy metals-concepts and applications. *Chemosphere* 91, 869–881. doi: 10.1016/j.chemosphere.2013.01.075

Ali, H., Naseer, M., and Sajad, M. A. (2012). Phytoremediation of heavy metals by *Trifolium alexandrinum. Int. J. Environ. Sci.* 2 1459–1469. doi: 10.6088/ijes.002020300031

Arshad, M., Saleem, M., and Hussain, S. (2007). Perspectives of bacterial ACC deaminase in phytoremediation. *Trends Biotechnol.* 25, 356–362. doi: 10.1016/j.tibtech.2007.05.005

Assunção, A. G. L., Da Costa Martins, P., De Folter, S., Vooijs, R., Schat, H., and Aarts, M. G. M. (2001). Elevated expression of metal transporter genes in three accessions of the metal hyperaccumulator *Thlaspi caerulescens. Plant, Cell Environ.* 24, 217–226.

Aybar, M., Bilgin, A., and Sağlam, B. (2015). Removing heavy metals from the soil with phytoremediation. *Artvin Çoruh Univ. J. Nat. Hazards Environ.* 1, 59–65.

Baderna, D., Caloni, F., Benfenati, E. (2019). Investigating landfill leachate toxicity in vitro: a review of cell models and endpoints. *Environ. Int.* 122, 21–30.

Baker, A., and Brooks, R. (1989). Terrestrial higher plants which hyperaccumulate metallic elements. a review of their distribution, ecology and phytochemistry. *Biorecovery* 1, 81–126. doi: 10.1080/01904168109362867

Banuelos, G., Cardon, G., Mackey, B., Ben–Asher, J., Wu, L., Beuselinck, P., et al. (1993). Boron and selenium removal in boron-laden soils by four sprinkler irrigated plant species. *J. Environ. Qual.* 22, 786–792. doi: 10.2134/jeq1993.00472425002200040021x

Banuelos, G., and Meek, D. (1990). Accumulation of selenium in plants grown on selenium-treated soil. *J. Environ. Qual.* 19, 772–777. https://doi.org/10.2134/jeq1990.0047242 5001900040023x

Bastow, I. D., Booth, A. D., Corti, G., Keir, D., Magee, C., Jackson, C. A. L., et al. (2018). The development of late-stage continental breakup: seismic reflection and borehole evidence from the Danakil Depression, Ethiopia. *Tectonics* 37, 2848–2862. doi: 10.1029/2017TC004798

Becerra-Castro, C., Prieto-Fernández, Á., Álvarez-López, V., Monterroso, C., Cabello-Conejo, M., Acea, M., et al. (2011). Nickel solubilizing capacity and characterization of rhizobacteria isolated from hyperaccumulating and non-hyperaccumulating subspecies of Alyssum serpyllifolium. *Int. J. Phytoremediat.* 13, 229–244. doi: 10.1080/15226514.2011.568545

Benfenati, E., Maggioni, S., Campagnola, G., Senese, V., Lodi, M., Testa, S., Schramm, K. W. (2007). "A protocol to evaluate organic compounds present in a landfill," in *Landfill Research Trends*, ed. A.A. Velinni (Nova Science), pp. 141–166.

Berken, A., Mulholland, M. M., Leduc, D. L., and Terry, N. (2002). Genetic engineering of plants to enhance selenium phytoremediation. *Crit. Rev. Plant Sci.* 21, 567–582. doi: 10.1080/0735-260291044368

Bhargavi, V. L. N., and Sudha, P. N. (2015). Removal of heavy metal ions from soil by electro-kinetic assisted phytoremediation method. *Int. J. ChemTech Res.* 8, 192–202.

Bishop, J. E. (1995). Pollution fighters hope a humble weed will help reclaim contaminated soil. *AP NEWS.* Online. Available at https://apnews.com/article/70ada5a18a3150d1524ef08d4d4747ff

Bizily, S. P., Rugh, C. L., and Meagher, R. B. (2000). Phytodetoxification of hazardous organomercurials by genetically engineered plants. *Nat. Biotechnol.* 18, 213–217. doi: 10.1038/72678

Blaylock, M., and Huang, J. (2000). "Phytoextraction of metals," in *Phytoremediation of Toxic Metals: Using Plants to Clean-up the Environment*, eds I. Raskin and B. D. Ensley (New York: Wiley), 303.

Braud, A., Jézéquel, K., Bazot, S., and Lebeau, T. (2009). Enhanced phytoextraction of an agricultural Cr- and Pb-contaminated soil by bioaugmentation with siderophore-producing bacteria. *Chemosphere* 74, 280–286. doi: 10.1016/j.chemosphere.2008.09.013

Brewer, E. P., Saunders, J. A., Angle, J. S., Chaney, R. L., and Mcintosh, M. S. (1999). Somatic hybridization between the zinc accumulator *Thlaspi caerulescens* and *Brassica napus*. *Theor. Appl. Genet.* 99, 761–771. doi: 10.1007/s001220051295

Broadley, M. R., White, P. J., Hammond, J. P., Zelko, I., and Lux, A. (2007). Zinc in plants. *New Phytol.* 173(4), 677–702. doi: 10.1111/j.1469-8137.2007.01996.x. PMID: 17286818.

Burges, A., Alkorta, I., Epelde, L., and Garbisu, C. (2018). From phytoremediation of soil contaminants to phytomanagement of ecosystem services in metal contaminated sites. *Int. J. Phytoremediat.* 20, 384–397. doi: 10.1080/15226514.2017.1365340

Cailliatte R., Schikora A., Briat J. F., Mari S., and Curie C. (2010). High-affinity manganese uptake by the metal transporter NRAMP1 is essential for Arabidopsis growth in low manganese conditions. *Plant Cell* 22, 904–917.

Cang, L., Zhou D. M., Wang Q. Y., and Fan, G. P. (2012). Impact of electrokinetic-assisted phytoremediation of heavy metal contaminated soil on its physicochemical properties, enzymatic and microbial activities. *Electrochim. Acta 86.* https://doi.org/10.1016/j.electacta.2012.04.112

Cang, L., Zhou, D. M., Wang, Q. Y., and Wu, D. Y. (2009). Effect of electrokinetic treatment of a heavy metal contaminated soil on soil enzyme activities. *J. Hazard. Mater.* 172, 1602–1607. doi.org/10.1016/j.jhazmat.2009.08.033

Cao, X. (2004). Antioxidative responses to arsenic in the arsenic-hyperaccumulator Chinese brake fern (*Pteris vittata* L.). *Environ. Pollut.*128, 317–325.

Chandra, R., Dubey, N.K., and Kumar, V. (Eds.) (2017). *Phytoremediation of Environmental Pollutants* (1st ed.). CRC Press. https://doi.org/10.1201/9781315161549

Chaney, R. L., Broadhurst, C. L., and Centofanti, T. (2010). "Phytoremediation of soil trace elements," in *Trace Elements in Soils*, ed. P. S. Hooda (Chichester: Wiley), 311–352.

Chen, B. D., Li, X. L., Tao, H. Q., Christie, P., and Wong, M. H. (2003). The role of *Arbuscular mycorrhiza* in zinc uptake by red clover growing in a calcareous soil spiked with various quantities of zinc. *Chemosphere* 50, 839–846. doi: 10.1016/S0045-6535(02)00228-X

Chen, B., Stein, A.F., Castell, N., Gonzalez-Castanedo, Y., De La Campa, A.S., and De La Rosa, J. (2016). Modeling and evaluation of urban pollution events of atmospheric heavy metals from a large Cu-smelter. *Sci. Total. Environ.* 539, 17–25. doi: 10.1016/j.scitotenv.2015.08.117

Cisneros B. J. (2011). Safe Sanitation in Low Economic Development Areas. *Treatise Water Sci.* 2011, 147–200. doi: 10.1016/B978-0-444-53199-5.00082-8. Epub 2011 Jan 24. PMCID: PMC7158271.

Clemens, S. (2006). Toxic metal accumulation, responses to exposure and mechanisms of tolerance in plants. *Biochimie* 88, 1707–1719. doi: 10.1016/j.biochi.2006.07.003

Clemens, S., Palmgren, M. G., and Krämer, U. (2002). A long way ahead: understanding and engineering plant metal accumulation. *Trends Plant Sci* 7, 309–315.

Confalonieri UEC (2003). Variabilidade climática, vulnerabilidade social e saúde no Brasil. *São Paulo Terra Livre* 19-I(20), 193–204.

DalCorso, G., Fasani, E., Manara, A., Visioli, G., and Furini, A. (2019). Heavy metal pollutions: state of the art and innovation in phytoremediation. *Int. J. Mol. Sci.* 20, 3412. doi: 10.3390/ijms20143412

Dalvi, A. A., and Bhalerao, S. A. (2013). Response of plants towards heavy metal toxicity: an overview of avoidance, tolerance and uptake mechanism. *Ann. Plant Sci.* 2, 362–368.

de-Bashan, L.E., Hernandez, J.P., Bashan, Y. (2012). The potential contribution of plant growth-promoting bacteria to reduce environmental degradation—a comprehensive evaluation. *Appl. Soil Ecol.* 61, 171–189. doi: https://doi.org/10.1016/j.apsoil.2011.09.003.

Desbrosses-Fonrouge, A.-G., Voigt, K., Schröder, A., Arrivault, S., Thomine, S., and Krämer, U. (2005). *Arabidopsis thaliana* MTP1 is a Zn transporter in the vacuolar membrane which mediates Zn detoxification and drives leaf Zn accumulation. *FEBS Lett.* 579, 4165–4174.

de Souza, M. P., Lytle, C. M., Mulholland, M. M., Otte, M. L., and Terry, N. (2000). Selenium assimilation and volatilization from dimethylselenoniopropionate by Indian mustard. *Plant Physiol.* 122, 1281–1288. doi: 10.1104/pp.122.4.1281

Dhanwal, P., Kumar, A., Dudeja, S., Chhokar, V., and Beniwal, V. (2017). "Recent advances in phytoremediation technology," in *Advances in Environmental Biotechnology*, eds R. Kumar, A. K. Sharma, and S. S. Ahluwalia (Singapore: Springer), 227–241. doi: 10.1007/978-981-10-4041-2_14

Dix, M. E., Klopfenstein, N. B., Zhang, J. W., Workman, S. W., Kim, M. S. (1997). Potential use of Populus for phytoremediation of environmental pollution in riparian zones. *Micropropag., Genet. Eng. Mol. Biol. Populus* 297, 206–211.

Domínguez–Solís, J. R., López–Martín, M. C., Ager, F. J., Ynsa, M. D., Romero, L. C., and Gotor, C. (2004). Increased cysteine availability is essential for cadmium tolerance and accumulation in Arabidopsis thaliana. *Plant Biotechnol. J.* 2, 469–476. doi: 10.1111/j.1467-7652.2004.00092.x

Dowling, David N., and Doty, S. L. (2009). Improving phytoremediation through biotechnology. *Curr. Opin. Biotechnol.* 20(2), 204–206.

Dushenkov, S. (2003). Trends in phytoremediation of radionuclides. *Plant Soil* 249, 167–175. doi: 10.1023/A:1022527207359

Eapen, S., Suseelan, K. N., Tivarekar, S., Kotwal, S. A., Mitra, R. (2003). Potential for rhizofiltration of uranium using hairy root cultures of *Brassica juncea* and *Chenopodium amaranticolor*. *Environ. Res.* 91, 127–133.

EPA (Environmental Protection Agency) (2000). *Introduction to Phytoremediation, EPA/600/R-99/107*. Ohio, USA: National Risk Management Research Laboratory Office of Research and Development U.S. Environmental Protection Agency Cincinnati.

Ernst, W. H., Verkleij, J., and Schat, H. (1992). Metal tolerance in plants. *Acta Bot. Neerl.* 41, 229–248. doi: 10.1111/j.1438-8677.1992.tb01332.x

Faucon, M. P., Shutcha, N., and Meerts, P. (2007). Revisiting copper and cobalt concentrations in supposed hyperaccumulators from SC Africa: influence of washing and metal concentrations in soil. *Plant Soil* 301, 29–36.

Foyer, C. H., Noctor, G., Buchanan, B., Dietz, K. J., Pfannschmidt, T. (2009). Redox regulation in photosynthetic organisms: signaling, acclimation and practical implications. *Antioxid. Redox Signal* 11 (4), 861–905.

Foyer, C. H., and Noctor, G. (2005). Oxidant and antioxidant signalling in plants: a re-evaluation of the concept of oxidative stress in a physiological context. *Plant, Cell Environ.* 28(8), 1056–1071.

Frutos, F. J., Perez, R., Escolano, O., Rubio, A., Gimeno, A., Fernandez, M. D., Carbonell, G., Perucha, C., Laguna, J. (2012). Remediation trials for hydrocarbon-contaminated sludge from a soil washing process: evaluation of bioremediation technologies. *J. Hazard. Mater.* 199–200, 262–271. https://doi.org/10.1016/j.jhazmat.2011.11.017

Gabor, T. S., North, A. K., Ross, L. C. M., Murkin, H. R., Anderson, J. S., and Turner, M. A. (2001). *Beyond the Pipe: The Importance of Wetlands and Upland Conservation Practises Practices in Watershed Management: Function and Values for Water Quality and Quantity.* Ducks Unlimited Canada, 55.

Gerhardt, K. E., Gerwing, P. D., and Greenberg, B. M. (2017). Opinion: taking phytoremediation from proven technology to accepted practice. *Plant Sci.* 256, 170–185. doi: 10.1016/j.plantsci.2016.11.016

Ghosh, M., and Singh, S. P. (2005). A review on phytoremediation of heavy metals and utilization of it's by products. *Asian J. Energy Environ.* 6(4), 18.

Ginn, B. R. G., Szymanowski, J. S., and Fein, J. B. (2008). Metal and proton binding onto the roots of *Fescue rubra. Chem. Geol.* 253(3–4), 130–135. https://doi.org/10.1016/j.chem geo.2008.05.001

Ginneken, L. V., Meers, E., Guisson, G., Ruttens, A., Elst, K., Tack, F. M., Vangronsveld, J., Ludo Diels, L., and Dejonghe, W. (2007). Phytoemediation for heavy metal-contaminated soil combined with bioenergy production. *J. Environ. Eng. Landsc. Manage.* 15 (4), 227–236.

Glick, B. R. (2014). Bacteria with ACC deaminase can promote plant growth and help to feed the world. *Microbiol. Res.* 169, 30–39. doi: 10.1016/j.micres.2013.09.009

Göhre, V., and Paszkowski, U. (2006). Contribution of the *Arbuscular mycorrhizal* symbiosis to heavy metal phytoremediation. *Planta* 223, 1115–1122. doi: 10.1007/s00425-006-0225-0

Gupta, D. K., Huang, H. G., and Corpas, F. J. (2013). Lead tolerance in plants: strategies for phytoremediation. *Environ. Sci. Pollut. R* 20, 2150–2161. doi: 10.1007/s11356-013-1485-4

Gupta, D. K., Srivastava, A., and Singh, V. P. (2008). EDTA enhances lead uptake and facilitates phytoremediation by vetiver grass. *J. Environ. Biol.* 29, 903–906.

Gustin, J. L., M.E. Loureiro., M. E., D. Kim., D., M. Tikhonova, M., and D.E. Salt, D.E. (2009). MTP1-dependent Zn sequestration into shoot vacuoles suggests dual roles in Zn tolerance and accumulation in Zn-hyperaccumulating plants. *Plant J.* 57, 1116–1127.

Hall, J. (2002). Cellular mechanisms for heavy metal detoxification and tolerance. *J. Exp. Bot.* 53, 1–11. doi: 10.1093/jexbot/53.366.1

Hamzah, A., Hapsari, R. I., and Wisnubroto, E. I. (2016). Phytoremediation of Cadmium-contaminated agricultural land using indigenous plants. *Int. J. Environ. Agric. Res.* 2, 8–14.

Herschbach, C., and Rennenberg, H. (1997). Sulfur nutrition of conifers and deciduous trees,". in *Trees—Contributions to Modern Tree Physiology*, eds Rennenberg, H., Eschrich, W., and Ziegler, H. (Leiden, The Netherlands: Backhuys Publishers), 293–311.

Herzig, R., Nehnevajova, E., Pfistner, C., Schwitzguebel, J.-P., Ricci, A., and Keller, C. (2014). Feasibility of labile Zn phytoextraction using enhanced tobacco and sunflower: results of five-and one-year field-scale experiments in Switzerland. *Int. J. Phytoremediat.* 16, 735–754. doi: 10.1080/15226514.2013.856846

Hooda, V. (2007). Phytoremediation of toxic metals from soil and waste water. *J. Environ. Biol.* 28(2 Suppl.), 367–376.

Huang, D., Xu, Q., Cheng, J., Lu, X., and Zhang, H. (2012). Electrokinetic remediation and its combined technologies for removal of organic pollutants from contaminated soils. *Int. J. Electrochem. Sci.* 7, 4528–4544.

Ignatius, A., Arunbabu, V., Neethu, J., and Ramasamy, E.V (2014). Rhizofiltration of lead using an aromatic medicinal plant *Plectranthus amboinicus* cultured in a hydroponic nutrient film technique (NFT) system. *Environ. Sci. Pollut. Res.* 21 (22), 13007–13016.

INFOSAN. (2011). *Information on Nuclear Accidents and Radioactive Contamination of Foods International Food Safety Authorities Network*. (Geneva: World Health Organization).

Javed, M. T., Tanwir, K., Akram, M. S., Shahid, M., Niazi, N. K., and Lindberg, S. (2019). "Chapter 20—Phytoremediation of cadmium-polluted water/sediment by aquatic macrophytes: role of plant-induced pH changes," in *Cadmium Toxicity and Tolerance in Plants*, eds M. Hasanuzzaman, M. N. V. Prasad, and M. Fujita (London: Academic Press), 495–529. doi: 10.1016/B978-0-12-814864-8.00020-6

Kalkan, H., Oraman, Ş., and ve Kaplan, M. (2011). Kirlenmiş Arazilerin Islah Edilmesinde Fitoremidasyon Tekniği. *Selçuk Üniversitesi Selçuk Tarım ve Gıda Bilimleri Dergis. i.* 25(4), 103–108.

Karaman, M. R., Adiloğlu, A., Brohi, R., Güneş, A., İnal, A., Kaplan, M., Katkat, V., Korkmaz, A., Okur, N., Ortaş, İ., Saltalıl, K., Taban, S., Turan, M., Tüfenkçi, Ş., Eraslan, F., and ve Zengin, M. (2012). *Bitki Besleme*. ISBN 978-605-87103-2-0 Dumat Ofset. Şti., Ankara: Matbacılık San. Tic. Ltd..

Khan, F. I., Husain, T., and Hejazi, R. (2004). An overview and analysis of site remediation technologies. *J. Environ. Manag.* 71, 95–122. doi: 10.1016/j.jenvman.2004.02.003.

Koprivova, A., Meyer, A., Schween, G., Herschbach, C., Reski, R., Kopriva, S. (2002a). Functional knockout of the adenosine 5′ phosphosulfate reductase gene in *Physcomitrella patens* revives an old route of sulfate assimilation. *J. Biol. Chem.* 277, 32195–32201.

Kostal, Jan, Yang, R., Wu, C.H., Mulchandani, A., and Chen, W. (2004). Enhanced arsenic accumulation in engineered bacterial cells expressing ArsR. *Appl. Environ. Microbiol.* 70 (8), 4582–4587.

Koźmińska, A., Wiszniewska, A., Hanus-Fajerska, E., and Muszyńska, E. (2018). Recent strategies of increasing metal tolerance and phytoremediation potential using genetic transformation of plants. *Plant Biotechnol. Rep.* 12, 1–14. doi: 10.1007/s11816-017-0467-2

Kubiak, J. J., Khankhane, P. J., Kleingeld, P. J., and Lima, A. T. (2012). An attempt to electrically enhance phytoremediation of arsenic contaminated water. *Chemosphere* 87, 259–64. doi: 10.1016/j.chemosphere.2011.12.048

Kumpiene, J., Fitts, J. P., and Mench, M. (2012). Arsenic fractionation in mine spoils 10 years after aided phytostabilization. *Environ. Pollut.* 166, 82–88. https://doi.org/10.1016/J.ENVPOL.2012.02.016

Lasat, M. (1999). Phytoextraction of metals from contaminated soil: a review of plant/soil/metal interaction and assessment of pertinent agronomic issues. *J. Hazard Subst. Res.* 2, 5. doi: 10.4148/1090-7025.1015

Lawal, A. T. (2017). Polycyclic aromatic hydrocarbons. A review, *Cogent Environ. Sci.* 3 (1), 1339841. doi:https://doi.org/10.1080/23311843.2017.1339841

Lee, J. H. (2013). An overview of phytoremediation as a potentially promising technology for environmental pollution control. *Biotechnol. Bioproc.* E 18, 431–439. https://doi.org/10.1007/s12257-013-0193-8.

Lee, J., Reeves, R. D., Brooks, R. R., and Jaffré, T. (1977). Isolation and identification of a citrato-complex of nickel from nickel-accumulating plants. *Phytochemistry* 16, 1503–1505. doi: 10.1016/0031-9422(77)84010-7

Lee, J., and Sung, K. (2014). Effects of chelates on soil microbial properties, plant growth and heavy metal accumulation in plants. *Ecol. Eng.* 73, 386–394. doi: 10.1016/j.ecoleng.2014.09.053

Leist, Michael, R.J. Casey, and Caridi, D. (2000). The management of arsenic wastes: problems and prospects. *J. Hazard. Mater.* 76 (1), 125–138.

Li, Y. M., Chaney, R., Brewer, E., Roseberg, R., Angle, J. S., Baker, A., et al. (2003). Development of a technology for commercial phytoextraction of nickel: economic and technical considerations. *Plant Soil* 249, 107–115. doi: 10.1023/a:1022527330401

Lim, J. M., Salido, A. L., and Butcher, D. J. (2004). Phytoremediation of lead using Indian mustard (*Brassica juncea*) with EDTA and electrodics. *Microchem. J.* 76, 3–9. doi: 10.1016/j.microc.2003.10.002

Lingua, G., Bona, E., Todeschini, V., Cattaneo, C., Marsano, F., et al. (2012). Effects of heavy metals and *Arbuscular mycorrhiza* on the leaf proteome of a selected poplar clone: a time course analysis. *PLoS ONE* 7(6), e38662. doi:10.1371/journal.pone.0038662

Liphadzi, M., Kirkham, M., Mankin, K., and Paulsen, G. (2003). EDTA-assisted heavy-metal uptake by poplar and sunflower grown at a long-term sewage-sludge farm. *Plant Soil* 257, 171–182. https://doi.org/10.1023/A:1026294830323

Lynch, J. M., and Moffat, A. J. (2005). Bioremediation-prospects for the future application of innovative applied biological research. *Ann. Appl. Biol.* 146, 217. doi: 10.1111/j.1744-7348.2005.040115.x

Ma, Y., Prasad, M., Rajkumar, M., and Freitas, H. (2011). Plant growth promoting rhizobacteria and endophytes accelerate phytoremediation of metalliferous soils. *Biotechnol. Adv.* 29, 248–258. doi: 10.1016/j.biotechadv.2010.12.001

Macnair, M. R. (2003). The hyperaccumulation of metals by plants. *Adv. Bot. Res.* 40, 63–105.

Mahar, A., Wang, P., Ali, A., Awasthi, M. K., Lahori, A. H., Wang, Q., et al. (2016). Challenges and opportunities in the phytoremediation of heavy metals contaminated soils: a review. *Ecotox. Environ. Safe.* 126, 111–121. doi: 10.1016/j.ecoenv.2015.12.023

Maila M. P, and Colete T. E. (2004). Bioremediation of petroleum hydrocarbons through land farming: are simplicity and cost-effectiveness the only advantages? *Rev. Environ. Sci. Bio/Technol.* 3, 349–360.

Malik, R. N., Husain, S. Z., and Nazir, I. (2010). Heavy metal contamination and accumulation in soil and wild plant species from industrial area of Islamabad, Pakistan. *Pak. J. Bot.* 42, 291–301.

Manara, A. (2012). "Plant responses to heavy metal toxicity," in *Plants and Heavy Metals*, ed. A. Furini (Dordrecht: Springer), 27–53. doi: 10.1007/978-94-007-4441-7_2

Mani, D., and Kumar, C. (2014). Biotechnological advances in bioremediation of heavy metals contaminated ecosystems: an overview with special reference to phytoremediation. *Int. J. Environ. Sci. Technol.* 11, 843–872. doi: 10.1007/s13762-013-0299-8

Mao, X. Y., Han, F. X. X., Shao, X. H., Guo, K., McComb, J., Arslan, Z., and Zhang, Z. Y. (2016). Electro-kinetic remediation coupled with phytoremediation to remove lead, arsenic and cesium from contaminated paddy soil. *Ecotoxicol. Environ. Saf.* 125, 16–24.

Marques, A. P., Rangel, A. O., and Castro, P. M. (2009). Remediation of heavy metal contaminated soils: phytoremediation as a potentially promising clean-up technology. *Crit. Rev. Env. Sci. Technol.* 39, 622–654. doi: 10.1080/10643380701798272

Martin, T. A., and Ruby, M. V. (2004). Review of in situ remediation technologies for lead, zinc, and cadmium in soil. *Remed. J.* 14, 35–53. doi: 10.1002/rem.20011

Memon, A. R., Aktoprakhgil, D., Özdemir, A., and Vertii, A. (2000). *Heavy Metal Accumulation and Detoxification Mechanisms in Plants. TÜBITAK MAM.* (Kocaeli, Turkey: Institute for genetic Engineering and Biotechnology).

Memon, A. R., and Schröder, P. (2009). Implications of metal accumulation mechanisms to phytoremediation. *Environ. Sci. Pollut. R* 16, 162–175. doi: 10.1007/s11356-008-0079-z

Mengoni, A., Baker, A. J. M., Bazzicalupo, M., Reeves, R. D., Adigüzel, N., Chianni, E., Galardi, F., Gabbrielli, R., and Gonnelli, C. (2003). Evolutionary dynamics of nickel hyperaccumulation in Alyssum revealed by ITS nrDNA analysis. *New Phytol.* 159, 691–699.

Milner, M. J., and Kochian, L. V. (2008). Investigating heavy-metal hyperaccumulation using *Thlaspi caerulescens* as a model system. *Ann. Bot.* 102, 3–13.

Morah, F. N. I.. (2007). *Medicinal Plants and Health Care Delivery, 45th Inaugural Lecture of the University of Calabar.* Available at (2007) Retrieved from www.unical.edu.ng/inaug ral/0022.pdf (August 22, 2014).

Muradoglu, F., Gundogdu, M., Ercisli, S., Encu, T., Balta, F., Jaafar, H. Z., et al. (2015). Cadmium toxicity affects chlorophyll a and b content, antioxidant enzyme activities and mineral nutrient accumulation in strawberry. *Biol. Res.* 48, 11. doi: 10.1186/S40659-015-0001-3

Neff, J., Lee, K., and Deblois, E. M. (2011). "Produced water: overview of composition, fates, and effects," in *Produced Water: Environmental Risks and Advances in Mitigation Technologies*, eds K. Lee and J. Neff (New York: Springer), 3–54. doi: 10.1007/978-1-4614-0046-2_1

Nehnevajova, E., Herzig, R., Federer, G., Erismann, K.-H., and Schwitzguébel, J.-P. (2007). Chemical mutagenesis—a promising technique to increase metal concentration and extraction in sunflowers. *Int. J. Phytoremediat.* 9, 149–165. doi: 10.1080/15226510701232880

Newman, L., Strand, S., Duffy, J., Ekuan, G., Raszaj, M., Shurtleff, B., Wilmoth, J., Heilman, P., and Gordon, M. (1997). Uptake and biotransformation of trichloroethylene by hybrid poplars. *Environ. Sci. Technol.* 31, 1062–1067.

Noctor, G., Arisi, A., Jouanin, L., Kunert, K., Rennenberg, H., and Foyer, C. (1998). Glutathione: biosynthesis, metabolism and relationship to stress tolerance explored in transformed plants. *J. Exp. Bot.* 49, 623–647.

Oni, O. M., Isola, G. A., Oni, F. G. O., and Sowole, O. (2011). Natural activity concentrations and assessment of radiological dose Equivalents in Medicinal Plants Around Oil and Gas Facilities in Ughelli and Environs, Nigeria, *Environ. Nat. Resour. Res.* 1, 201–206.

Omondi, D., and Navalia, A. (2020). "Constructed wetlands in wastewater treatment and challenges of emerging resistant genes filtration and reloading," in *Inland Waters: Dynamics and Ecology*, eds A. Devlin, J. Pan, M. Shah (London: IntechOpen). 10.5772/intechopen.93293

Ow, D. W. (1996) Heavy metal tolerance genes: prospective tools for bioremediation resources. *Conserv. Recycl* 18, 135–149.

Padmavathiamma, P. K., and Li, L. Y. (2012). Rhizosphere influence and seasonal impact on phytostabilisation of metals—a field study. *Water Air Soil Pollut.* 223, 107–124. doi: 10.1007/s11270-011-0843-4

Peer, W. A., Baxter, I. R., Richards, E. L., Freeman, J. L., and Murphy, A. S. (2005). "Phytoremediation and hyperaccumulator plants," in *Molecular Biology of Metal Homeostasis and Detoxification*, eds M. J. Tamas and E. Martinoia (Berlin: Springer), 299–340. doi: 10.1007/4735_100

Persans, M. W., Nieman, K., and Salt, D. E. (2001). Identification of a novel family of metal transporters uniquely expressed in metal hyperaccumulating *Thlaspi goesingense* (Ha'la ʹcsy). *Proc. Natl. Acad. Sci. USA* 98, 9995–10000.

Perttu, K. L., and Kowalik, P. J. (1997). Salixvegetation filters for purification of waters and soils. *Biomass Bioenergy* 12, 9–19.

Pichtel, J. (2016). Oil and gas production wastewater: soil contamination and pollution prevention. *Appl. Environ. Soil Sci.* 2016 (8), 1–24. doi: 10.1155/2016/2707989

Pivetz, B. E. (2001). *Ground Water Issue: Phytoremediation of Contaminated Soil and Ground Water at Hazardous Waste Sites*. United States Environmental Protection Agency, EPA, 540/S-01/500, 36.

Prasad, M. N. V. (2003). Phytoremediation of metal-polluted ecosystems: hype for commercialization. *Russ. J. Plant Physiol.* 50, 686–701. doi: 10.1023/A:1025604627496

Rafique, N., and Tariq, S. R. (2016). Distribution and source apportionment studies of heavy metals in soil of cotton/wheat fields. *Environ. Monit. Assess.* 188, 309. doi: 10.1007/s10661-016-5309-0

Rai, V. (2002). Role of amino acids in plant responses to stresses. *Biol. Plantarum* 45, 481–487. doi: 10.1023/A:1022308229759

Rajurkar, N. S., and Pardesh, B. M. (1996). *Elemental Analysis of Some Herbal Plant used in Control Diabetes using Neutron Activation Analysis (NAA) and Atomic Absorption Spectroscopy (AAS)*. Pune, India: University of Pune..

Ribé, V., Aulenius, E., Nehrenheim, E., Martell, U., and Odlare, M. (2012). Applying the triad method in a risk assessment of a former surface treatment. *J. Hazard. Mater.*, 207–208, 15–20.

Riddell-Black D. (1994). Heavy metal uptake by fast growing willow species," in *Willow Vegetation Filters for Municipalwastewaters and Sludges. A Biological Purification System*, eds P. Aronsson and K. Perttu (Uppsala: Swedish University of Agricultural Sciences), 145–151.

Rosestolato, D., Bagatinbergio, B., and Ferro, S. (2015). Electrokinetic remediation of soils polluted by heavy metals (mercury in particular). *Chem. Eng. J.* 264(15), 16–23.

Roy, M., Giri, A. K., Dutta, S., and Mukherjee, P. (2015). Integrated phytobial remediation for sustainable management of arsenic in soil and water. *Environ. Internat.* 75, 180–198. doi: 10.1016/j.envint.2014.11.010

Ryan, R.P., Ryan, D., and Dowling, D. N. (2007). Plant protection by the recombinant, root-colonizing Pseudomonas fluorescens F113rifPCB strain expressing arsenic resistance: improving rhizoremediation. *Lett. Appl. Microbiol.* 45 (6), 668–674.

Roy, S. B., and Bera, A. (2002). Individual and combined effect of mercury and manganese on phenol and proline content in leaf and stem of mungbean seedlings. *J. Environ. Biol.* 23, 433–435.

Santos, J. A. G., Gonzaga, M. I. S., Ma, L. Q., and Srivastava, M. (2008). Timing of phosphate application affects arsenic phytoextraction by *P. vittata* L. of different ages. *Environ. Pollut.* 154, 306–311. doi: 10.1016/j.envpol.2007.10.012

Sarwar, N., Imran, M., Shaheen, M. R., Ishaque, W., Kamran, M. A., Matloob, A., et al. (2017). Phytoremediation strategies for soils contaminated with heavy metals: modifications and future perspectives. *Chemosphere* 171, 710–721. doi: 10.1016/j.chemosphere.2016.12.116

Schalk, I. J., Hannauer, M., and Braud, A. (2011). New roles for bacterial siderophores in metal transport and tolerance. *Environ. Microbiol.* 13, 2844–2854. doi: 10.1111/j.1462-2920.2011.02556.x

Serfor-Armah, Y., Akaho, E. H. K., Nyarko, B. J. B., Kyere, A. W. K., and Oppon-Boachie, K. (2003). Application of instrumental neutron activation analysis to plant medicine in Ghana: a review. *J. Radioanal. Nuclear Chem.* 257 (1), 125–128.

Selvi A., Rajasekar, A., Theerthagiri, J., Ananthaselvam, A., Sathishkumar, K, Madhavan, J., Rahman. P. K. S. M. (2019). Integrated remediation processes toward heavy metal removal/recovery from various environments—a review. *Front Environ. Sci.* 7, 66

Seth, C. S. (2012). A review on mechanisms of plant tolerance and role of transgenic plants in environmental clean-up. *Bot. Rev.* 78, 32–62. doi: 10.1007/s12229-011-9092-x

Sheoran, V., Sheoran, A., and Poonia, P. (2011). Role of hyperaccumulators in phytoextraction of metals from contaminated mining sites: a review. *Crit. Rev. Env. Sci. Technol.* 41, 168–214. doi: 10.1080/10643380902718418

Singh, J., Kumar, V., Kumar, P., Kumar, P., Yadav, K.K., Cabral-Pinto, M. M. S., Kamyab, H., and Chelliapan, S. (2021). An experimental investigation on phytoremediation performance of water lettuce (*Pistia stratiotes* L.) for pollutants removal from paper mill effluent. *Water Environ. Res.* 93(9), 1543–1553.

Singh, J., Upadhyay, A. K., Bahadur, A., Singh, B., Singh, K. P. and Rai, M. (2006). Antioxidant phytochemicals in cabbage (*Brassica oleracea* L. var. capitata). *Sci. Hortic.* 108, 233–237.

Singh, N., Kaur, K., and Katnoria, J.K. (2017). Analysis on bioaccumulation of metals in aquatic environment of Beas River Basin: a case study from Kanjli wetland. *GeoHealth* 1(3), 93–105.

Souza, L. A., Piotto, F. A., Nogueirol, R. C., and Azevedo, R. A. (2013). Use of non-hyperaccumulator plant species for the phytoextraction of heavy metals using chelating agents. *Sci. Agric.* 70, 290–295. doi: 10.1016/j.chemosphere.2008.11.007

Suman, J., Uhlik, O., Viktorova, J., and Macek, T. (2018). Phytoextraction of heavy metals: a promising tool for clean-up of polluted environment? *Front Plant Sci.* 9, 1476. doi: 10.3389/fpls.2018.01476

Sun, R.-L., Zhou, Q.-X., and Jin, C.-X. (2006). Cadmium accumulation in relation to organic acids in leaves of *Solanum nigrum* L. as a newly found cadmium hyperaccumulator. *Plant Soil* 285, 125–134. doi: 10.1007/s11104-006-0064-6

Terry, N., Carlson, C., Raab, T. K., and Zayed, A. (1992). Rates of selenium volatilization among crop species. *J. Environ. Qual.* 21, 341–344.

Terry, N., Zayed, A. M., De Souza, M. P., and Tarun, A. S. (2000). Selenium in higher plants. *Annu. Rev. Plant Phys.* 51, 401–432. doi: 10.1146/annurev.arplant.51.1.401

Thakur, S., Singh, L., Wahid, Z. A., Siddiqui, M. F., Atnaw, S. M., and Din, M. F. M. (2016). Plant-driven removal of heavy metals from soil: uptake, translocation, tolerance mechanism, challenges, and future perspectives. *Environ. Monit. Assess.* 188, 206. doi: 10.1007/s10661-016-5211-9

Thangavel, P., and Subbhuraam, C. (2004). Phytoextraction: role of hyperaccumulators in metal contaminated soils. *Proc. Indian Nat. Sci. Acad. B* 70, 109–130.

Thijs, S., Sillen, W., Rineau, F., Weyens, N., and Vangronsveld, J. (2016). Towards an enhanced understanding of plant–microbiome interactions to improve phytoremediation: engineering the metaorganism. *Front. Microbiol.* 7, 341.

Tong, Y.-P., Kneer, R., and Zhu, Y.-G. (2004). Vacuolar compartmentalization: a second-generation approach to engineering plants for phytoremediation. *Trends Plant Sci.* 9, 7–9. doi: 10.1016/j.tplants.2003.11.009

Vamerali, T., Bandiera, M., and Mosca, G. (2010). Field crops for phytoremediation of metal-contaminated land. A review. *Environ. Chem. Lett.* 8, 1–17. doi: 10.1007/s10311-009-0268-0

Van Aken, B. (2009). Transgenic plants for enhanced phytoremediation of toxic explosives. *Curr. Opin. Biotechnol.* 20, 231–236.

van der Ent, A., Baker, A. J., Reeves, R. D., Pollard, A. J., and Schat, H. (2013). Hyperaccumulators of metal and metalloid trace elements: facts and fiction. *Plant Soil* 362, 319–334. doi: 10.1007/s11104-012-1287-3

Vangronsveld, J., Herzig, R., Weyens, N., Boulet, J., Adriaensen, K., Ruttens, A., et al. (2009). Phytoremediation of contaminated soils and groundwater: lessons from the field. *Environ. Sci. Pollut. R* 16, 765–794. doi: 10.1007/s11356-009-0213-6

Vanlı, I., Ö. (2007). Pb, Cd, B Elementlerinin Topraklardan Şelat Destekli Fitoremediasyon Yöntemiyle Giderilmesi. ABD: İTÜ, Fen Bil. *Enst. Çevre Müh, Yüksek Lisans Tezi.* 2007, 80–88.

Vaverková, M., and Adamcová, D. (2014). Heavy metals uptake by select plant species in the landfill area of Štěpánovice, Czech Republic. *Pol. J. Environ. Stud.* 23 (6), 2265–2269.

Verret, F., Gravot, A., Auroy, P., Leonhardt, N., David, P., Nussaume, L., Vavasseur, A., and Richaud, P. (2004). Overexpression of AtHMA4 enhances root-to-shoot translocation of zinc and cadmium and plant metal tolerance. *FEBS Lett.* 576, 306–312.

Wang, A. S., Angle, J. S., Chaney, R. L., Delorme, T. A., and Reeves, R. D. (2006). Soil pH effects on uptake of Cd and Zn by *Thlaspi caerulescens*. *Plant Soil* 281, 325–337. doi: 10.1007/s11104-005-4642-9

Wang, Y. P., Shi, J. Y., Wang, H., Lin, O., Chen, X. C., and Chen, Y. X. (2007). The influence of soil heavy metals pollution on soil microbial biomass, enzyme activity, and community composition near a copper smelter. *Ecotoxicol. Environ. Saf.* 67, 75–81

Wang, Y. P., Xu, C. X., and Wang, S. M. (2005). Correlation between chemical element contents in tree rings and those in soils near tree roots in the southern suburbs of Beijing, China. *Geol.Bull. China* 24(10–11), 952–956.

Wang, Z., Xu, Y., Zhao, J., Li, F., Gao, D., and Xing, B. (2011). Remediation of petroleum contaminated soils through composting and rhizosphere degradation. *J. Hazard Mater.* 190, 677–685. doi:10.1016/j.jhazmat.2011.03.103

Watson, C. (2002). *The Phytoremediation Potential of Salix: Studies of the Inter-action of Heavy Metals and Willows.* PhD thesis, University of Glasgow.

Williams, L. E., and Mills, R. F. (2005). P1B-ATPases—an ancient family of transition metal pumps with diverse functions in plants. *Trends Plant Sci.* 10: 491–502.

WNA (2014). Naturally-occurring radioactive materials (NORMs) World Nuclear Association [online]. Retrieved from https://world-nuclear.org/information-library/safety-and-security/radiation-and-health/naturally-occurring-radioactive-materials-norm.aspx/ (July 1, 2014)

Wu, G., Kang, H., Zhang, X., Shao, H., Chu, L., and Ruan, C. (2010). A critical review on the bio-removal of hazardous heavy metals from contaminated soils: issues, progress, eco-environmental concerns and opportunities. *J. Hazard Mater.* 174, 1–8. doi: 10.1016/j.jhazmat.2009.09.113

Wu, X., Abbondanza, C., Altamimi, Z., Chin, T.M., Collilieux, X., Gross, R.S., Heflin, M.B., Jiang, Y., and Parker, J.W. (2015). KALREF—A Kalman filter and time series approach to the International Terrestrial Reference Frame realization, *J. Geophys. Res. Solid Earth* 120, 3775– 3802, doi:10.1002/2014JB011622

Wu, Y., Wang, X., Li, Y., and Ma, Y. (1995). Compound pollution of Cd, Pb, Cu, Zn and As in plant soil system and its prevention. *J. Environ. Sci.* 8(4), 474–482.

Wuana, R. A., and Okieimen, F. E. (2011). Heavy metals in contaminated soils: a review of sources, chemistry, risks and best available strategies for remediation. *ISRNsrn Ecol.* 2011, 402647. doi: 10.5402/2011/402647

Wong, M.H. (2003). Ecological restoration of mine degraded soils, with emphasis on metal contaminated soils. *Chemosphere* 50(6), 775–780. https://doi.org/10.1016/s0045-6535(02)00232-1

Yan, A., Wang, A., Tan, S.N., Yusof, M.L.M., Ghosh, S., and Chen, Z. (2020) Phytoremediation: a promising approach for revegetation of heavy metal-polluted land. *Front. Plant Sci.* 11, 359.

Yancheshmeh, J. B., Khavazi, K., Pazira, E., and Solhi, M. (2011). Evaluation of inoculation of plant growth-promoting rhizobacteria on cadmium uptake by canola and barley. *Afr. J. Microbiol. Res.* 5, 1747–1754. doi:https://doi.org/10.5897/AJMR10.625.

Zhu, Y. L., Pilon-Smits, E. A. H., Jouanin, L., and Terry, T. (1999). Overexpression of glutathione synthetase in *Brassica juncea* enhances cadmium accumulation and tolerance. *Plant Physiol.* 119, 73–79.

8 Green Chemistry and Its Applications in Waste Management

Neha Sharma[1]* and Sanjay K. Sharma[2]*
[1]Department of Biotechnology, Maharishi Markandeshwar
(Deemed to be University), Mullana, Ambala
(Haryana), India
[2]Green Chemistry and Sustainability Research Group,
Department of Chemistry, JECRC University, Jaipur
(Rajasthan), India
*Corresponding authors: Neha Sharma,
nehamicrobiologist@ gmail.com; Sanjay K. Sharma,
sk.sharmaa@outlook.com

CONTENTS

8.1   INTRODUCTION

The U.S. environmental law, The Pollution Prevention Act of 1990, proposed that pollution can be prevented by designing industrial processes that do not lead to waste generation, in the first place. This concept has led to inception of *green chemistry*, the term coined by Paul Anastas in 1991. According to Environmental Protection Agency (EPA), green chemistry is defined as "the design of chemical products and processes that reduce or completely exterminate the use or generation of hazardous substances". This constitutes diminished waste products that are rendered nontoxic, thereby improving the efficiency of a manufacturing process (USEPA 2006). It

DOI: 10.1201/9781003132349-9

mainly emphasizes the design of both safer chemicals and processes that may replace utilization of hazardous substances.

8.2   GREEN CHEMISTRY: HISTORICAL BACKGROUND

Breakthrough sensitization regarding environmental issues stringently affecting industrial activities began in the 1940s when the industrial practices initiated a leap from conventional manufacturing to sustained and resilient practices through capacity building programs related to chemical research and industrial ecology. It was in the 1990s that 12 principles of green chemistry were postulated, accounting for a shift in industrial practices leading to sustenance, economic viability, reduced waste generation, and pollution abatement, to name a few (Figure 8.1). This necessarily adds to the value proposition of environmentally favorable actions from product planning, use of environmentally benign raw material, synthesis, manufacturing, and source destination (Anastas 1999).

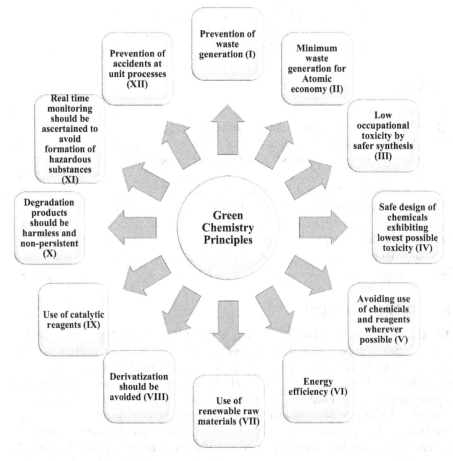

FIGURE 8.1   Principles of green chemistry.

The primary objective is to reduce occupational hazards indigenous to industrial activities (Prado 2003). Furthermore, these principles were utilized in new methods aimed to reduce environmental impacts (Anastas 1999). Co-relating green chemistry with economic benefits in industrial settings has been sought in the recent past, specifically dealing with attributes of waste management and efficient effluent treatment and storage systems (Prado 2003).

8.3 AMALGAMATION OF WASTE WITH GREEN CHEMISTRY

Chemical production processes are less efficient in the proper utilization of feed stocks and raw materials, thereby generating enormous volumes of wastes. The mitigation strategy could be accomplished by increasing atom economy wherein all the atoms in reactants should undergo transformation to yield desirable products. An alternative to eliminate or minimize waste generation is to integrate different reactions and processes in which the by-product in one reaction is the feedstock of another at each subsequent step.

Waste generation and its consequent eco-toxicological implications in addition to health adversities have exerted a detrimental impact on low-income countries. A recent report has opined that annually 11.2 billion tons of solid waste is generated where more than 90% of waste is openly dumped or incinerated (UNEP 2020). Attributed to paucity of proper waste management facilities, in Southeast Asian countries such as Vietnam, Malaysia, the Philippines, and Thailand, waste-related issues are further worsened by plastic and electronic waste imported from industrially advanced nations (Kaza et al. 2018; Dell 2019; Sukanan 2020). Subsequently, this "trend" of waste accumulation has instigated the development, execution, and strengthening of policy framework to mitigate the waste-assisted environmental challenge (Figure 8.2).

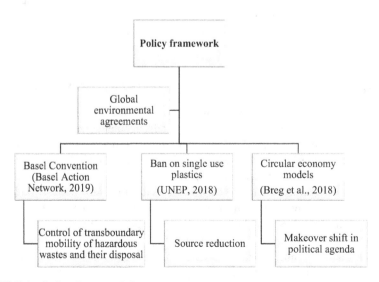

FIGURE 8.2   Policy framework for waste management.

The strategic implementation of these policies encounters significant technical, economic, environmental, and social constraints for putting in place "closed-loop management" of processed materials at a commercial scale. Noteworthy is the fact that recycling of plastic and metals may involve use of hazardous substances for extraction and purification, which leads to both health and environmental adversities (Kral et al. 2019). Besides, a realistic projection of the practical viability of recycling and repurposing strategies has a direct linkage with monetary value of end products. This has led to emergence of "clean and green manufacturing" processes to develop "value-added" commodities from waste materials, at the base of which lies the support system of green chemistry. This improvisation can be scaled up to synergize both circularity and sustainability goals (Ferronato and Torretta 2019).

8.4   CONVERGENCE OF SOLID WASTE WITH INDUSTRIAL ECOLOGY AND GREEN CHEMISTRY

Municipal solid waste (MSW) is an entity generated regularly by households and over a period of time tends to bio-accumulate and bio-magnify, leading to unprecedented environmental challenges, thereby altering bio-geochemical cycling and ecological balance. The composition of MSW is highly variable in nature and changes significantly with time (Chhipa et al. 2013). Incidentally, MSW does not include wastes that are generated from agricultural, industrial, health care, and radioactive materials. Broadly, solid wastes are categorized as shown in Figure 8.3. A typical waste management plan constitutes the following interlinked verticals under idealistic projections (Figure 8.4). The economic layout constituting the core deliverables is far from viability in terms of losses incurred for the practices. MSW has been explicitly considered as a feedstock for biotransformation of waste-derived products inspired by industrial ecology practices (Smith et al. 2015) (Figure 8.5). This encompasses symbiotic and sequential processing of feedstock (waste) to generate a value-added product (e.g., waste to energy). Until recently, the term waste has been replaced by *sustainable materials management* (USEPA 2014).

8.5   GREEN ROUTE OF WASTE-DERIVED INDUSTRIAL ECOLOGY

A concept pertaining to waste to wealth through interventions of green chemistry involves addressing issues like technological advancement, feedstock supply, and commercial viability of the product, which are integral to scaled up production. The basic approach in chemical processes is the transformation of low-cost feedstock into a value-added product. For instance, a pelletized material produced from organic fraction of MSW (OFMSW) could be transformed into refuse-derived fuel (RDF). The term would then be a RDF-OFMSW spread. In case of biogas or ethanol, so one could have biogas-OFMSW and ethanol-OFMSW spreads, respectively. The benefit of a product feedstock spread is that people familiar with the processes involved develop an understanding of when specific conversion processes will be profitable, because they also know the other costs for the process.

The fundamentals of green chemistry have been applied to produce chemical precursors from food wastes as alternatives to petroleum derivatives. Wine industry

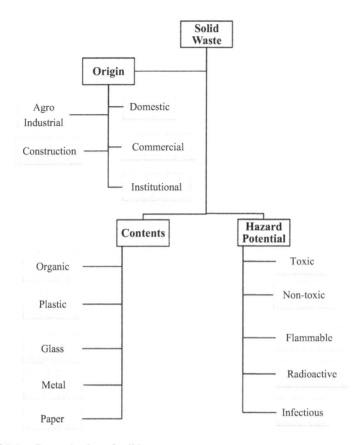

FIGURE 8.3   Categorization of solid waste.

produces a substantial amount of grape seeds used as bio-fuel for energy production, impacting both gastronomic and economic interests (Fiori et al. 2012). Value addition of waste-derived products is a foundation of green chemistry. Looking into multifaceted implications of grape seeds, it has found utility as a precursor of grape seed oil. Furthermore, double bonds undergo chemical transformation into epoxy and hydroxyl groups. Recently, rigid polyurethane foam has been developed from winery waste (de Haro et al. 2018). The end product was found to possess improved flame-retardant properties and was presented as an insulating material in buildings. Their results confirm the possibility of using the phosphorylated biopoliols obtained from the grape seed oils as an alternative to petroleum-derived polyester polyols.

8.6   GREEN SYNTHESIS OF WASTE-DERIVED NANOPARTICLES AND THEIR ENVIRONMENTAL APPLICATIONS

Research showcasing the utility of nanoparticles for widespread applications has gained momentum in recent past spanning across energy transformation and storage (Hussein 2015; Verma et al. 2016; Sonawane et al. 2017; Yang et al. 2019), monitoring of pollution (Baruah and Dutta 2009), precision agriculture and controlled delivery

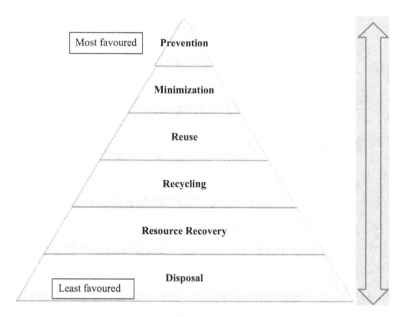

FIGURE 8.4 Waste management practices in an ideal scenario.

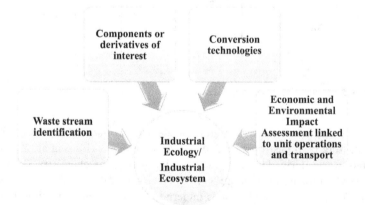

FIGURE 8.5 Industrial ecology based on green chemistry.

of food ingredients (Bindraban et al. 2015; Peters et al. 2016), membrane technology (Goh et al. 2015), water treatment (Westerhoff et al. 2013), drug delivery and diagnostics (Turcheniuk and Mochalin 2017), and bone and tissue engineering (Shadjou and Hasanzadeh 2016), among others. Within a purview of environmental implications, a ceaseless effort to minimize potential health hazards associated with the production, use, and disposal of novel nanomaterials, approaches to attain sustainability, frameworks, and metrics are being considered (Dhingra et al. 2010; Mata et al. 2015; Falinski et al. 2018).

Industrial wastes such as discarded batteries, tires, effluents, and sludges have been explored as potential, cost-effective precursors for nanoparticle synthesis after suitable physicochemical pretreatment. Chemical pretreatment is carried out to decontaminate waste sample by treating it with strong acids (e.g., H_2SO_4, HNO_3, HCl) (Samaddar et al. 2018).

The common route to synthesize waste-derived nanoparticles is through sodium borohydride ($NaBH_4$) treatment following its excessive removal with ultrapure water and absolute alcohol. Iron nanoparticles have been formulated from pickling waste derived from a steel plant. Another widely used method is the solvent thermal method using a closed autoclave reactor involving high-throughput temperatures (Dubin et al. 2010). Another approach is voltage driven, which catalyzes chemical reactions in aqueous solutions for nanoparticle synthesis. This method has been applied to produce nanowires, nanoporous materials, and nanocylinders. In order to ascertain environmental benignity and address sustainability goals, the greener route of nanoparticle synthesis has been extensively explored by using biocompatible materials through plants and microbiological interventions (Jeevanandam et al. 2016; Abdelbasir et al. 2020) (Figure 8.6).

The recovery of metal nanoparticles (e.g., Pb, Hg, Cu, Fe, Au, Ag, Pd, Pt, and Rh) and polymers can be achieved from electronic wastes such as computer circuit boards, cellular phones, laptops, automobiles, and super-capacitors with diversified applications including pollution monitoring and health care (Xiu and Zhang 2012; Singh and Lee 2016; Vermisoglou et al. 2016). Waste printed circuit boards (WPCBs) have also been recycled to obtain metals such as Cu, Pb, Fe, Au, and Hg (Calgaro et al. 2015; Chen et al. 2015 a, b; Chu et al. 2015; Fogarasi et al. 2015; Hadi et al. 2015; Cui and Anderson 2016). The recovery of copper nanoparticles from recovered metals in WPCBs has been reported (Yousef et al. 2018; El-Nasr et al. 2020). Additionally, waste-derived functionalized nanoparticles have been successfully used for health

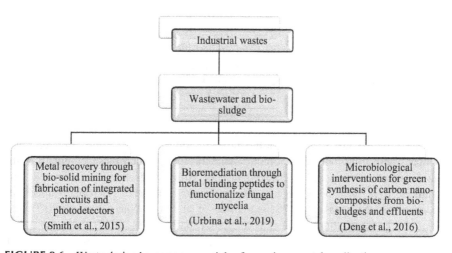

FIGURE 8.6 Waste derived green nanoparticles for environmental applications.

TABLE 8.1
Environmental Applications of Waste-Derived Reusable Nanoparticles

No	Nanoparticle	Source	Application	Reference
	Wastewater treatment and remediation			
1	Iron oxide nanoparticles	Mill scale	Dye removal (adsorption)	Arifin et al. 2017
2	Magnetite (Fe_3O_4)	Iron ore tailings		Giri et al. 2011
3	Graphene	Polyethylene terephthalate (PET)		El Essawy et al. 2017
4	CaCO3	Eggshell	Lead (Pb^{2+}) adsorption	Wang et al. 2018
5	Porous aerogels	Paper, cotton textiles, and plastic bottles	Oil adsorption	Thai et al. 2019
6	$NiFe_2O_4$/ ZnCuCr-LDH composite	Saccharin wastewater	Dye removal (adsorption)	Zhang et al. 2020
7	Silica nanoparticles (SiO_2 NPs)	Sugarcane waste ash		Rovani et al. 2018
8	Silica nanoparticles (MW– $nSiO_2$, $nSiO_2$)	Corn husk waste		Peres et al. 2018
9	Nanocomposite of ZnO and CuO	Printed circuit boards	Dye removal (photocatalysit)	Nayak et al. 2019
10	Metal-doped ZnO (M-ZnO)	Fabric filter dust		Wu et al. 2014
11	Zero-valent iron nanoparticles	Pickling line of a steel plant	Nitrobenzene removal	Lee et al. 2015
	Monitoring of pollutants in water			
1	Carbon nanoparticles	Pomelo peels	Detection of mercury (Hg^{2+})	Lu et al. 2012
2	Nano-cuprous oxide	Electrical waste	Dopamine and mercury	Abdelbasir et al. 2018
	Capture of air pollutants			
1	Porous silica nanoparticles (PSNs)	Rice husks	CO_2 capture	Zeng and Bai 2016

care application to capture toxins from spiked blood plasma samples followed by glycine application to free up the nanoparticles after their initial use (Hassanain et al. 2017). Recycled waste batteries and electronics that have been extensively used to recover nanomaterials has provided insights into environmental research (Dutta et al. 2017).

Industrial effluents constitute a wide array of recalcitrant pollutants unique to the industries that create them. One of the most common examples is the textile effluents influxed with multitude of xenobiotic synthetic dyes, which bio-accumulate in the environment exerting adverse health effects. Effluent streams contaminated with synthetic dyes have been targeted for dye removal through microbially mediated mechanisms (Sharma et al. 2013; Sharma et al. 2019; Sharma et al. 2020) and through bio-nanotechnological interventions (Sharma et al. 2017) to ensure public and environmental health (Arslan et al. 2016). Table 8.1 exemplifies environmental applications of waste derived nanoparticles derived by a conglomerate of green chemistry principles (Abdelbasir et al. 2020).

8.7   CHAPTER SUMMARY

All industrial practices by virtue of their intersectoral manufacturing practices generate enormous volumes of wastes. Ever since the inception of green chemistry for sustainable industrial practices, much of the impetus had been imparted to transforming wastes into value-added products. A multitude of policies had been in place for waste management culminating into a newer concept of *circular economy*, its fundamental backbone being industrial ecology. Despite these efforts, a comprehensive visualization needs to be done for economic viability of applying green chemistry to industrial processes. A concrete understanding of cost analysis versus risk–benefit analysis needs to be channelized to provide converging solutions developed from waste.

That said, one of the classical examples of wastes as feedstock comprise nanoparticles derived through functionalities of green chemistry. A multitude of possibilities for using waste materials as feedstocks of nanoparticles had existed and are being researched upon in emerging areas of targeted drug delivery systems and other molecular models of importance in health care settings. Nanoparticles derived through green route have found immense application in different environmental applications including management of municipal solid waste. Contrastingly, this concept seems to be technologically sound as a full-circle approach at laboratory scale but underlying knowledge gaps for technology transfer are seemingly bleak in the domain of energy use, generation of secondary wastes, fate and transport behavior, routes of exposure in different habitats, and levels of toxicity across different trophic levels by processes of bioaccumulation, bio-magnification, and bio-augmentation. There isn't a uniform cyclic process to overcome negative implications of current "end-of-life" waste management systems. A critical analysis of life cycle assessments and subsequent environmental impact assessment remains a daunting task during the development of new technologies and processes, and also, to account for aging transformations and potential release of nanomaterials and by-products into the environment.

ACKNOWLEDGMENTS

The authors would like to thank Maharishi Markandeshwar (Deemed to be University), Mullana and JECRC University Jaipur for providing necessary administrative and allied assistance.

REFERENCES

Abdelbasir, S. M., S. M. El-Sheikh, V. L. Morgan, H. Schmidt, L. M. Casso-Hartmann, D. C. Vanegas, I. Velez-Torres, and E. S. Mc Lamore. 2018. "Graphene-anchored cuprous oxide nanoparticles from waste electric cables for electrochemical sensing." *ACS Sustainable Chemistry and Engineering* 6: 12176–12186. doi: 10.1021/acssuschemeng.8b02510

Abdelbasir, S. M., K. M. McCourt, C. M. Lee, and D. C. Vanegas. 2020. "Waste-derived nanoparticles: synthesis approaches, environmental applications, and sustainability considerations." *Frontiers in Chemistry* 782: 1–18. doi: 10.3389/fchem.2020.00782

Anastas, P. T. 1999. "Green chemistry and the role of analytical methodology development." *Critical Reviews in Analytical Chemistry* 29 (3): 167–175. doi: 0.1080/10408349891199356

Arifin, S. A. R. A., I. Ismail, A. H. Abdullah, F. Nabilah Shafiee, R. Nazlan, and I. Riati Ibrahim. 2017. "Iron oxide nanoparticles derived from mill scale waste as potential scavenging agent in dye wastewater treatment for batik industry." *Solid State Phenomena* 268: 393–398. doi: 10.4028/www.scientific.net/SSP.268.393

Arslan, S., M. Eyvaz, E. Gürbulak, and E. Yüksel. 2016. "A review of state-of-the-art technologies in dye-containing wastewater treatment—the textile industry case." *Textile Wastewater Treatment*, 1–29. doi: 10.5772/64140

Baruah, S., and J. Dutta. 2009. "Nanotechnology applications in pollution sensing and degradation in agriculture." *Environmental Chemistry Letters* 7: 191–204. doi: 10.1007/s10311-009-0228-8

Bindraban, P. S., C. Dimkpa, L. Nagarajan, A. Roy, and R. Rabbinge. 2015. "Revisiting fertilisers and fertilisation strategies for improved nutrient uptake by plants." *Biology and Fertility of Soils* 51: 897–911. doi: 10.1007/s00374-015-1039-7

Calgaro, C. O., D. F. Schlemmer, M. D. C. R. Da Silva, E. V. Maziero, E. H. Tanabe, and D. A. Bertuol. 2015. "Fast copper extraction from printed circuit boards using supercritical carbon dioxide." *Waste Management* 45: 289–297. doi: 10.1016/j.wasman.2015.05.017

Chen, M., J. Huang, O. A. Ogunseitan, N. Zhu, and Y. Wang. 2015a. "Comparative study on copper leaching from waste printed circuit boards by typical ionic liquid acids." *Waste Management* 41: 142–147. doi: 10.1016/j.wasman.2015.03.037

Chen, M., S. Zhang, J. Huang, and H. Chen. 2015b. "Lead during the leaching process of copper from waste printed circuit boards by five typical ionic liquid acids." *Journal of Clean Production* 95: 142–147. doi: 10.1016/j.jclepro.2015.02.045

Chhipa, M. R. N., V. P. Jatakiya, P. A. Gediya, S. M. Patel, and D. J. Sen. 2013. "Green chemistry: a unique relationship between waste and recycling." *International Journal of Advances in Pharmaceutical Research* 4 (7): 2000–2008.

Chu, Y., M. Chen, S. Chen, B. Wang, K. Fu, and H. Chen. 2015. "Micro copper powders recovered from waste printed circuit boards by electrolysis." *Hydrometallurgy* 156: 152–157. doi: 10.1016/j.hydromet.2015.06.006

Cui, H., and C. G. Anderson. 2016. "Literature review of hydrometallurgical recycling of printed circuit boards (PCBs)." *Journal of Advanced Chemical Engineering* 6: 1–11. doi: 10.4172/2090-4568.1000142

de Haro, J. C., D. López-Pedrajas, A. Pérez, J. F. Rodríguez, and M. Carmona. 2018. "Synthesis of rigid polyurethane foams from phosphorylated biopolyols." *Environmental Science and Pollution Research* 26 (4): 3174–3183. https://doi.org/10.1007/ s11356-017-9765-z

Dell, J. 2019. "157,000 Shipping Containers of U.S. Plastic Waste Exported to Countries with Poor Waste Management in 2018—Plastic Pollution Coalition". Available online at: www.plasticpollutioncoalition.org/blog/2019/3/ 6/157000-shipping-containers-of-us-plastic-waste-exported-to-countrieswith-poor-waste-management-in-2018

Dhingra, R., S. Naidu, G. Upreti, and R. Sawhney. 2010. "Sustainable nanotechnology: through green methods and life-cycle thinking." *Sustainability* 2: 3323–3338. doi: 10.3390/su2103323

Dubin, S., S. Gilje, K. Wang, V. C. Tung, K. Cha, A. S. Hall, J. Farrar, R. Varshneya, Y. Yang, and R. B. Kaner. 2010. "A one-step, solvothermal reduction method for producing reduced graphene oxide dispersions in organic solvents." *ACS Nano* 4: 3845–3852. doi: 10.1021/nn100511a

Dutta, T., K. H. Kim, A. Deep, J. E. Szulejko, K. Vellingiri, S. Kumar, E. E. Kwon, and S. T. Yun. 2017. "Recovery of nanomaterials from battery and electronic wastes: a new paradigm of environmental waste management." *Renewable and Sustainable Energy Reviews* 82: 3694–3704. doi: 10.1016/j.rser.2017.10.094

El Essawy, N. A., S. M. Ali, H. A. Farag, A. H. Konsowa, M. Elnouby, and H. A. Hamad. 2017. "Green synthesis of graphene from recycled PET bottle wastes for use in the adsorption of dyes in aqueous solution." *Ecotoxicology and Environmental Safety* 145: 57–68. doi: 10.1016/j.ecoenv.2017.07.014

El-Nasr, S. R., S. M. Abdelbasir, A. H. Kamel, and S. S. M. Hassan. 2020. "Environmentally friendly synthesis of copper nanoparticles from waste printed circuit boards." *Separation and Purification Technology* 230: 15860. doi: 10.1016/j.seppur.2019.115860

Falinski, M. M., D. L. Plata, S. S. Chopra, T. L. Theis, L. M. Gilbertson, and J. B. Zimmerman. 2018. "A framework for sustainable nanomaterial selection and design based on performance, hazard, and economic considerations." *Nature Nanotechnology* 13: 708–714. doi: 10.1038/s41565-018-0120-4

Ferronato, N., and V. Torretta. 2019. "Waste mismanagement in developing countries: a review of global issues." *International Journal of Environmental Research and Public Health* 16: 1060. doi: 10.3390/ijerph16061060

Fiori, L, M. Valbusa, D. Lorenzi, and L. Fambri. 2012. "Modeling of the devolatilization kinetics during pyrolysis of grape residues." *Bioresource Technology* 103: 389–397.

Fogarasi, S., F. Imre-Lucaci, A. Egedy, Á. Imre-Lucaci, and P. Ilea. 2015. "Eco-friendly copper recovery process from waste printed circuit boards using Fe^{3+}/Fe^{2+} redox system." *Waste Management* 40: 136–143. doi: 10.1016/j.wasman.2015.02.030

Giri, S. K., N. N. Das, and G. C. Pradhan. 2011. "Synthesis and characterization of magnetite nanoparticles using waste iron ore tailings for adsorptive removal of dyes from aqueous solution." *Colloids Surfaces A: Physicochemical Engineering Aspects* 389: 43–49. doi: 10.1016/j.colsurfa.2011.08.052

Goh, P. S., B. C. Ng, W. J. Lau, and A. F. Ismail. 2015. "Inorganic nanomaterials in polymeric ultrafiltration membranes for water treatment." *Separation and Purification Reviews* 44: 216–249. doi: 10.1080/15422119.2014.926274

Hadi, P., M. Xu, C. S. K. Lin, C. W. Hui, and G. McKay. 2015. "Waste printed circuit board recycling techniques and product utilization." *Journal of Hazardous Materials* 283: 234–243. doi: 10.1016/j.jhazmat.2014.09.032

Hassanain, W. A., E. L. Izake, M. S. Schmidt, and G. A. Ayoko. 2017. "Gold nanomaterials for the selective capturing and SERS diagnosis of toxins in aqueous and biological fluids." *Biosensors and Bioelectronics* 91: 664–672.

Hussein, A. K. 2015. "Applications of nanotechnology in renewable energies—a comprehensive overview and understanding." *Renewable and Sustainable Energy Reviews* 42: 460–476. doi: 10.1016/j.rser.2014.10.027

Jeevanandam, J., Y. S. Chan, and M. K. Danquah. 2016. "Biosynthesis of metal and metal oxide nanoparticles." *Chem Bio Eng Reviews* 3: 55–67. doi: 10.1002/cben.201500018

Kaza, S., L. Yao, P. Bhada-Tata, F. VanWoerden, and K. Ionkova. 2018. *What a Waste 2.0: A Global Snapshot of Solid Waste Management to 2050*. Washington, DC: World Bank. doi: 10.1596/978-1-4648-1329-0

Kral, U., L. S. Morf, D. Vyzinkarova, and P. H. Brunner. 2019. "Cycles and sinks: two key elements of a circular economy." *Journal of Material Cycles and Waste Management* 21: 1–9. doi: 10.1007/s10163-018-0786-6

Lee, H., B. H. Kim, Y. K. Park, S. J. Kim, and S. C. Jung. 2015. "Application of recycled zero-valent iron nanoparticle." *Journal of Nanomaterials* 16 (1): 392537. doi: 10.1155/2015/392537

Lu, W., X. Qin, S. Liu, G. Chang, Y. Zhang, Y. Luo, A. M. Asiri, A. O. Al-Youbi, and X. Sun. 2012. "Economical, green synthesis of fluorescent carbon nanoparticles and their use as probes for sensitive and selective detection of mercury (II) ions." *Analytical Chemistry* 84 (12): 5351–5357. doi: 10.1021/ac3007939

Mata, M. T., A. A. Martins, A. V. C. Costa, and K. S. Sikdar. 2015. "Nanotechnology and sustainability—current status and future challenges." In: A. Vaseashta (Ed.), *Life Cycle Analysis of Nanoparticles: Risk, Assessment and Sustainability*. Lancaster, PA: DEStech Publications, pp. 271–306.

Nayak, P., S. Kumar, I. Sinha, and K. K. Singh. 2019. "ZnO/CuO nanocomposites from recycled printed circuit board: preparation and photocatalytic properties." *Environmental Science and Pollution Research* 26: 16279–16288. doi: 10.1007/s11356-019-04986-6

Peres, E. C., J. C. Slaviero, A. M. Cunha, A. Hosseini-Bandegharaei, and G. L. Dotto. 2018. "Microwave synthesis of silica nanoparticles and its application for methylene blue adsorption." *Journal of Energy, Environmental and Chemical Engineering* 6: 649–659. doi: 10.1016/j.jece.2017.12.062

Peters, R. J., H. Bouwmeester, S. Gottardo, V. Amenta, M. Arena, P. Brandhoff, H. J. Marvin, A. Mech, F. B. Moniz, L. Q. Pesudo, and H. Rauscher. 2016. "Nanomaterials for products and application in agriculture, feed and food." *Trends in Food Science and Technology* 54: 155–164. doi: 10.1016/j.tifs.2016.06.008

Prado, A. G. S. 2003. "Green chemistry, the chemical challenges of the new millennium." *Quimica Nova* 26 (5): 738–744.

Rovani, S., J. J. Santos, P. Corio, and D. A. Fungaro. 2018. "Highly pure silica nanoparticles with high adsorption capacity obtained from sugarcane waste ash." *ACS Omega* 3: 2618–2627. doi: 10.1021/acsomega.8b00092

Samaddar, P., Y. S. Ok, K. H. Kim, E. E. Kwon, and D. C. W. Tsang. 2018. "Synthesis of nanomaterials from various wastes and their new age applications." *Journal of Cleaner Production* 197: 1190–1209. doi: 10.1016/j.jclepro.2018.06.262

Shadjou, N., and M. Hasanzadeh. 2016. "Graphene and its nanostructure derivatives for use in bone tissue engineering: recent advances." *Journal of Biomaterials Materials Research* 104: 1250–1275. doi: 10.1002/jbm.a.35645

Sharma, N., H. Bhagwani, N. Yadav, and N. Chahar. 2020. "Biodegradation of textile waste water by naturally attenuated *Enterbacter* sp." *Nature Environment and Pollution Technology* 19 (2): 845–850.

Sharma, N., P. Bhatnagar, S. Chatterjee, P. J. John, and I. P. Soni. 2017. "Bio nanotechnological intervention: a sustainable alternative to treat dye bearing waste waters." *Indian Journal of Pharmaceutical and Biological Research* 5: 17–24.

Sharma, N., S. Chatterjee, and P. Bhatnagar. 2013. "Assessment of physicochemical properties of textile wastewaters and screening of bacterial strains for dye decolourisation." *Universal Journal of Environmental Research and Technology* 3: 345–355.

Sharma, N., S. Chatterjee, and P. Bhatnagar. 2019. "Degradation of Direct Red 28 by Alcaligenes sp. TEX S6 isolated from aeration tank of common effluent treatment plant (CETP), Pali, Rajasthan." *Nature Environment and Pollution Technology* 18: 9–20.

Singh, J., and B. K. Lee. 2016. "Recovery of precious metals from low grade automobile shredder residue: a novel approach for the recovery of nano zerovalent copper particles." *Waste Management* 48: 53–365. doi: 10.1016/j.wasman.2015.10.019

Smith, K. S., P. L. Hageman, G. S. Plumlee, J. R. Budahn, and D. I. Bleiwas. 2015. "Potential metal recovery from waste streams." International Applied Geochemistry Symposium (IAGS), Tucson, AZ.

Sonawane, J. M., A. Yadav, P. C. Ghosh, and S. B. Adeloju. 2017. "Recent advances in the development and utilization of modern anode materials for high performance microbial fuel cells." *Biosensors and Bioelectronics* 90: 558–576. doi: 10.1016/j.bios.2016.10.014

Sukanan, D. 2020. "Waste Imports Are Flooding Asian Countries Like Thailand." Available online at: www.sustainability-times.com/environmental-protection/asian-countries-like-thailand-are-flooded-by-waste-imports/ (accessed May 10, 2020).

Thai, Q. B., D. K. Le, T. P. Luu, N. Hoang, D. Nguyen, and H. M. Duong. 2019. "Aerogels from wastes and their applications." *Juniper Online Journal Material Science* 5: 555663. doi: 10.19080/JOJMS.2019.05.555663

Turcheniuk, K., and V. N. Mochalin. 2017. "Biomedical applications of nano diamond (Review)." *Nanotechnology* 28 (2017): 252001. doi: 10.1088/1361 6528/aa6ae4

UNEP. 2020 Solid Waste Management I UNEP—UN Environment Programme. Available online at: www.unenvironment.org/explore-topics/resourceefficiency

USEPA. 2006, June 28. *Green Chem.* United States Environmental Protection Agency. Available online at: www.epa.gov/greenchemistry

USEPA. 2014. Sustainable Materials Management. Available online at: www.epa.gov/smm/ (accessed May 27, 2014).

Verma, M. L., M. Puri, and C. J. Barrow. 2016. "Recent trends in nanomaterials immobilised enzymes for biofuel production." *Critical Reviews in Biotechnology* 36: 108–119. doi: 10.3109/07388551.2014.928811

Vermisoglou, E. C., M. Giannouri, N. Todorova, T. Giannakopoulou, C. Lekakou, and C. Trapalis. 2016. "Recycling of typical supercapacitor materials." *Waste Management and Research* 34: 337–344. doi: 10.1177/0734242X15625373

Wang, H., B. Gao, J. Fang, Y. S. Ok, Y. Xue, K. Yang, and X. Cao. 2018. "Engineered biochar derived from eggshell-treated biomass for removal of aqueous lead." *Ecological Engineering* 121: 124–129. doi: 10.1016/j.ecoleng.2017.06.029

Westerhoff, P. K., A. Kiser, and K. Hristovski. 2013. "Nanomaterial removal and transformation during biological wastewater treatment." *Environmental Engineering Science Journal* 30: 109–117. doi: 10.1089/ees.2012.0340

Wu, Z. J., W. Huang, K. K. Cui, Z. F. Gao, and P. Wang. 2014. "Sustainable synthesis of metals-doped ZnO nanoparticles from zinc-bearing dust for photo-degradation of phenol." *Journal of Hazardous Materials* 278: 1–99. doi: 10.1016/j.jhazmat.2014.06.001

Xiu, F. R., and F. S. Zhang. 2012. "Size-controlled preparation of Cu_2O nanoparticles from waste printed circuit boards by supercritical water combined with electro-kinetic process." *Journal of Hazardous Materials* 233 (234): 200–206. doi: 10.1016/j.jhazmat.2012.07.019

Yang, Z., J. Tian, Z. Yin, C. Cui, W. Qian, and F. Wei. 2019. "Carbon nanotube- and graphene-based nanomaterials and applications in high-voltage super-capacitor: a review." *Carbon N. Y.* 141: 467–480. doi: 10.1016/j.carbon.2018.10.010

Yousef, S., M. Tatariants, V. Makarevičius, S. I. Lukošiute, R. Bendikiene, and G. Denafas. 2018. "A strategy for synthesis of copper nanoparticles from recovered metal of waste printed circuit boards." *Journal of Cleaner Production* 185: 653–664. doi: 10.1016/j.jclepro.2018.03.036

Zeng, W., and H. Bai. 2016. "High-performance CO_2 capture on amine-functionalized hierarchically porous silica nanoparticles prepared by a simple template-free method." *Adsorption* 22: 117–127. doi: 10.1007/s10450-015-9698-0

Zhang, H., B. Xia, P. Wang, Y. Wang, Z. Li, Y. Wang, L. Feng, X. Li, and S. Du. 2020. "From waste to waste treatment: mesoporous magnetic $NiFe_2O_4$/ZnCuCr-layered double hydroxide composite for wastewater treatment." *Journal of Alloys and Compounds* 819: 153053. doi: 10.1016/j.jallcom.2019.153053

9 Bioenergy and Biofuels from Wastes

Perspectives from Developing Nations

Eustace Fernando[1*] and Shalini Lalanthika Rajakaruna[2]
[1]Rajarata University, Faculty of Applied Sciences, Mihintale, Sri Lanka
[2]National Institute of Fundamental Studies, Hantana, Sri Lanka
*Corresponding author: Eustace Fernando, eustace6192@as.rjt.ac.lk

CONTENTS

DOI: 10.1201/9781003132349-10

9.1   INTRODUCTION

The recovery of energy from biomass and waste material has been around for millennia. The most basic and primitive form of this process is the burning of lignocellulosic biomass such as firewood and other lignified plant material for heat and light energy. This has now manifested in several more advanced forms such as the burning of waste woody materials such as wood shavings, wood chips, and saw-dust from various industries for recovering energy. Throughout history, the use of lignocellulosic biomass including firewood, agricultural residue such as hay, stover and stalks of field crops, and other naturally available lignocellulosics such as reeds for heat and energy needs is well documented (Cai et al., 2017; Hassan et al., 2018). This has remained unchanged for hundreds of years and forms a major part of energy generation in different parts of the world even to the current day. The most basic forms of energy recovery from lignocellulosic matter have most often been employed in the developing parts of the world such as parts of Asia (India, Sri Lanka, Bangladesh, Nepal, and Philippines), many sub-Saharan African nations (Nigeria, Ghana, Uganda, Kenya, Somalia, and Ethiopia), and many nations in the South American continent (Brazil, Chile, Ecuador, Peru, and Uruguay). It is apparent that the most basic use of lignocellulosic biomass for energy recovery (agricultural lignocellulosic residue and firewood burning) is the most inefficient form of biomass biofuel usage. The best estimates by the US Environmental Protection Agency (EPA) and many other environmental protection authorities around the world put the recoverable energy content of properly dried firewood within the range of 14–16.9 MJ/kg (US EPA and Sustainable Energy Development Office, Australia). The US EPA estimates that by using best industrial practices, the energy efficiency obtained from firewood is about 70% (10 MJ/kg) if a 14 MJ/kg value is considered. Apart from this, there are many other ways in which usable energy from waste biomass can be recovered. All such methods possess their strengths and weaknesses. This chapter discusses the utility of select waste types used for bioenergy generation. It places a special emphasis on the waste-to-energy nexus technologies used in developing countries, strengths and limitations of such methods, and the future outlook of these waste to energy conversion methods for their sustainable development.

9.1.1   ENERGY RECOVERY FROM BIOLOGICAL WASTES: PERSPECTIVES FROM THE DEVELOPING WORLD

Many developing nations such as India, Pakistan, Sri Lanka, parts of the Middle East such as modern-day Iraq and Iran, parts of Africa such as Egypt, Libya, and Sudan and Mesoamerica such as Mexico and Nicaragua and South America such as Peru, Guatemala, and Brazil all had ancient civilizations that primarily relied on lignocellulosic biomass and lignocellulosic residue for energy requirements (Pecha and Garcia-Perz, 2020). Therefore, it is accurate to state that entire civilizations were fueled by biomass and waste residue biomass in the ancient past. The lignocellulosic biomass and its residues still fulfill a large part of the energy requirements of developing nations.

9.1.2 Agricultural and Forestry Residue for Bioenergy Generation

Apart from direct combustion of lignocellulosics containing biomass for bioenergy generation, several other methods are in use for converting the energy content of such materials into useful energy. Attempts have been made to convert the chemical energy contained within cellulosic residue material into other forms of liquid and gaseous fuel types. Biomass energy conversion technologies are inherently flexible because a variety of technological options are available to obtain useful energy out of waste biomass (Van Meerbeek et al., 2019). It can be directly burned, co-fired with other fuel types, or can be subjected to other intermediate conversion processes such as gasification and electrical energy production by combined heat and power (CHP) processes.

9.1.3 Conversion of Cellulosic Biomass into Liquid Biofuels

The polymeric cellulose content of cellulosic and lignocellulosic biomass can be saccharified (depolymerized into sugar residues) either by chemical or biological means. These simple sugars are then converted into liquid biofuels using processes such as ethanol fermentation and solventogenic fermentations such as acetone–ethanol–butanol (ABE) processes (Su et al., 2020). This involves the conversion of the carbohydrate components of the cellulosic matter found in such feedstocks into combustible alcohols such as ethanol and butanol by means of microbial fermentation (Mahmood et al., 2019).

Feedstock types for fuel ethanol production indicate a considerable variety. Primary among them are the sugar-rich or starchy feedstock types such as sugarcane or corn (Ibarra-Gonzalez., 2019). This is because the fermentable substrates for the ethanol fermentation reactions are readily available in such feedstocks. Unfortunately, the readily fermentable substrates happen to be a big part of the global food supply (Zhu et al., 2020). In other words, fuel ethanol production on a massive scale using readily fermentable feedstock types will interfere with the global food production in a very adverse way. For example, widespread use of feedstocks such as corn, rice, and wheat for fuel ethanol production will directly compete with food supply and land use for agriculture in places such as Africa. The end result of such practices will be catastrophic food shortages in the regions of the world that are already strained by food insecurity.

One of the main alternatives for circumventing this problem is to use the cellulose content of the lignocellulosic biomass for saccharification and fermentation reactions of ethanol production (Figure 9.1). In fact, the biggest natural reservoir of fermentable sugars is postulated to be locked-up in the lignocellulosic materials (Zhu et al., 2020). The liberation of fermentable sugars from this reservoir, however, is a difficult task due to the presence of lignin (Moodley et al., 2020). This can be achieved by several

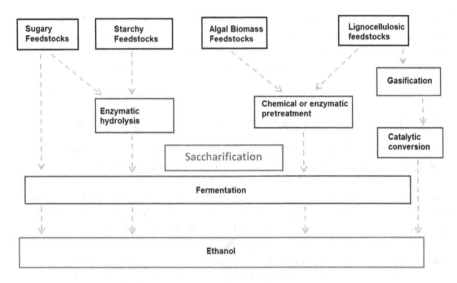

FIGURE 9.1   Feedstock utilization and the different industrial workflows used in fuel ethanol production.

means including physicochemical methods such mechanical maceration followed by acid or alkali pretreatment (Aftab et al., 2017) or biological pretreatment such as enzymatic hydrolysis by lignin peroxidases, manganese peroxidases, and laccases (De La Torre., 2017). With regard to fuel ethanol production from readily fermentable feedstock such as corn starch and cane sugar, microbial fermentation reactions are considered as the rate-limiting step. On the contrary, for fuel ethanol production from lignocellulosic material, the pretreatment of the feedstock leading to saccharification is considered as the rate-limiting step (Fatma et al., 2018). Depending on the feedstock types used for fuel ethanol production, a basic classification is made. If sugary feedstocks are utilized, the resulting fuel ethanol is classified as "first generation". If the feedstock utilized is lignocellulosics, the fuel ethanol produced is coined "second generation" and if the feedstock is algal biomass, the resulting fuel ethanol is termed "third generation". Additionally, if the carbon for fuel ethanol originates from waste CO_2 and catalytic chemical conversion is utilized for ethanol production, the resulting biofuel is termed "fourth generation" fuel ethanol (Figure 9.2). Out of these, all of the generations that follow the "first generation" ethanol are regarded as sustainable and considered to be interfering minimally with food supply (Aftab et al., 2017).

Agricultural residues such as corn fiber, corn stalks and cobs, sugarcane bagasse, and other industrial residue such as hardwood residue, sawdust, wood chips, and wood pellets have been successfully used in numerous fuel ethanol plants as suitable lignocellulosic feedstock in developing countries. In addition to this, forestry residue such as woody residue, switchgrass, grass seeds, and grass straw has also been used as raw material for large-scale fuel ethanol production (Lamichhane et al., 2021). Algal biomass, one of the more desirable feedstock types for fuel ethanol production, has been used much less frequently when compared to other feedstock types. This is

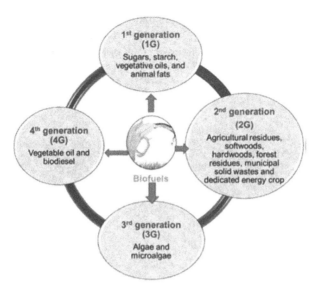

FIGURE 9.2 Classification of biofuels based on substrate utilization (Reproduced with permission from Zabed et al., 2019).

partly owing to the technical challenges and costly nature of cultivating and downstream processing of such biomass in dedicated photobioreactors, natural or artificial ponds, and marine environments for fuel ethanol production (Kumar et al., 2018).

9.2 FUEL ETHANOL PRODUCTION IN THE DEVELOPING WORLD

The biggest global producer of fuel ethanol 2020 was the United States with an annual production of 63 billion liters and this constitutes approximately 53% of the global fuel ethanol production (global total fuel ethanol production in 2020 being approximately 117 billion liters). The next biggest annual fuel ethanol output of 36 billion liters in 2020 came from Brazil, a developing country (30% of the total global production). Among other developing nations that made it into the top ten producers of fuel ethanol in 2020, China stands at the fourth position, producing approximately 4 billion liters (3% of global production) and India closely follows at the sixth position, producing 2.3 billion liters (2% of the global production). Other developing nations such as Thailand have also ramped up fuel ethanol production since 2015 and it had a 1.5% share (1.8 billion liters) of global production in 2020. The only other developing nation that has made its way into this list is Argentina, with an annual global production share of approximately 0.9% (1 billion liters) in 2020 (data from US Energy Information Administration). Apart from the aforesaid countries, the fuel ethanol production in the rest of the world (excluding the European Union) contributed a mere 2% (2.25 billion liters) to the global share in 2020. Both China and India can be considered to have a strong industrial production base for massive-scale production of fuel ethanol and other biofuels, whereas other developing nations only possess emerging levels of infrastructure for biofuel production. Apart from India, many other fuel ethanol-producing developing nations have not shown a steady rise in production in the past five years (2015–2020) (Figure 9.3A). In fact, fuel ethanol production in some countries was either stagnant or has indicated a slight downward trend over the same time period (Figure 9.3B).

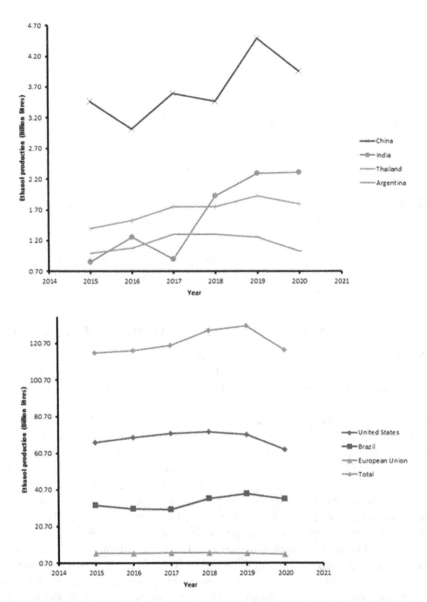

FIGURE 9.3 Global trends in fuel ethanol production of (A) developing nations excluding Brazil and (B) temporal comparison of fuel ethanol production of top-three producers compared to the global total fuel ethanol output (Raw data from the U.S. Department of Agriculture).

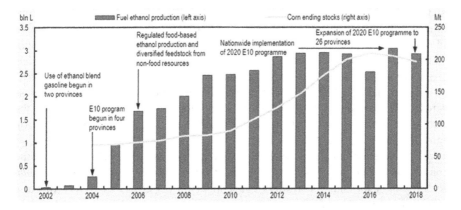

FIGURE 9.4   The timeline of Chinese fuel ethanol program and the utilization of excess grain stocks for the nationwide implementation of the nationwide E10 drive (Raw data from OECD).

9.3   CHINESE FUEL ETHANOL PROGRAM

China is expected to make a big impact in the fuel ethanol market share in the coming decades. The Chinese government imposed an official policy of using an E10 blend (10% fuel ethanol in petrol) in some petrol outlets in the country since 2002. This was done to deal with air pollution, ensure energy security, and to deal with excess grain stocks (mainly excess maize stocks). Between 2007 and 2015, a system of government subsidies and improvements to storage system stimulated maize production in China. This, however, drove up the excess maize stocks in the country from 82 Mt in 2008 to 209 Mt in 2016 (estimated values). The Chinese government then commissioned a nationwide utilization drive for excess corn to boost fuel ethanol production (Figure 9.4). This was also synonymous with the Chinese government policy of implementing a nationwide E10 fuel ethanol mixture in all petrol stations in the country (data from the Organisation for Economic Co-operation and Development (OECD)).

Chinese fuel ethanol production topped 3.7 billion liters in 2020 and the expected fuel ethanol output by 2025 is 13.6 billion liters (OECD). With the current outlook and future policy shifts, China is set to make a sizable contribution to the global fuel ethanol production. However, the challenge remains in making most of that contribution from lignocellulosic residue and waste material rather than relying on corn or any other type of grain-based feedstock.

9.4   SOLVENTOGENIC FERMENTATIONS FOR LIQUID BIOFUELS PRODUCTION

Solventogenic fermentations for the simultaneous fermentative production of three volatile products—acetone, ethanol, and butanol (known as ABE fermentations)—have been known since the 1860s and it has been used on a very large scale during World War I to produce many of the precursors of explosives (Lee et al., 2016). With respect to biofuel production, ABE fermentations offers several benefits over conventional ethanol fermentations. Firstly, it can produce two liquid biofuels, butanol and ethanol, in a single fermentation run (produces acetone, butanol, and ethanol in a ratio of 3:6:1) (Mayank et al., 2013). Secondly, butanol as a liquid biofuel has longer carbon chain-length compared to ethanol (two-carbon chain for ethanol versus four-carbon chain for butanol) and therefore has a bigger energy density. It is less hygroscopic, less corrosive to the engine, and butanol more closely resembles the combustion profile of petrol than ethanol when burned in an internal combustion engine (Mahapatra et al., 2017).

ABE fermentations are primarily driven by bacteria belonging to the genus *Clostridium* and species such as *Clostridium beijerinckii*, *Clostridium saccharobutylicum*, and *Clostridium acetobutylicum* (Li et al., 2020). All of these organisms are strict anaerobes and are relatively slow growing compared to aerobic organisms and facultative aerobes. Therefore, ABE fermentations essentially are anaerobic processes. This places ABE fermentations at a distinct disadvantage because anaerobic bioprocesses are inherently slow processes (Patakova et al., 2013). Therefore, the kinetic conversion rates of even the well-optimized ABE fermentations are lower compared to ethanol fermentations driven by the facultative aerobe *Saccharomyces cerevisiae*.

In addition to the utilization of lignocellulosic waste materials as feedstock for ABE fermentations, other industrial residue such as molasses and rapeseed cake can also be used (Gu et al., 2014). Moreover, there are instances where large-scale ABE fermentations are in operation utilizing algal biomass as the feedstock (Ellis et al., 2012). The total global biobutalol production volume yielding from ABE fermentation process is projected to be about 6.9 billion liters in 2023 (US Department of Energy (DoE)). Although biobutanol production from the ABE process is a promising prospect for liquid biofuels, the US DoE projects its future growth and near-term outlook as negative. The US DoE and US EPA also state that the global market for biobutanol is small and intermittent. Close to a 90% of biobutanol produced from the ABE process comes from the two countries, the United States and Brazil. This is partly because the preferred substrate for ABE fermentations, molasses, is widely available in these two countries (Mansur et al., 2010).

9.5 ANAEROBIC CONVERSION OF HIGH ORGANIC STRENGTH WASTEWATER AND WASTE MATERIAL INTO BIOENERGY

High organic strength wastewater, usually exceeding a chemical oxygen demand (COD) value of 3000 mg/L, originating from many different industries can be used in anaerobic digestion processes and biogas energy production (Liew et al., 2020). These industries primarily include dairy industry, confectionary industry, slaughterhouses, rice mills, and other food processing industries. Animal manure such as cow, pig, and chicken manure also produces a high organic strength raw material that can be used for anaerobic digestion and biogas production (Kong et al., 2019). In addition to these, the excess biological sludge resulting from the return activated sludge (RAS) streams of municipal activated sludge wastewater treatment plants also produce large amounts of digestible material that can be easily converted to biogas energy.

Historically, developed nations, especially the ones in Europe such as Germany, the Netherlands, Belgium, the United Kingdom, and Denmark, as well as others like the United States, Japan, and Canada have been at the forefront of this process where biogas is harnessed efficiently from wastewater. In such places, anaerobic digesters are an integral part of the wastewater treatment plant (WWTP) design where biogas is directly utilized for the energy needs of the WWTP or biogas is given into the national gas grid (grid injection) for heating and electricity generation purposes (Figure 9.5).

This technology has now proliferated into considerably large-scale anaerobic digestion processes where large quantities of energy-rich products such as biomethane and biohydrogen can be generated, even in developing nations (Kumar et al., 2020). Developing nations that possess a strong industrial base such as India, China, Malaysia, Indonesia, and Thailand are emerging as strong players of biogas recovery from waste materials. Back in the 1980s, 1990s through to early 2000s, developing nations such as India, Bangladesh, Vietnam, and China made it an official government policy to support the operation of community-based small- to medium-scale biogas reactors (Bharatiraja et al., 2018). While keeping this policy and the infrastructure intact, now these countries are making a departure into industry-operated, very large-scale anaerobic digesters and biogas plants. The generation of high-COD waste streams therefore is consequential. Industries such as dairy processing, confectionaries, and

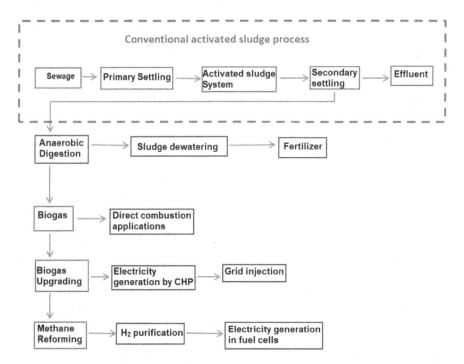

FIGURE 9.5 The link between process flows of conventional activated sludge wastewater treatment process and the biogas production from high organic load waste products from the activated sludge process.

slaughterhouses produce very high organic strength wastewater and they are plentiful in developing nations. In addition to these, countries such as Indonesia operate extensive palm oil extraction and processing industries. Such industries are known to produce wastewater streams that are ultra-high (>10,000 mg COD/L) in organic strength (Choong et al., 2018). Treatment of such wastewaters using conventional wastewater treatment systems such as activated sludge systems is not feasible due to the high organic load in such wastewater. Therefore, one of the very few options available for dealing with wastewater streams of such strengths is to rely on anaerobic digestion and biogas production as a by-product of that process.

Bio-wastes that are co-digested in anaerobic digesters for biomethane include aforementioned industrial waste residues, sewage sludge, the organic fraction of the municipal solid waste, silage, microbial biomass, manure (cow, pig, and chicken), and renewable lignocellulosic biomass (Solé-Bundó et al., 2019).

Connected to the same biogas generation process is the process of biogas upgrading. Biogas is a complex mixture of methane gas, hydrogen gas, carbon dioxide, hydrogen sulfide, nitrogen gas, and water vapor. The useful gases for combustion and energy production out of these are methane and biohydrogen. Carbon dioxide does not aid in combustion and others such as hydrogen sulfide are corrosive to mechanical components. All the other gaseous components apart from methane and biohydrogen are undesirable with regards to energy recovery. The removal of these undesirable gaseous components and the refinement and separation of biomethane and biohydrogen is known as biogas upgrading. Upgraded biomethane can be directly burned in automobile internal combustion engines, in CHP processes for electricity generation, in heating and cooking applications, and can be fed into the central gas grid as a biogas grid injection. Recovered biohydrogen can be used in fuel cells for clean electricity generation and in fuel cell-based clean public transportation services.

9.6   MUNICIPAL SOLID WASTE MANAGEMENT AND BIOENERGY GENERATION

The management of disposal of solid waste generated from urban municipalities in the developing world is one of the most pressing environmental problems faced by developing countries. The lack of technological capacity, poor land use patterns, and laxity of environmental regulations have caused the landfills in such countries to fill up with municipal solid waste. One of the best examples of problems caused by such scenarios is the tragic garbage dump landslide that took place at Meethotamulla, Sri Lanka, in 2017, where nearly three dozen lives were lost and damage in the order of several millions of US dollars to property occurred (Basha and Raviteja, 2018; Jayaweera et al., 2019). This and many such incidents around the developing world highlight the magnitude of this problem faced by developing countries. Such poorly managed landfill dumps lead to uncontrolled build-up of methane gas, causing landfill gas explosions that significantly contribute to global warming.

Many of the solid organic waste types that end up in landfills can be used for bioenergy generation. However, non-biodegradable organic materials such as plastics, polyethylene and their derivatives, other synthetic polymers, and fabrics cannot be used in conventional bioenergy generation and biofuel production methods. Other disposal methods such as salvage and recycling, incineration, and co-combustion with other materials are utilized for the management of such material. Biodegradable material such as biodegradable packaging materials, woody material food waste, and other biodegradable solids are successfully subjected to anaerobic co-digestion (sometimes subsequent to pretreatment) and gasification/liquefaction followed by catalytic conversion to other combustible biofuels.

9.7   WASTE COOKING OIL IN BIODIESEL PRODUCTION

Waste cooking oil (WCO) or used cooking oil (UCO) has long been considered an attractive starting material for biodiesel production for several reasons (Azeem et al., 2016, Canesin et al., 2014; Chhetri et al., 2008). The first advantage is that it provides a stable source of raw material for biodiesel manufacture by transesterification reactions. Secondly, it provides an attractive recycling solution for used cooking oil that has reached its end of product cycle. Used waste cooking oil has limited use and poses potential disposal problems once it reaches the end of its product cycle (Canesin et al., 2014). Converting large amounts of waste cooking oil into biodiesel that can be burned in internal combustion engines presents an attractive way of averting some of the potential environmental problems that improper disposal of waste cooking oil may create (Chhetri et al., 2008).

In addition to WCO from domestic and food industry usage, other substrates that fit in this category include slaughterhouse wastage, tallow, waste products of fish oil industry, and poultry fat (Thamsiriroj and Murphy, 2011). Major producers of WCOs in large quantities except for domestic production include food and restaurant industry (especially when deep fat fryers are employed), hotels, and canteens (Hatzisymeon et al., 2019). The current contribution of WCOs to global biodiesel output is estimated to be about 23% (Figure 9.6) (OECD-FAO Agricultural Outlook Report 2020–2029).

All the European Union (EU) member states are mandated to cutting emissions and employing renewable energy sources under the EU Renewable Energy Directive (RED). Under this scheme, the EU provides technical assistance and subsidies to biofuel companies for developing and utilizing WCO to biodiesel conversion technologies (Chiaramonti and Goumas, 2019). A key aspect of this approach is to have a strong, systematic, and widespread WCO feedstock collection network. Therefore, such a network is currently being laid out in many EU member states to channel an uninterrupted supply of WCO feedstock to be converted to biodiesel (Chiaramonti and Goumas, 2019 and Braungardt et al., 2019). In addition to this scheme however, WCO is imported from developing nations such as Indonesia, Malaysia, and China. The origin for most of this WCO is palm oil-based cooking oils (Figure 9.7).

However, this importation drive for WCO, especially from the developing nations such as China, Indonesia, and Malaysia, have raised serious environmental concerns. This is because when the demand for WCO for developed nations is increasing, the lucrative nature of WCO exports for the exporter countries drives them to produce more WCO, primarily from sources such as palm oil, rapeseed oil, and many other

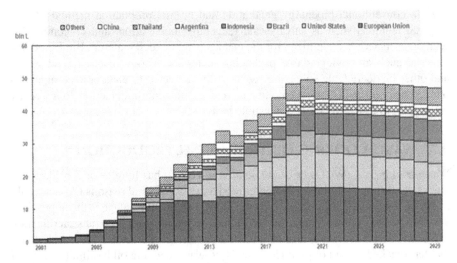

FIGURE 9.6 Past, current, and projected trends of global biodiesel production in major biodiesel producers. Major contributors to the global supply have largely remained unchanged (Raw data from OECD-FAO agricultural outlook report 2020–2029).

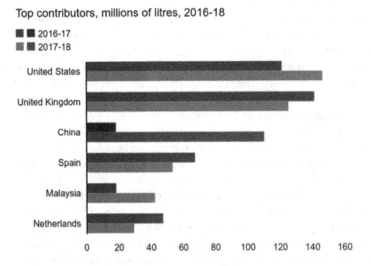

FIGURE 9.7 Top contributors of WCO to the UK for production of biodiesel during the period of 2016 to 2018, indicates a growing WCO supply from developing countries such as Malaysia and China (Data from the Department of Transport, UK).

types of oil seeds (Table 9.1). Consequently, this leads to increased deforestation in developing nations that export WCO. Moreover, this practice reduces the WCO stocks available for the developing nations to produce more biodiesel for their own consumption and to reduce their own greenhouse gas emissions.

TABLE 9.1
Feedstock Types Used for Biofuel Production in Major Biofuel-Producing Developing Nations

Developing Nation	Feedstock Type	Type of Biofuel Produced	Reference
Brazil	Sugarcane, corn, lignocellulosic biomass	Bioethanol	Martins et al., 2019
Brazil	Waste cooking oil, soybean, beef tallow and cottonseed oil	Biodiesel	Lima et al., 2017
China	Maize, corn and lignocellulosic biomass	Bioethanol	Fu et al., 2021
China	Waste cooking oil, rapeseed oil, soybean oil	Biodiesel	Chen et al., 2020
Thailand	Cassava, molasses, sugarcane, agricultural residue (corn stover) and lignocellulosic biomass.	Bioethanol	Silalertruska and Gheewala., 2010, Haputta et al.,2020
Thailand	Crude palm oil and refined palm oil, coconut oil, palm stearin, and waste cooking oil	Biodiesel	Napueng et al., 2018, Heo and Choi., 2019
Indonesia	Cassava, sugarcane, sweet sorghum, potato, starch, corn, and lignocellulosic biomass	Bioethanol	Khatiwada and Silveira., 2017, Darmayanti et al., 2019.
Indonesia	Palm oil, animal fat, Jatropha oil, and waste cooking oil	Biodiesel	Harahap et al., 2019, Syafuddin et al., 2020
Argentina	Soybean oil and waste cooking oil	Biodiesel	Rouhani and Montgomery, 2019
India	Agricultural residue, domestic and industrial food waste, manure, and farm residue	Biomethane and biohydrogen	Kalyanasundaram et al., 2020

9.8 INVASIVE PLANT BIOMASS IN DEVELOPING NATIONS AS FEEDSTOCK FOR BIOFUELS

9.8.1 WHY USE "INVASIVE PLANTS" AS FEEDSTOCK FOR BIOFUELS?

As a result of a decline in fossil fuel sources and increasing global warming, the need of a viable energy alternative to fossil fuels has emerged. The research and discussion about energy derived from biomass or plant material have taken place for more than 40 years in the Western world (Lewandowski et al., 2003). Unlike fossil fuels, biomass is renewable and eco-friendly due to its lower contribution to climate change compared to fossil fuels. A variety of plant species including crops and wild plants are being considered for the use of biofuel generation. However, this topic has been controversial for some reasons. One of them is that most of the researched plant material for bioenergy generation is focused on food crops. The plant materials used for bioenergy generation are converted to energy through direct burning and conversion to gas or to ethanol. In the United States, many crops have been tested as feedstock for bioenergy generation including the food crops such as corn (*Zea mays* L.)

and sorghum grain, by converting to ethanol through fermentation, and soybean and canola by transforming into fatty acid methyl esters by reaction with alcohol (Demirbas 2007). The cultivation of food crops for bioenergy generation requires a large amount of land, which requires the conversion of forests in to monocultivated lands, deteriorating biodiversity. Considering the cost of cultivation of food crops for biofuel, the economic benefits also would be negligible. As an alternative, perennial grasses were taken into consideration for bioenergy generation due to their nature of continuous growth in large quantities of biomass and especially their noncrop status. Whether food crops or grasses, earlier the cellulosic bioenergy research was mainly focused on cultivated biomass while paying very little attention to the existing non-cultivated plants, which are widespread in non-agricultural lands like open grasslands, riparian habitats, and lakes (Young et al., 2011). However, most of the non-crop species for bioenergy generation are exotic species with the potential to become invasive species or already identified as invasive species (Van Meerbeek et al., 2015). There is a major concern among researchers about the potential of these plant species being cultivated for biofuel generation, to escape from the managed systems and introduce and invade the natural ecosystems (US Department of Agriculture, 2021). However, less attention is given to invasive species that are already present in the environment and need to be removed from their ecosystems, to be used as a source of biomass for bioenergy generation (Van Meerbeek et al., 2015). Using invasive plant species as feedstock for biofuel can also be used as a management tool for naturally occurring invasive species and several success stories of invasive species management by using the biomass for bioenergy can also be found from different parts of the world. It is estimated that the impact and management costs associated with invasive species globally amounts to $1.5 trillion annually or 5% of annual global domestic product (GDP) (Pimentel, 2001).

9.8.2 INVASIVE PLANTS AS BIOMASS FOR BIOENERGY GENERATION FROM THE DEVELOPED WORLD

The research on bioenergy crops have shown that plant species used for the process should possess properties such as high competitiveness, high yields, rapid establishment, tolerance to harsh environmental conditions, and being maximally productive with less recourses being used (Barney, 2014). Therefore, invasive plant species standout from other crop species with the aforementioned properties, making them the most suitable as biomass for energy generation. There are many control mechanisms used to mitigate the problem of invasive plants in different parts of the world. Burning, application of herbicides, and mechanical removal of plants have been accepted as the most effective control measures for most of the invasive species so far. However, in aquatic bodies, burning and herbicide application have not been successful as neither of these methods address the root cause of invasive plant spread in aquatic bodies—the nutrient enrichment in water (Carson et al., 2018). Therefore, in water bodies mechanical removal of invasive plants is practiced as a control strategy. However, the mechanically removed plant materials have to be discarded properly to prevent them from being re-introduced to the ecosystem Carson et al. (2018) showed that the removal of the single growing season biomass of the three most dominant invasive plant species in the Great Lakes coastal wetlands of the United States—invasive cattail, common reed, and reed canary grass—could generate 1.8 million barrels of oil if combusted, or 0.9 million barrels of oil if converted to biogas in an anaerobic digester. A study conducted in Iberian Peninsula in Europe used *Cortaderia selloana* (CS), which is an invasive alien species to the region, to see the possibility of conversion the biomass in it to bioenergy through pyrolysis technology and showed successful results (Perez et al., 2021). Eucalyptus (*Eucalyptus globulus Labill.*) is a fast-growing exotic species in Portugal and is grown for paper industry for over 800,000 ha over the country and harvest residues are removed from the sites to bioenergy production (Barreiro and Tomé, 2012 in Carneiro et al., 2014). In countries such as France, Spain, and Portugal, Australian species of the genus *Acacia* Mill. have become problematic invasive plants and therefore Portugal has imposed laws to restrict the cultivation of these species in plantations and it has been recommended to use the biomass for bioenergy and bio products (Carneiro et al., 2014).

9.8.3 Invasive Plants as Biomass for Bioenergy Generation from the Developing World

Invasive species is a great ecological threat to developing countries as it deteriorates biodiversity and threatens food security. The cost of invasive species management is huge, having an adverse impact on the economy. An assessment of case studies done in early 1990s in India showed that the energy needs of India's rural population could be fulfilled using bioenergy and it could be achieved by using invasive plant biomass also as feedstock, reducing dependence on forests and the national electricity grid and leading the nation toward sustainable development (Ravindranath and Hall, 1995). The analysis is applicable to the majority of developing countries that are battling with the energy crisis as well as the spread of invasive species. Aquatic invasive plants that grow in inland water bodies could affect the nutrient dynamics and water quality of the water bodies, causing decline in fish population and it could directly impact the inland fisheries (Kovalenko et al., 2010). For instance, *Eichhornia crassipes* is a fast-growing invasive species that can be found in tropical water bodies, and an individual can produce more than 140 million plants annually covering a 1.4 km^2 area in a water body (Tao et al., 2016). A case study in Sri Lanka showed *E. crassipes* was the most dominant aquatic plant in some of the natural inland water bodies with high nutrient contents, which are vital to inland fisheries in the North Central Province of the island (Rajakaruna et al., 2017). There are few other examples from different parts of Asia like Indonesia, India, and China, where expansive growth of *E. crassipes* has become a huge environmental problem (Sukarni et al., 2019; Das et al., 2016; Tao et al., 2016). However, promising results have been obtained by several researchers on the use of *E. crassipes* as a material for bioenergy generation (Sukarni et al., 2019; Das et al., 2016; Tao et al., 2016). Tao et al., 2016 claim that *E. crassipesis* is the best alternative cellulose source with 58.6% combined cellulose and hemicellulose content and low lignin content.

In Africa, particularly in Kenya and South Africa. the use of *Prosopis* species—*P. glandulosa* Torrey and *P. velutina* Wooton, *P. juliflora* Swartz DC, and *P. pallida*—for biofuel generation is been practiced as a control mechanism for these invasive species

(Witt, 2010). The introduction of these species have caused some typical problems of invasive plants in Africa by invading more than 4 million hectares. However, it is recommended to use the plant material for biofuel generation should be along with the other control measures in order to control the spread of the plant (Witt, 2010).

Another study from Brazil showed that *Panicum maximum*, which is a widespread weed in the tropical countries, could be effectively used for biofuel production. In Brazil, using *P. purpureum* for bioenergy is already in practice, producing high yields of biomass (Jank et al., 2013).

In Asia and Africa, *Jatropha curcas* was also a very popular species used for biofuel production; however, the cultivation of this species has pros and cons in economic and ecological perspectives (Low et al., 2010).

Although there are many experiments carried out on the possibilities of using invasive plant biomass as feedstock for bioenergy generation, large-scale industrial bioenergy production using invasive plants is not yet common in developing countries.

9.9 ECOLOGICAL AND LAND-USE ASPECTS OF WASTE TO BIOENERGY PROCESS IN THE DEVELOPING WORLD

9.9.1 LAND-USE CRISIS IN THE DEVELOPING WORLD AND ITS IMPLICATIONS FOR BIOFUEL PRODUCTION

In the developing world, population growth and the demand for land has become a major economic concern in the process of development of nations. Growth of the population is directly linked with land needed for expansion of agriculture, industries, and for residence. Studies around the world have emphasized the fact that people make decisions about land use relying on the economic benefits that are designed by the market and economic policies of the countries (Schreinemachers and Berger, 2006). Therefore, it is vital that the land-use policies and the economic policies are interconnected in order to protect both environment and people. However, the lack of balance between these two has led to many ecological and land-use problems in the developing world such as habitat destruction, deforestation for agriculture, and spread of alien invasive species (Shah et al., 2020). To meet the demand for food, land has to be used for agriculture. However, not every available land in the world is suitable for agriculture while most of the suitable land for agriculture is already being used and the rest remains in the forests (Tinker, 1997). Apart from the land-use and waste management problems, the struggle for sustainable energy has also become a crisis in the developing world. As the world in total has moved to 33% of the energy consumption with renewable energy, with the decline of fossil fuel resources and climate change, the time has come for the developing world also to move toward bioenergy generation using waste and plant biomass (Rajmohan et al., 2019).

9.9.2 ECOLOGICAL BENEFITS AND CONCERNS IN THE WASTE TO BIOENERGY PROCESS

Traditional biofuels such as firewood and charcoal are now replaced by biodiesel, ethanol, or biogas derived from the combustion of natural vegetation, grown crops, or agricultural and municipal waste (Von Braun and Pachauri, 2006). These newer sources of biofuel are considered cleaner as they are produced and used in a way that reduces carbon emission to the atmosphere. The reduction of greenhouse gas emission creates a lot of environmental and ecological benefits as the world's flora and fauna are sensitive to climate change, and global warming has contributed to loss of biodiversity throughout the world. It is demonstrated that forest sustainability can be significantly achieved through agricultural waste management strategies (Kammen, 2006). Crop residues have a greater potential in acting as a feedstock for bioenergy, and this potential could be used in the sustainable management of agricultural waste; for example, large amount of sugarcane residues is used in ethanol plants in Brazil (Dale et al., 2010). In Africa and other coffee-growing countries, coffee pulp is the most available agricultural waste obtained by wet process of coffee. Research conducted in Ethiopia has demonstrated the possibility of converting coffee pulp into bioethanol for energy generation (Kefale et al., 2012). In that way, the use of food crops (like corn, maize, and sorghum) for biofuel, production which may lead to food insecurity, could be prevented by using alternative nonedible agricultural waste (Kefale et al., 2012). Not only crop residues but also municipal solid waste, sludge, yard waste, and dairy and cattle manure are researched to be used as biomass for bioenergy production (Champagne, 2008). When municipal waste is collected in landfills without any treatment, many environmental problems like greenhouse gas emissions and contamination of the soil and water bodies, releasing of leachate and odors arise, and this scenario is common in most densely populated areas of developing countries. Contamination of water bodies leads to a major ecological and global pollution problem known as eutrophication; excessive plant growth resulting from nutrient enrichment in water by waste pollutants Smith and Schindler, 2009). Therefore, bioenergy generation through waste, without releasing them to water bodies, can help gain a huge ecological benefit by minimizing eutrophication and by using waste as feedstock for biofuel without letting them accumulate in landfills, land and air pollution could be reduced.

The spread of invasive plant species is considered as one of the greatest ecological threats globally. Although the mechanical removal of invasive plants is widely used as a successful mitigation method, discarding of removed plant materials remains a problem (Carson et al., 2018). However, the successful use of invasive plant biomass for bioenergy has been demonstrated, as was earlier discussed in this chapter (Section 9.8).

However, there is a great concern of ecologists about cultivating invasive plants to cater to the bioenergy industry. It is a great challenge for this industry to meet the demands of sustainability by using a minimum amount of land, less harmful by-products, and also controlling unintended invasive species spread. However, biomass production at the industrial scale aims at maximum yield with minimum management and therefore the industries tend to depend on self-sustaining species that regenerate with minimal cultural practices (Petri, L. 2019). Therefore, countries like the United States moved toward the cultivation of primarily exotic perennial grasses for bioenergy that possess many beneficial traits, making some of the introduced species invasive (Barney, 2014). Due to these practices, a discussion was prompted among environmentalists that development in the biofuel industry may lead to some environmental threats through land-use change and extensive introduction of potentially invasive species (Low and Booth, 2007). Considering these ecological threats, a lot of research has been conducted on the assessment of the invasiveness of new bioenergy crops and preventing new introductions (Van Meerbeek et al., 2015). However, the most appropriate and recommended practice is to use the existing invasive species growing in the natural ecosystems for bioenergy production in order to gain the maximum ecological benefits.

9.9.3 CHALLENGES AND IMPLICATIONS OF BIOFUELS POLICY IN THE DEVELOPING WORLD

Biofuel production and utilization offers large benefits to developing nations for cutting back on their relatively large carbon footprints compared to developed nations. The replacement of a portion of their carbon emissions originating from fossil fuels with renewable sources will ensure that some of the biggest global contributors to climate change (such as China and India) will act more responsibly and cut back on a significant proportion of carbon emissions. Embedding the production, use, and promotion of biofuels in the energy policy of developing nations will be beneficial to the world in general. Some developing nations such as Brazil are already world leaders in this respect. When coupled with waste to energy nexus and its defining feature, the five R principle (reduce, reuse, recycle, recovery, and restore) for bioenergy generation from waste material is an ideal solution for developing countries to deal with their waste material in a more meaningful manner (Sharma et al., 2020).

The biggest challenge pertaining to biofuels for developing nations, however, is to prevent biofuel feedstock supplies interfering with food production. If biofuels are to make a real impact, the feedstocks supply must be stable and sustained over long periods of time. Dwindling land area for biofuel feedstock production and the competition between food production and biofuel supply is one of the biggest problems faced by the developing nations. The use of non-arable land, the use of marine, lagoon, and mangrove environments for biofuel feedstock production, and the use of genetically modified organisms as biofuel feedstock can be explored as potential solutions for the problems discussed above. The utilization of waste materials as an alternative feedstock will have far-reaching positive implications in meeting these challenges. Globally, this trend has started to take shape both in developed and developing regions. However, a lot more needs to be done in order to promote waste material as the leading biomass feedstock type for biofuel production.

REFERENCES

Aftab, M.N., Iqbal, I., Riaz, F., Karadag, A., and Tabatabaei, M. 2019. Different pretreatment methods of lignocellulosic biomass for use in biofuel production. In: *Biomass for Bioenergy-Recent Trends and Future Challenges*, A. E.-F. Abomohra (ed). London: IntechOpen.

Aftab, A.A.R.I., Ismail, A.R., and Ibupoto, Z.H. (2017). Enhancing the rheological properties and shale inhibition behavior of water-based mud using nanosilica, multi-walled carbon nanotube, and graphene nanoplatelet. *Egyptian Journal of Petroleum* 26(2), 291–299. https://doi.org/10.1016/j.ejpe.2016.05.004

Azeem, M.W., Hanif, M.A., Al-Sabahi, J.N., Khan, A.A., Naz, S., and Ijaz, A. 2016. Production of biodiesel from low priced, renewable and abundant date seed oil. *Renewable Energy* 86, 124–132.

Barney, J.N. 2014. Bioenergy and invasive plants: quantifying and mitigating future risks. *Invasive Plant Science and Management* 7, 199–209.

Barreiro, S., and Tomé, M. 2012. Analysis of the impact of the use of eucalyptus biomass for energy on wood availability for eucalyptus forest in Portugal: a simulation study. *Ecology and Society* 17, 1–16.

Basha, B.M., and Raviteja, K. 2018. Meethotamulla landfill failure analysis: a probabilistic approach. In: *Geotechnics for Natural and Engineered Sustainable Technologies*, A. M. Krishna, A. Dey, and S. Sreedeep (eds), pp. 341–351 Singapore: Springer. https://doi.org/10.1007/978-981-10-7721-0_20

Bharathiraja, B., Sudharsana, T., Jayamuthunagai, J., Praveenkumar, R., Chozhavendhan, S., and Iyyappan, J. 2018. Biogas production—a review on composition, fuel properties, feed stock and principles of anaerobic digestion. *Renewable and Sustainable Energy Reviews* 90, 570–582.

Braungardt, S., Bürger, V., Zieger, J., and Bosselaar, L. 2019. How to include cooling in the EU Renewable Energy Directive? Strategies and policy implications. *Energy Policy* 129, 260–267.

Cai, J., He, Y., Yu, X., Banks, S.W., Yang, Y., Zhang, X., Yu, Y., Liu, R., and Bridgwater, A.V. 2017. Review of physicochemical properties and analytical characterization of lignocellulosic biomass. *Renewable and Sustainable Energy Reviews* 76, 309–322.

Canesin, E.A., de Oliveira, C.C., Matsushita, M., Dias, L.F., Pedrão, M.R., and de Souza, N.E. 2014. Characterization of residual oils for biodiesel production. *Electronic Journal of Biotechnology* 17, 39–45.

Carneiro, M., Moreira, R., Gominho, J., and Fabião, A. 2014. Could control of invasive Acacias be a source of biomass for energy under Mediterranean conditions? *Chemical Engineering* 37, 187–192.

Carson, B.D., Lishawa, S.C., Tuchman, N.C., Monks, A.M., Lawrence, B.A., and Albert, D.A. 2018. Harvesting invasive plants to reduce nutrient loads and produce bioenergy: an assessment of Great Lakes coastal wetlands. *Ecosphere* 9, e02320.

Champagne, P. 2008. Bioethanol from agricultural waste residues. *Environmental Progress* 27, 51–57.

Chen, H., Ding, M., Li, Y., Xu, H., Li, Y., and Wei, Z. 2020. Feedstocks, environmental effects and development suggestions for biodiesel in China. *Journal of Traffic and Transportation Engineering* (English Edition) 7(6), 791–807.

Chhetri, A.B., Watts, K.C., and Islam, M.R. 2008. Waste cooking oil as an alternate feedstock for biodiesel production. *Energies* 1, 3–18.

Chiaramonti, D., and Goumas, T. 2019. Impacts on industrial-scale market deployment of advanced biofuels and recycled carbon fuels from the EU Renewable Energy Directive II. *Applied Energy* 251, 113351.

Choong, Y.Y., Chou, K.W., and Norli, I. 2018. Strategies for improving biogas production of palm oil mill effluent (POME) anaerobic digestion: a critical review. *Renewable and Sustainable Energy Reviews* 82, 2993–3006.

Dale, V.H., Kline, K.L., Wiens, J., and Fargione, J. 2010. *Biofuels: Implications for Land Use and Biodiversity*. Washington, DC: Ecological Society of America.

Darmayanti, R.F., Amini, H.W., Rizkiana, M.F., Setiawan, F.A., Palupi, B., Rahmawati, I., Susanti, A., and Fachri, B.A. 2019. Lignocellulosic material from main indonesian plantation commodity as the feedstock for fermentable sugar in biofuel production. *ARPN Journal of Engineering and Applied Sciences* 14, 3524–3534.

Das, S.P., Gupta, A., Das, D., and Goyal, A. 2016. Enhanced bioethanol production from water hyacinth (Eichhornia crassipes) by statistical optimization of fermentation process parameters using Taguchi orthogonal array design. *International Biodeterioration & Biodegradation* 109, 174–184.

De La Torre, M., Martín-Sampedro, R., Fillat, Ú., Eugenio, M.E., Blánquez, A., Hernández, M., Arias, M.E., and Ibarra, D. 2017. Comparison of the efficiency of bacterial and fungal laccases in delignification and detoxification of steam-pretreated lignocellulosic biomass for bioethanol production. *Journal of Industrial Microbiology and Biotechnology* 44, 1561–1573.

Demirbas, A. 2007. Progress and recent trends in biofuels. *Progress in Energy and Combustion Science* 33, 1–18.

Ellis, J.T., Hengge, N.N., Sims, R.C., and Miller, C.D. 2012. Acetone, butanol, and ethanol production from wastewater algae. *Bioresource Technology* 111, 491–495.

Fatma, S., Hameed, A., Noman, M., Ahmed, T., Shahid, M., Tariq, M., Sohail, I., and Tabassum, R. 2018. Lignocellulosic biomass: a sustainable bioenergy source for the future. *Protein and Peptide Letters* 25, 148–163.

Fu, J., Du, J., Lin, G., and Jiang, D. 2021. Analysis of yield potential and regional distribution for bioethanol in China. *Energies* 14, 4554.

Gu, Y., Jiang, Y., Yang, S., and Jiang, W. 2014. Utilization of economical substrate-derived carbohydrates by solventogenic clostridia: pathway dissection, regulation and engineering. *Current Opinion in Biotechnology* 29, 124–131.

Haputta, P., Puttanapong, N., Silalertruksa, T., Bangviwat, A., Prapaspongsa, T., and Gheewala, S.H. 2020. Sustainability analysis of bioethanol promotion in Thailand using a cost-benefit approach. *Journal of Cleaner Production* 251, 119756.

Harahap, F., Silveira, S., and Khatiwada, D. 2019. Cost competitiveness of palm oil biodiesel production in Indonesia. *Energy* 170, 62–72.

Hassan, S.S., Williams, G.A., and Jaiswal, A.K. 2018. Emerging technologies for the pretreatment of lignocellulosic biomass. *Bioresource Technology* 262, 310–318.

Hatzisymeon, M., Kamenopoulos, S., and Tsoutsos, T. 2019. Risk assessment of the life-cycle of the Used Cooking Oil-to-biodiesel supply chain. *Journal of Cleaner Production* 217, 836–843.

Heo, S., and Choi, J.W. 2019. Potential and environmental impacts of liquid biofuel from agricultural residues in Thailand. *Sustainability* 11, 1502.

Ibarra-Gonzalez, P., and Rong, B.-G. 2019. A review of the current state of biofuels production from lignocellulosic biomass using thermochemical conversion routes. *Chinese Journal of Chemical Engineering* 27, 1523–1535.

Jank, L., de Lima, E.A., Simeão, R.M., and Andrade, R.C. 2013. Potential of Panicum maximum as a source of energy. *Tropical Grasslands-Forrajes Tropicales* 1, 92–94.

Jayaweera, M., Gunawardana, B., Gunawardana, M., Karunawardena, A., Dias, V., Premasiri, S., Dissanayake, J., Manatunge, J., Wijeratne, N., and Karunarathne, D. 2019. Management of municipal solid waste open dumps immediately after the collapse: an integrated approach from Meethotamulla open dump, Sri Lanka. *Waste Management* 95, 227–240.

Kalyanasundaram, G.T., Ramasamy, A., Godwin, B., Desikan, R., and Subburamu, K. 2020. Prospects and challenges in biogas technology: Indian scenario. In: *Biogas Production*, N. Balagurusamy and A.K. Chandel (eds), pp. 19–37. Cham: Springer. https://doi.org/10.1007/978-3-030-58827-4_2

Kammen, D.M. 2006. Bioenergy and agriculture: Promises and challenges. Bioenergy in developing countries: Experiences and prospects. https://lib.icimod.org/record/12537/files/4385.pdf

Kefale, A., Redi, M., and Asfaw, A. 2012. Potential of bioethanol production and optimization test from agricultural waste: the case of wet coffee processing waste (pulp). *International Journal of Renewable Energy Research* 2, 446–450.

Khatiwada, D., and Silveira, S. 2017. Scenarios for bioethanol production in Indonesia: how can we meet mandatory blending targets? *Energy* 119, 351–361.

Kong, Z., Li, L., Xue, Y., Yang, M., and Li, Y.-Y. 2019. Challenges and prospects for the anaerobic treatment of chemical-industrial organic wastewater: a review. *Journal of Cleaner Production* 231, 913–927.

Kovalenko, K.E., Dibble, E.D., Agostinho, A.A., Cantanhêde, G., and Fugi, R. 2010. Direct and indirect effects of an introduced piscivore, *Cichla kelberi* and their modification by aquatic plants. *Hydrobiologia* 638(1), 245–253. https://doi.org/10.1007/s10 750-009-0049-6

Kumar, V., Nanda, M., Joshi, H.C., Singh, A., Sharma, S., and Verma, M. 2018. Production of biodiesel and bioethanol using algal biomass harvested from fresh water river. *Renewable Energy* 116, 606–612.

Kumar, A., and Samadder, S.R. 2020. Performance evaluation of anaerobic digestion technology for energy recovery from organic fraction of municipal solid waste: a review. *Energy* 197, 117253.

Lamichhane, G., Acharya, A., Poudel, D.K., Aryal, B., Gyawali, N., Niraula, P., Phuyal, S.R., Budhathoki, P., Bk, G., and Parajuli, N. 2021. Recent advances in bioethanol production from Lignocellulosic biomass. *International Journal of Green Energy* 18, 731–744.

Lee, S.-H., Yun, E.J., Kim, J., Lee, S.J., Um, Y., and Kim, K.H. 2016. Biomass, strain engineering, and fermentation processes for butanol production by solventogenic clostridia. *Applied Microbiology and Biotechnology* 100, 8255–8271.

Lewandowski, I., Scurlock, J.M.O., Lindvall, E., and Christou, M. 2003. The development and current status of perennial rhizomatous grasses as energy crops in the US and Europe. *Biomass and Bioenergy* 25, 335–361.

Li, S., Huang, L., Ke, C., Pang, Z., and Liu, L. 2020. Pathway dissection, regulation, engineering and application: lessons learned from biobutanol production by solventogenic clostridia. *Biotechnology for Biofuels* 13, 1–25.

Liew, Y.X., Chan, Y.J., Manickam, S., Chong, M.F., Chong, S., Tiong, T.J., Lim, J.W., and Pan, G.-T. 2020. Enzymatic pretreatment to enhance anaerobic bioconversion of high strength wastewater to biogas: a review. *Science of the Total Environment* 713, 136373.

Lima, M.A., Linhares, F.G., Mothe, G.A., Castro, M.P.P., and Sthel, M.S. 2017. Study of gaseous emissions derived from the combustion of diesel/beef tallow biodiesel blends. *Sustainability in Environment* 2, 210–222.

Low, T., Booth, C., and Council, I. 2007. *The Weedy Truth About Biofuels*. Melbourne: Invasive Species Council.

Low, T., Booth, C., and Sheppard, A. 2011. Weedy biofuels: what can be done? *Current Opinion in Environmental Sustainability* 3(1–2), 55–59.

Mahapatra, M.K., and Kumar, A. 2017. A short review on biobutanol: a second generation biofuel production from lignocellulosic biomass. *Journal of Clean Energy Technology* 5, 27–30.

Mahmood, H., Moniruzzaman, M., Iqbal, T., and Khan, M.J. 2019. Recent advances in the pretreatment of lignocellulosic biomass for biofuels and value-added products. *Current Opinion in Green and Sustainable Chemistry* 20, 18–24.

Mansur, M.C., O'Donnell, M.K., Rehmann, M.S., and Zohaib, M. 2010. *ABE Fermentation of Sugar in Brazil*. Senior Design Reports, Department of Chemical and Biomolecular Engineering, University of Pennsylvania, Philadelphia.

Martins, L.H.D.S., Neto, J.M., Gomes, P.W.P., De Oliveira, J.A.R., Penteado, E.D., and Komesu, A. 2019. Potential feedstocks for second-generation ethanol production in Brazil. In: *Sustainable Biofuel and Biomass*, A. Kuila (ed), pp. 145–166. Apple Academic Press. https://doi.org/10.1201/9780429265099

Mayank, R., Ranjan, A., and Moholkar, V.S. 2013. Mathematical models of ABE fermentation: review and analysis. *Critical Reviews in Biotechnology* 33, 419–447.

Moodley, P., Sewsynker-Sukai, Y., and Kana, E.B.G. 2020. Progress in the development of alkali and metal salt catalysed lignocellulosic pretreatment regimes: potential for bioethanol production. *Bioresource Technology* 310, 123372.

Nupueng, S., Oosterveer, P., and Mol, A.P.J. 2018. Implementing a palm oil-based biodiesel policy: the case of Thailand. *Energy Science & Engineering* 6, 643–657.

OECD-FAO Agricultural Outlook 2020–2029. www.oecd.org/publications/oecd-fao-agricultu ral-outlook-19991142.htm [Accessed on June 1, 2021].

Patakova, P., Linhova, M., Rychtera, M., Paulova, L., and Melzoch, K. 2013. Novel and neglected issues of acetone–butanol–ethanol (ABE) fermentation by clostridia: *Clostridium* metabolic diversity, tools for process mapping and continuous fermentation systems. *Biotechnology Advances* 31, 58–67.

Pecha, M.B., and Garcia-Perez, M. 2020. Pyrolysis of lignocellulosic biomass: oil, char, and gas. In: *Bioenergy*, Anju Dahiya (ed), pp. 581–619. Elsevier. https://doi.org/10.1016/ B978-0-12-815497-7.00029-4

Pérez, A., Ruiz, B., Fuente, E., Calvo, L.F., and Paniagua, S. 2021. Pyrolysis technology for *Cortaderia selloana* invasive species. Prospects in the biomass energy sector. *Renewable Energy* 169, 178–190.

Petri, L. 2019. Invitation. ESA Annual Meeting (August 11–16), ESA.

Pimentel, D. 2001. Economic and environmental impacts of invasive species and their management. *Pesticides and You* 21, 10–11.

Rajakaruna, S.L., Ranawana, K.B., Gunarathne, A., and Madawala, H. 2017. Impacts of river regulation and other anthropogenic activities on floodplain vegetation: a case study from Sri Lanka. *Community Ecology* 18, 203–213.

Rajmohan, K.V.S., Ramya, C., Viswanathan, M.R., and Varjani, S. 2019. Plastic pollutants: effective waste management for pollution control and abatement. *Current Opinion in Environmental Science & Health* 12, 72–84. https://doi.org/10.1016/ j.coesh.2019.08.006

Ravindranath, N.H., and Hall, D.O. 1995. *Biomass, Energy and Environment: A Developing Country Perspective from India*. Oxford: Oxford University Press.

Rouhany, M., and Montgomery, H. 2019. Global biodiesel production: the state of the art and impact on climate change. *Biodiesel* 8, 1–14.

Schreinemachers, P., and Berger, T. 2006. Land use decisions in developing countries and their representation in multi-agent systems. *Journal of Land Use Science* 1, 29–44.

Shah, K.K., Tiwari, I., Tripathi, S., Subedi, S., and Shrestha, J. 2020. Invasive alien plant species: a threat to biodiversity and agriculture in Nepal. *Agriways* 8, 62–73.

Sharma, S., Basu, S., Shetti, N.P., Kamali, M., Walvekar, P., and Aminabhavi, T.M. 2020. Waste-to-energy nexus: a sustainable development. *Environmental Pollution* 267, 115501.

Silalertruksa, T., and Gheewala, S.H. 2010. Security of feedstocks supply for future bio-ethanol production in Thailand. *Energy Policy* 38, 7476–7486.

Smith, V.H., and D.W. Schindler. 2009. Eutrophication science: where do we go from here? *Trends in Ecology and Evolution* 24(4), 201–207. https://doi.org/10.1016/ j.tree.2008.11.009

Solé-Bundó, M., Passos, F., Romero-Güiza, M.S., Ferrer, I., and Astals, S. 2019. Co-digestion strategies to enhance microalgae anaerobic digestion: a review. *Renewable and Sustainable Energy Reviews* 112, 471–482.

Su, T., Zhao, D., Khodadadi, M., and Len, C. 2020. Lignocellulosic biomass for bioethanol: recent advances, technology trends and barriers to industrial development. *Current Opinion in Green and Sustainable Chemistry* 24, 56–60.

Sukarni, S., Zakaria, Y., Sumarli, S., Wulandari, R., Permanasari, A.A., and Suhermanto, M. 2019, April. Physical and chemical properties of water hyacinth (eichhornia crassipes) as a sustainable biofuel feedstock. *IOP Conference Series: Materials Science and Engineering* 515(1), 12070.

Syafiuddin, A., Hao, C.J., Yuniarto, A., and Hadibarata, T. 2020. The current scenario and challenges of biodiesel production in Asian countries: a review. *Bioresource Technology Reports* 12, 100608.

Tao, L., Ahmad, A., de Roode, J.C., and Hunter, M.D. 2016. Arbuscular mycorrhizal fungi affect plant tolerance and chemical defences to herbivory through different mechanisms. *Journal of Ecology* 104(2), 561–571. https://doi.org/10.1111/1365-2745.12535

Thamsiriroj, T., and Murphy, J.D. 2011. The impact of the life cycle analysis methodology on whether biodiesel produced from residues can meet the EU sustainability criteria for biofuel facilities constructed after 2017. *Renewable Energy* 36, 50–63.

Tinker, P.B. 1997. The environmental implications of intensified land use in developing countries. *Philosophical Transactions of the Royal Society of London. Series B: Biological Sciences* 352, 1023–1033.

US Department of Agriculture. 2021. www.ers.usda.gov/data-products/us-bioenergy-statistics/ [Accessed on May 31, 2021].

US Energy Information Administration. 2021. *Renewable & Alternative Fuels*. www.eia.gov/renewable/data.php#alternative [Accessed on June 31, 2021].

Van Meerbeek, K., Appels, L., Dewil, R., Calmeyn, A., Lemmens, P., Muys, B., and Hermy, M. 2015. Biomass of invasive plant species as a potential feedstock for bioenergy production. *Biofuels, Bioproducts and Biorefining* 9(3), 273–282. https://doi.org/10.1002/bbb.1539

Van Meerbeek, K., Muys, B., and Hermy, M. 2019. Lignocellulosic biomass for bioenergy beyond intensive cropland and forests. *Renewable and Sustainable Energy Reviews* 102, 139–149.

Von Braun, J., and Pachauri, R.K. 2006. *The promises and challenges of biofuels for the poor in developing countries: IFPRI 2005–2006 Annual Report Essay*. Washington, DC: International Food Policy Research Institute.

Witt, A.B.R. 2010. Biofuels and invasive species from an African perspective—a review. *GCB Bioenergy* 2, 321–329.

Young, S.L., Gopalakrishnan, G., and Keshwani, D.R. 2011. Invasive plant species as potential bioenergy producers and carbon contributors. *Journal of Soil and Water Conservation* 66, 45A–50A.

Zabed, H.M., Akter, S., Yun, J., Zhang, G., Awad, F.N., Qi, X., and Sahu, J.N. 2019. Recent advances in biological pretreatment of microalgae and lignocellulosic biomass for biofuel production. *Renewable and Sustainable Energy Reviews* 105, 105–128.

Zhu, P., Abdelaziz, O.Y., Hulteberg, C.P., and Riisager, A. 2020. New synthetic approaches to biofuels from lignocellulosic biomass. *Current Opinion in Green and Sustainable Chemistry* 21, 16–21.

10 Upcycling

A New Perspective on Waste Management in a Circular Economy

*Randika Jayasinghe[1]*and P. H. L. Arachchige[2]*
[1]University of Sri Jayewardenepura, Sri Lanka
[2]Monash University, Australia
*Corresponding author: randika@sjp.ac.lk

CONTENTS

DOI: 10.1201/9781003132349-11

10.1   INTRODUCTION

Waste is a global problem that affects every aspect of our lives and the environment we live in. Disposing of waste—out of sight, out of mind—is not an option anymore as many urban cities around the world have reached their limits to capacity (Guerrero, Maas, and Hogland 2013). It is not only the volume of items we consume and dispose of that is problematic but the complexity of the products has also changed. Over the last century, products have shifted from simple natural items that are biodegradable to products that are complex and difficult to biodegrade. The products we use on a daily basis remind us of the resource-intensive nature of our everyday lives and the pressure it exerts on the ecosystem (Hawken 2018). The discarded materials generated due to this wasteful consumption and their haphazard disposal pollute our water sources and the ocean and contribute to air pollution and global warming, disrupting the ecological balance. During the past few decades, the unsustainable production and consumption of materials and products have led to widespread concerns, including resource depletion, scarcity of raw materials, environmental degradation, and anthropogenic climate change.

The world needs a transition to a circular economy (CE), where waste is reduced or eliminated through the development of new business models, eco-designs, and product life extension (Ellen MacArthur Foundation 2013). The CE concepts have seen significant growth in recent years, following an upsurge of interest in practices such as waste upcycling that encourage the manufacturing of value-added products from discarded materials. The economic burden of waste disposal and a growing consciousness of the environment also contribute to the increasing interest in alternative approaches to conventional waste management practices. Upcycling can be identified as reusing discarded objects or materials to create a product of higher quality or value than the original (Sung et al, 2018).

Upcycling, also known as creative reuse or creative recycling, transforms waste materials or unwanted products into new materials or products with greater quality and value (Caldera et al. 2020; Sung et al. 2019). It encourages us to see materials and products through a new lens—as having the potential to be something valuable. The term "upcycling" gained recognition after William McDonough and Michael Braungart incorporated the word in their book *Cradle to Cradle: Remaking the Way We Make Things*, initially published in 2002. The authors described upcycling as a process of converting a material that has completed its purpose into something of similar or greater value in its second life (McDonough and Braungart 2009). Upcycling contributes to reducing the consumption of new raw materials when creating products. This, in turn, will reduce energy usage, air and water pollution, and even anthropogenic greenhouse gas emissions associated with continuous extracting and processing of new raw materials (McDonough, Braungart, and Clinton 2013).

Moreover, upcycling is not a new approach. It has been practiced since prehistoric times. Upcycling has captured the attention of environmentally conscious individuals as a sustainable solution for a long time. However, mass production and easy access to goods and services that came about with the industrial revolution slowly detached people from the need to upcycle used products. The growing environmental concerns about resource availability, overconsumption and production, and increasing waste volumes have now motivated the industries and communities to follow the upcycling trend. Bridgens et al. (2018) identify it as a step toward a regenerative design culture. Upcycling has many benefits, including increased quality and lifetime of materials and products, reducing wastes, creating new employment opportunities, supporting new social enterprises, and encouraging sustainable consumption (Jagdeep Singh et al. 2019).

In developing countries, upcycling has the potential to drive the transition toward a circular economy by reducing the dependency on virgin raw materials and energy-intensive processes (Conlon, Jayasinghe, and Dasanayake 2019). Therefore, this chapter aims to explore waste upcycling—converting discarded material into value-added products in a developing context. Furthermore, this chapter will help to have a deeper reflection on applying simple solutions to managing different types of wastes, especially for developing countries. The chapter mainly focuses on low-tech, low-energy upcycling that is much more applicable to a developing country. This type of upcycling focuses on retaining the value of the waste material in its original state. However, a few examples of advanced resource-intensive upcycling methods are also mentioned to demonstrate the potential of upcycling on a commercial scale.

10.2 ADDING VALUE THROUGH DESIGN AND CREATIVITY

The tension caused by the population growth is heightening due to the improvement of modern standards of living. At a glance, these advanced standards may look like a positive impact on human well-being, while, in-depth, improvement of living standards and resource consumption have a proportionate relationship. Population growth and industrialization have been accelerating unsustainable resource extraction around the world. Recent studies have suggested that current global resource extraction rates are ten times more than the rates at the beginning of the last century, while global material stock accumulation has also increased 23-fold (Krausmann et al. 2017; Wiedenhofer et al. 2019). In 1972, Meadows and her team projected that the human–nature interacting system could start collapsing around the mid-21st century (Meadows, Meadows, and Randers 1972). Later in 2004, they substantiated their projection, revealing the data from 1970 to 2000 (Meadows, Randers, and Meadows 2004). This kind of revelations raise the dire question, "is collapse likely and imminent?"

10.2.1 VALUE RETENTION THROUGH CIRCULAR ECONOMY

It is clear that the current 'take-make-use-dispose' economic model is not sustainable. Generally, the resources extracted to be utilized for economic purposes are not literally "consumed". Instead, they generate waste residuals that do not disappear in the system, resulting in severe environmental damages and unpaid social costs (Ayres and Kneese 1969; Di Maio et al. 2015). Hence, we need alternative economic models comprised of regenerative and restorative approaches to replace resource-intensive models. The CE is an emerging concept built upon the idea of closed-loop systems, predominantly aiming to economically and efficiently circulate resources and materials, including technical and biological nutrients, as long as possible in the anthroposphere (Conlon, Jayasinghe, and Dasanayake 2019). The CE has the potential to replace the traditional economic system because it attempts to extract the value of the resources as much as possible through multiple life cycles (Figure 10.1).

In a CE, products that no longer serve their functions are also retained as resources in the economic system to be reused as inputs to enable further value creation. Hence, CE models surpass traditional linear models by creating more value from a unit of resources. Contemporary terminologies (end-user, supply chain, and value chain) embedded in industrial operations itself narrate a linear process where transiting to a more resource-efficient circular model will benefit the operations in several ways. Extensive research studies carried out in the European context highlight that transition to a circular economy model that could save up to $380 billion from annual net material costs, which can be advanced up to $630 billion in an optimal transition only considering a subset of manufacturing sectors (Ellen MacArthur Foundation 2013).

The CE concept concentrates on value retention through integrated resource management while shifting toward notions like industrial symbiosis. Industrial symbiosis is an offshoot of industrial ecology, which looks into how collocated industries can synergize and utilize energy and materials in the best possible way. Industrial symbiosis facilitates industries to work together to reduce material and energy intensity. In this scenario, the accumulation of waste residuals can be turned into a resource-generating opportunity by adding value to waste as a resource. This conceptualization renegotiates the perceptions of coupling waste with a lack of value in most traditional economic models (Camacho-Otero, Boks, and Pettersen 2018). Therefore, in the CE, waste acts as a repurposable resource or input in another system, redefining the waste as a resource generated at the "wrong place at the wrong time" (Conlon, Jayasinghe,

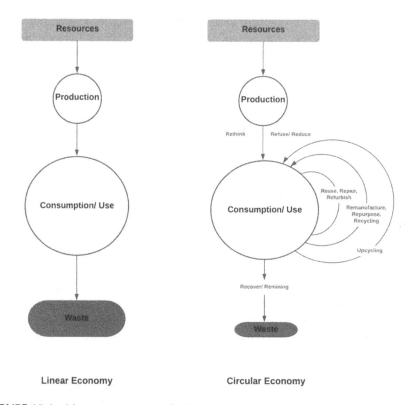

Linear Economy   **Circular Economy**

FIGURE 10.1 Linear economy to a circular economy.

and Dasanayake 2019). To explore how these waste sources could be redefined as inputs within the system or in another system, Reike et al. (2018) examined various value retention options (ten categories of recycling materials) that could be associated with the CE:

- The shortest loop of value retention—R0–R3: refuse, reduce, reuse, and repair;
- Medium length loop of value retention—R4–R6: refurbish, remanufacture, and repurpose; and
- The most resource-intensive way of value retention—R7–R9: recycling materials, recovering energy, and remining.

In value retention options R2–R5, the functionality of the product does not change. Under R6 (repurpose), a product can be used for a different purpose than originally intended, thus providing it with another lifecycle while, in some cases, only a part of the product could be repurposed. The last value retention option R9—remining route—considers the resources that are already present and growing in landfills. Options R0–R3 require significant behavioral change on the consumer side, and for any existing production unit, it calls for a thorough redesigning of the processes. Thus, these are the most difficult to implement. As we move up the ladder of retention options, R4–R6 are comparatively easier to adopt; however, they require more resources. R7–R9 options are the end of pipeline solutions, are easiest to implement, and are the most resource-intensive. Upcycling is positioned between the shortest loop of value retention (R0–R3: refuse, reduce, reuse, and repair) and the medium-length loop of value retention (R4–R6: refurbish, remanufacture, and repurpose). Thus, it can be argued that upcycling supports the transition from a linear to a circular economy by slowing down or closing the resource loops, designing out waste, and circulating resources in the production process.

10.2.2 Upcycling

Upcycling is the process of repurposing used or waste materials and discarded products to remake higher quality products than the original materials (Sung et al, 2019). Bridgens et al. (2018) identified upcycling as a creative process that connects people with the products and materials that they use and the environment they live in. Upcycling is also known as creative recycling. However, conventional recycling is considered a downcycling process due to the creation of lower-grade products or loss of quality of the material (Leal Filho et al. 2019). For instance, a study conducted by Mahpour in 2018 adopting a CE in construction and demolition waste management has recommended replacing recycling with reusing and upcycling practices to prevent unwanted recycling of micropollutants (Mahpour 2018). Moreover, recycling waste materials requires more energy and resources compared to upcycling (Steinhilper and Hieber 2001).

Waste management through upcycling brings a plurality of benefits to developing countries. It has the potential to promote a circular economy thinking at the national level (Dissanayake and Sinha 2015), support economic growth, promote the use of eco-friendly products, and support the livelihoods of many through entrepreneurial opportunities and job creation (Sung et al, 2019). From an environmental perspective, the conversion of waste into useful products supports sustainable production (Cumming 2016). Waste is given a second life where the lifetime and productivity of those materials are prolonged. This leads to less waste being dumped in open spaces, landfills, and waterways, reduce the requirement to extract virgin raw materials, and preserve natural resources (Ali et al, 2013).

Despite the benefits it creates and the availability of materials, upcycling is still considered a niche activity and requires scaling up to make a significant impact on the environment and society (Sung et al, 2017; 2019). Individuals, designers, and small-scale entrepreneurs are faced with many challenges in upcycling value chains. One of the main challenges is securing a continuous flow of quality materials (Singh et al. 2019). In addition, contaminated materials hinder the upcycling process and reduce the marketable value of the upcycled products (Bridgens et al. 2018). The availability of low-priced mass-produced products is also a challenge to secure a sustainable market for upcycled products. Moreover, the negative perception of using waste materials often discourages consumers from purchasing upcycled products. Therefore, creating awareness about the upcycling process and products, introducing a variety of upcycled products, and using an effective marketing strategy can generate a positive attitude toward upcycling (Singh et al. 2019).

10.3 UPCYCLING PROCESS

10.3.1 Process Flow

The upcycling process encourages individuals and designers to "think out of the box" and stimulate their creative and critical thinking. This leads to a unique outcome created by converting waste materials into creative products, contributing to a higher environmental value (Ali et al, 2013). The process, however, is not linear. There are many steps to follow in creating an upcycled product or a business. Upcycling is being practiced at the product and industrial material levels by individuals, designers, social enterprises, small and medium enterprises (SMEs), and industries (Sung et al, 2018). There are differences between small-scale upcycling by individuals, designers, local small business entrepreneurs to large-scale upcycling by industries. However, in both approaches, the process flow is the same (Figure 10.2).

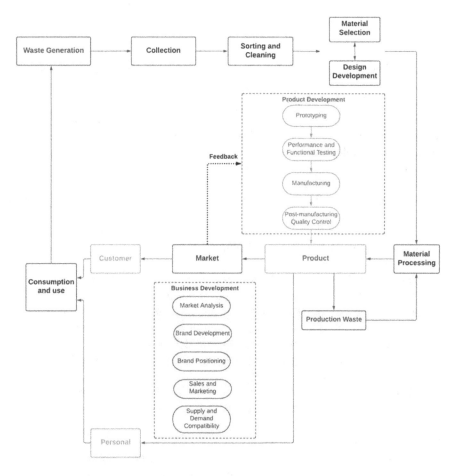

FIGURE 10.2 Upcycling process flow.

10.3.2 Critical Steps to Consider in the Upcycling Process

10.3.2.1 Material Flow Analysis and Sorting

The first step of the process is to carry out a material flow analysis to identify the types of materials and quantities available for upcycling. Identifying available waste materials from business enterprises and industries helps create long-term cooperation beneficial for both the industry and the designer. Then the industry will dispose of the waste materials, and the designer will gain access to clean waste materials. The distance to the source of materials, transport costs, and storage facilities at the production site are important factors to consider during this stage. After the materials are collected, they are sorted, cleaned, and stored for upcycling.

10.3.2.2 Design Development and Material Processing

In the design development stage, several designs are sketched based on the initial needs assessment. The type of materials required for production will be tentatively decided at this stage. Material selection begins with identifying and prioritizing critical design criteria or vice versa. For instance, readily available material can inspire a designer to develop a suitable product design. In individual or small-scale upcycling, materials are usually processed manually using basic tools and low-cost equipment. However, advanced resource-intensive methods are used in industrial upcycling processes.

10.3.2.3 Product Development

The product development process includes prototyping, testing, manufacturing, and quality control. Since upcycling is a learning process, the trial-and-error method in initial product development is essential to achieve a unique outcome at the end. At the beginning of the manufacturing process, several product prototypes are created to familiarize with the materials and the final design. Prototyping also helps to develop proper templates and molds useful for manufacturing products at a later stage. Then these prototyped products are tested for performance and functional testing. The selected designs and methods are perfected prior to starting the actual manufacturing process. Waste generated at the manufacturing stage is recirculated through the process to reduce waste generation.

It is essential to plan the production steps and pay particular attention to the quality of the product. When it comes to continuous production, the procedures used have to be standardized to ensure the quality is maintained throughout the production process. Further, post-manufacturing quality control of products is essential to identifying marketable or usable products and reject defective products. If a product is designed for personal use by an individual or a designer, quality control may not be that important as the product is not meant to be sold to a customer. According to Sung et al. (2014), upcycling is a journey for an individual in which the process is often more valued than the outcome.

10.3.2.4 Business Development

An individual can upcycle waste materials for personal use, such as furniture for the house, jewelry, artwork, or even an upcycled clothing item. However, there is an increasing interest to engage in upcycling as an entrepreneurial opportunity. An entrepreneur interested in an upcycling startup needs to consider specific criteria to develop a viable business model. These include conducting a detailed market analysis, creating a brand, positioning the brand, sales and marketing, and analyzing the supply and demand compatibility.

10.4 RECREATING VALUE-ADDED PRODUCTS FROM WASTE

Upcycling waste materials and products provide scope to create new objects with high monetary value. Moreover, it promotes concepts such as "small is beautiful" and "think global, buy local" (Wegener and Aakjær 2016). Upgrading small-scale upcycling to a commercial scale is still in its infancy in Sri Lanka. However, local examples identified in this chapter show potential in creating viable upcycling businesses that can be sustainable in developing countries.

10.4.1 PLASTICS

Plastics, often referred to as a wonder material, is essential in everything from simple household items, clothing, and automotive to medical equipment and electronics manufacturing (Shen and Worrell 2014). The worldwide generation of plastic waste increases daily and, once discarded, persists in the environment for a very long time. Open dumping of plastic waste leads to coastal and marine pollution, generation of micro and nano plastics, visual pollution, and breeding grounds for vector-borne diseases. Open burning of plastic waste is a major source of air pollution, releasing toxic gases including but not limited to dioxins, furans, and other harmful gases into the atmosphere (World Economic Forum, Ellen MacArthur Foundation 2016). Therefore, alternative practices of plastic waste management are urgently called for.

The plastic recycling process requires collecting, sorting, cleaning, melting, and manufacturing new products. However, the properties of plastics are significantly reduced after several recycling cycles (Kopnina 2018). Therefore, conventional recycling strategies are considered as downcycling in which the end products usually have a lower value than the original material (Shen and Worrell 2014). On the other hand, plastic waste can be converted to value-added materials using less energy and resources through upcycling. As such, upcycling has received attention as a sustainable alternative approach to recycling plastics. Upcycling requires clean waste materials. Therefore, collecting waste plastics, especially clean packaging materials, directly from industries is essential for upcycling. Contaminated plastic waste that does not fall under the hazardous waste category can be used if thoroughly washed and dried before use in the manufacturing process. Plastics can be upcycled into various products such as furniture, light fixtures, landscape and garden items (Figure 10.3), bags, storage containers, or artwork (Figure 10.4) using simple tools and manufacturing processes.

Plastics can also be upcycled into fuel, adhesives, chemicals, and even textile yarn using thermal and chemical processes (Celik et al. 2019). However, these methods would require more resources and energy. A successful example of this in Sri Lanka

FIGURE 10.3   Upcycled garden items.

FIGURE 10.4   Upcycled artwork.

is Eco Spindles, a local company operating a state-of-the-art polyester yarn plant in which yarn is produced from recycled polyethylene terephthalate (PET). The process creates yarn directly from PET flakes bypassing polymerization, where flakes are first converted to pellets and then to yarn (Figure 10.5). MAS Holdings, one of the leading textile manufacturers in the country, Eco Spindles, and the Sri Lankan Navy, has successfully upcycled waste plastics recovered from the beaches of Sri Lanka to produce the official jerseys of the Sri Lankan National Cricket Team for the ICC Cricket World Cup in 2019 (Figure 10.6).

10.4.2 TEXTILES

The textile industry is considered one of the most polluting industries globally, with extensive resource utilization and waste generation (Ellen MacArthur Foundation

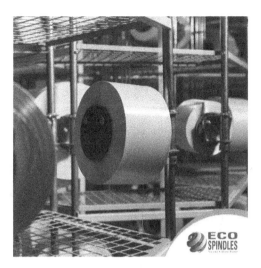

FIGURE 10.5   PET converted to polyester yarn.

FIGURE 10.6   Official jerseys of the Sri Lankan national cricket team upcycled from PET bottles collected from the beaches of Sri Lanka.

2017; Leal Filho et al. 2019). The need for sustainable, innovative solutions for managing the large volumes of waste associated with the textile industry is becoming urgent. Textile upcycling is considered an essential process within the sustainable fashion movement that actively seeks solutions to the industry's social and environmental issues (Dissanayake and Sinha 2015; Cuc and Tripa 2018; Jagdeep Singh et al. 2019).

Textile upcycling is widely practiced as a home-based industry in Sri Lanka. These small businesses are run mainly by women, especially single mothers, older women, and women who cannot find other work due to household responsibilities. They collect or buy textile pieces from various sources to make different upcycled products at home. These home-based businesses upcycle all types of textile waste— pre-consumer waste, production waste, and post-consumer waste. However, their

FIGURE 10.7   Upcycled home-based textile products.

design options are often decided by the limited fabric choices and the varying sizes of the fabric pieces collected. Upcycled textile products include new clothing items, jewelry, soft toys, bags, wallets, and rugs that are usually sold in the local market (Figure 10.7).

There are also successful local design brands such as *House of Lonali* who promotes upcycling textile waste. *House of Lonali* works closely with the rural cottage industry in Sri Lanka to convert pre-consumer textile waste and excess materials into unique fashionable clothing and accessories (Figure 10.8). This provides an opportunity for rural women to become a part of the high-end fashion world while generating a sustainable income.

10.4.3   Rubber Products

Rubber finds a wide field of application such as thermal insulation, flexible tubing, tires and tubes, adhesive, and hose pipes because of the material's beneficial properties. Millions of tires are discarded each year and end up in landfills. Tires have been proven to be excellent breeding grounds for mosquitoes and can also leach toxic chemicals into the environment (Walls 2013). Tires and other rubber products are recycled into crumbs and used in different applications such as crumb rubbers, playground covers, and sporting surfaces (Cerminara and Cossu 2018). However, rubber recycling requires specialized equipment and is not actively practiced everywhere. Therefore, upcycling rubber products, especially tires and inner tire tubes, into different products such as furniture, landscape designs, and accessories is a viable alternative (Figure 10.9).

FIGURE 10.8 Upcycled products by House of Lonali.

FIGURE 10.9 A landscape structure made out of tires.

10.4.4 PAPER AND CARDBOARD

Paper is a commodity widely used for printing magazines, newspapers, books, and as packaging material. Even amidst paperless campaigns, the demand for paper has not ceased and continues to this day. Paper comes in different forms such as packaging paper (case materials, carton board, wrapping papers, and other packaging papers),

FIGURE 10.10   Upcycled paper products by Earthbound Creations.

graphic paper (newsprint and other graphic paper), tissue paper (household and sanitary), and paper for technical and other purposes (Grossmann, Handke, and Brenner 2014). Any of these paper materials (except for used tissue paper and contaminated paper) can be used for upcycling. It is essential to source clean materials because, unlike plastics and glass, paper cannot be cleaned by washing prior to upcycling.

There are successful social entrepreneurs in Sri Lanka manufacturing upcycled paper and cardboard products for local and international markets. A company named Earthbound Creations manufactures a range of upcycled products from newspapers such as pencils, handicrafts, bowls and bins, and paper bags (Figure 10.10). Eco Maximus, another local company, manufactures value-added products from elephant dung and wastepaper. They use leftover paper bought from waste collectors and remains from their paper manufacturing process. Elephant dung and wastepaper are first recycled back to sheets of paper and converted to various products such as pen holders, notebooks, jewelry, and ornaments (Figure 10.11). Another company uses wastepaper pulp mixed with fiber materials such as banana skins and pineapple leaf fiber to make paper and value-added products.

10.4.5   Wood and Timber

Wood and timber are abundantly available natural resources used mainly in the construction industry. However, widespread deforestation and increasing climate change issues have prompted timber suppliers, manufacturers, and consumers to look for more sustainable timber sources to cater to their demands (Dodoo, Gustavsson, and Sathre 2014). Reusing, recycling, and upcycling wood reduce the need to cut down trees. Wood and timber materials can be reused as new materials for manufacturing different items, recycled into wood chips for various uses, or used for energy production. Wood and timber-based products that have come to the end of the product life can also be upcycled to give them a second life.

FIGURE 10.11   Upcycled paper products by Eco Maximus.

FIGURE 10.12   Furniture made out of upcycled wooden pallets.

Any wood-based material that is not contaminated with hazardous chemicals such as wooden windows or door frames, furniture, pallets, crates, and wood from demolition sites can be upcycled. The main methods for wood upcycling include repainting, repurposing, and refurbishing wood products. For example, furniture made from salvaged materials and then reassembled into new furniture with unique designs have good demand. One of the wood waste materials that is often upcycled in Sri Lanka is wooden pallets. Pallets can be used as a platform or flooring, displaying racks on the wall or furniture (Figure 10.12). Many local small-scale manufacturers use cheap wooden materials such as pallets and crates to make trendy furniture pieces that are sold at a much higher price.

FIGURE 10.13   Upcycled bottle light fixture.

10.4.6 Glass

Glass has made its way into every aspect of human activity, from packaging, vehicles, housewares, electrical equipment, insulation products to advanced science and technology applications. It is an infinitely recyclable material and can be remelted and reformed into products without loss in quality (Dyer 2014). Glass is a material that is reusable in many of its forms. For example, bottles and other glass jars can be used as containers long after their original use is over. Reusing and upcycling glass containers can generate significant savings in energy and resources. Upcycling glass containers is a practice that has become well established in many parts of the world. Many entrepreneurs are interested in upcycling glassware as a viable alternative to costly and resource-intensive glass recycling.

In upcycling, glass containers are first segregated according to quality, followed by a rigorous cleanup. These glass bottles and containers are then converted to products with a different purpose or use. Glass bottles and containers can be upcycled into many products such as light fixtures and lampshades (Figure 10.13), furniture, flower pots and vases, and jewelry. In some cases, the bottles are cut, edges are buffed to soften the rim, and polished, followed by a final cleaning to convert them to artwork or interior design pieces (Figure 10.14). Glass bottles are also used as building material and for outdoor landscape applications.

Glass upcycling can also be done using a technique called kiln forming (Adjei, Opoku-Bonsu, and Asiamah 2016). A kiln is an oven where the heat changes the solid glass into its molten state. At this stage, the glass is molded again into the desired shape. The kiln-forming process is expensive because of the high cost of the equipment and high electricity consumption. However, some glass manufacturing industries mentioned using their kilns to make upcycled products from defective

FIGURE 10.14   An interior design piece made out of glass pieces.

pieces rejected during the manufacturing process. Defective products will fetch a lower price in the market due to their poor quality. But the upcycled products can be sold at a much higher price due to their artistic value.

10.4.7   Metal Products

Unlike plastics, metals can be recycled repeatedly without altering their properties. Hence, metal recycling has been regarded as a profitable business. In Sri Lanka, metal fetches the highest price in the recycling market. However, there is a growing trend to use old metal items to produce scrap metal artwork, primarily for interior design applications. Metal upcycling requires specialized tools for cutting, polishing, and welding different parts to make a final product. Many metal waste products such as aluminum cans, barrels, automotive parts, and construction and demolition waste are upcycled to create artwork (Figure 10.15), sculptures, and furniture. For example, a local company called Barrel Chairs is manufacturing furniture using old oil drums (Figure 10.16). Further, old discarded electronic items can also be upcycled after removing the circuits and the wires (Figure 10.17).

10.5   CASE STUDY: THE WASTE FOR LIFE SRI LANKA PROJECT

10.5.1   Australia–Sri Lanka University Partnerships to Develop Community-Based Waste Upcycling Businesses in Sri Lanka

Complex issues surrounding sustainability have evolved in Sri Lanka more recently, particularly due to increased accumulation and poor waste management. The recent tragedies like the collapse of the Meethotamulla open dump in 2017 and the accelerated

FIGURE 10.15   Furniture made out of upcycled wooden pallets.

FIGURE 10.16   Sofa set made out of old oil drums by Barrel Chairs.

loss of wild animals due to plastic and polyethylene ingestion are clear signs of irresponsible waste management in Sri Lanka. The dominating attitude toward disposal arises from the perception of "waste as an unwanted material with no intrinsic value". A comprehensive longitudinal feasibility study conducted in Sri Lanka from 2011 to 2014 revealed the availability of opportunities to promote local economic development through upcycling waste resources (Jayasinghe 2015). It pointed out the existence of a multilayered network of stakeholders comprising individuals, recyclers,

FIGURE 10.17 Upcycled old electronic products.

social enterprises, and community-based organizations who rely on waste as their livelihoods. Their study also highlighted that the lack of technical knowledge, skills, affordability of appropriate machinery, proper designing protocols, and management strategies had hindered the quality and the effectiveness of the outcomes. Enabling and facilitating these skills are crucial to overcoming the challenges associated with existing small-scale ventures (Thamae and Baillie 2009).

Based on the findings of this feasibility study, the University of Western Australia, together with the not-for-profit organization Waste for Life, collaborated with three Sri Lankan universities (the University of Moratuwa, University of Jaffna, and the University of Sri Jayewardenepura) aiming to develop waste-based upcycling programs to educate and support community-based waste recycling and manufacturing avenues. Besides educational and training motives, the project also aimed to positively impact local economies to reduce poverty and promote environmental health. The project linked universities and communities by creating a hub that effectively integrates formal and informal education platforms to disseminate applied knowledge to the general public. Deviating from the traditional broad-focus education and training models, this project intended to equip participants with knowledge and training to instigate startup social enterprise by the end of the program. Initially, the community groups had one-on-one training sessions to provide hands-on experience on materials and product development. Later, after carefully analyzing the market and availability of resources, the project team worked with the groups to develop unique products and executed tailor-made brand positioning strategies. The project successfully established two design centers and three community projects at the end of the project in 2017 in the Northern and Western provinces.

10.5.2 Needs Assessment and the Feasibility Studies

Often, the efficacy of projects heavily depends on the informed decisions that emerged from feasibility studies and contextual needs. Feasibility studies evaluate the technical, socioeconomic, and environmental practicality or viability of a project

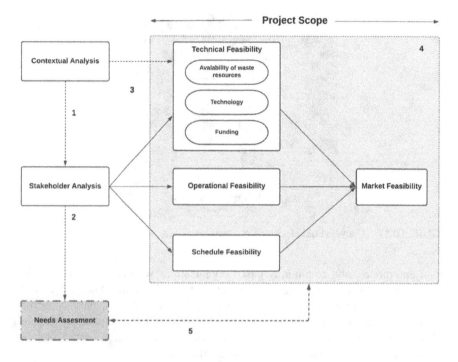

FIGURE 10.18   The participatory needs and feasibility analysis.

Note: This figure illustrates the aspects considered before implementing the project in Sri Lanka.

to make a conversant decision about the possibility of project execution (Jayasinghe and Baillie 2017; Karagiannidis et al. 2009). There is no one established structure to conduct a feasibility study; instead, they are specific and unique to each project. Accordingly, this project also followed a distinctive feasibility study (Table 10.1) to assess the practicability of waste-based upcycling in the Sri Lankan context. This project's need and feasibility study considered several aspects based on the project focus and contextual attributes (see Figure 10.18). The assessment comprised seven elements: (1) contextual analysis, (2) stakeholder analysis, (3) technical feasibility, (4) operational feasibility, (5) schedule feasibility, (6) market feasibility, and (7) needs assessment.

10.5.2.1   Contextual Analysis

This analysis was the initial step of the feasibility study. It aimed to identify and map the contextual factors that might affect the implementation and sustainability of the project (Nilsen and Bernhardsson 2019). A careful consideration of social, economic, political, and environmental factors enabled identifying opportunities and challenges entrenched to upcycling in the Sri Lankan context. In addition to the above, it was essential to create a complete "waste" profile of the local context to understand the nature of waste generation, accumulation, management, attitudes toward waste, and

TABLE 10.1
Summary of the Feasibility Study

Features Elements	Aim	Challenges	Recommendations
Contextual analysis	To analyze the holistic view of the context.	Spans over a broader range of disciplines and require augmented risk analysis. Requires time and experts who are familiar with the different disciplines and contexts.	Longitudinal PESTLE and SWOT analysis. Extending the project collaborations among different organization and institutions.
Stakeholder analysis	To discover stakeholders, their roles and how they are affected.	Difficulties in predicting stakeholder resistance toward change. The complexity of stakeholder politics. Hindered efforts of potential stakeholders due to the experience of previously failed waste management projects.	Approach via stakeholders and experts who already involved in the field. Launch community education programs.
Technical feasibility	To assess the availability of waste resources, accessibility to technology, and the obtainability of funding sources.	Lack of up-to-date databases on waste generation in developing countries. Barriers in accessing locally manufactured technological equipment. Lack of continual funding options for novel projects.	Mapping the waste profile of the context during the contextual analysis stage. Approach and encourage local technological providers. Collaborate with universities to promote waste-based R&D.
Operational feasibility	To assess the capacity to utilize, support, and perform the project tasks.	Unpredictability of prospect organizational conflicts and policies. Social acceptability of novel waste management approaches. Inadequacy of government regulations on proper waste collection. Lack of community–university knowledge-sharing channels.	Conduct monthly stakeholder meetings to brainstorm and to receive feedback on the current and desired status of the project. Collaborate with stakeholders in government institutions. Use of social media platforms and forums to disseminate awareness on upcycling.

(continued)

TABLE 10.1 (Continued)
Summary of the Feasibility Study

Features Elements	Aim	Challenges	Recommendations
Schedule feasibility	To evaluate the possibility of the project completion across the scheduled timeline.	Not having previous data related to similar projects. Complexity of quantifying the consistency of technology and human resource availability throughout the project timeline.	Use of decision science tools to navigate via project stages and tasks. Invest in training the trainers.
Market feasibility	To identify market opportunities and challenges.	Novelty of the concept and not being imitated for similar projects in the market. Price-conscious buying behaviors and attitudes in developing countries. Uncertainty of market timeliness.	Use crowdsourcing platforms to collect data. Analysis of buying behaviors and trends via similar product markets. Conduct test marketing campaigns.
Needs assessment	To analyze the gaps between current and desired conditions.	Inconsistency of needs and wants over time. Availability of trained personnel for longitudinal need assessments.	Open public communication channels to share knowledge between the general public and the collaborating institutions. Identify training needs and develop training plans and performance profiling.

possible alternative approaches. Even though contextual analysis was a pre-project stage, new factors often emerged during the project's functioning stage.

10.5.2.2 Stakeholder Analysis

Stakeholders either can influence or be affected by a particular issue or an action of a project (Gregory et al. 2020). The stakeholder analysis focused on discovering these actors, their roles, and how they are affected. The project team identified different stakeholders who supported or expressed their interest in building an alliance with the proposed project. There are three categories: primary stakeholders, secondary stakeholders, and external stakeholders (Da Camara 2006). These stakeholders later mapped into a network (Table 10.2) as an additional aid to explore their relationships with each other and, more importantly, to analyze the material flow inside this network. The key stakeholders involved in the upcycling system were identified as social enterprises, informal waste sector, local authorities, not-for-profit organizations, private companies that generate waste, retail businesses that sell upcycled products and

TABLE 10.2
Network of Stakeholders

Network of Stakeholders	Types of Entities
Primary stakeholders	Informal waste workers
	Communities in low-income settlements
	Women
	Local universities
Secondary stakeholders	Non-for-profit organizations
	Government and semi-government authorities
	Private organizations
	• Private organizations with foundations working on environment-related activities
	• Private waste management companies
	• Retail and B2B businesses
	Waste generators
	• Households and commercial establishments
External stakeholders	Research and development organizations

customers—both individuals and organizations who prefer to purchase upcycled products.

10.5.2.3 Technical Feasibility

Technical feasibility assessed the availability of waste resources, accessibility to technology, and the obtainability of funding sources. The "waste profile" developed in the contextual analysis further expanded, aiming to measure the resource availability in terms of quality, quantity, usability, and accessibility. The assessment of technology involved identifying the local technology standards as local knowledge was a solid indicator to determine whether machinery could be built locally. This evaluation further explored existing machinery-purchasing channels to ensure the accessibility to appropriate technology. It was evident that informal waste workers were already incorporating formal and informal technology development systems for their recycling businesses. The assessment highlighted the importance of supporting local small-scale businesses by commissioning them to manufacture the equipment, thereby enhancing their autonomy. Investigating the funding obtainability evaluated the accessibility to funds and the fund management as funding and financial management are sensitive topics in development work (Hoque 2017).

10.5.2.4 Operational Feasibility

Operational feasibility assesses how well the prospect ecosystem of the project addresses its difficulties, how it takes advantage of the opportunities identified during scope definition, and also how it gratifies the requisites specified in the contextual and technical feasibility studies (Bause et al. 2014). This project's operational feasibility also reviewed the availability and capacities of potential experts, educators, trainers, researchers, and administrative personnel and their willingness to assist the project. This was the most difficult during the feasibility assessment to gauge as it involved determining the continuity of the commitment offered by both voluntary and nonvoluntary project staff.

10.5.2.5 Schedule Feasibility

Schedule feasibility evaluates the possibility of the project completion across the scheduled timeline (Goodman 1988). A project's schedule feasibility is appraised as high when the chance of the project being completed on time is higher. Launching a novel project in an entirely new context has its own complexities and unforeseeable circumstances. Hence, the project team had to carefully strategize the management of uncertainties in advance during the assessment of schedule feasibility.

10.5.2.6 Market Feasibility

Generally, market feasibility is a well-known terminology in most fields. As described earlier, this project had distinctive means of assessing the market opportunities and challenges. The project team conducted a longitudinal market analysis to identify the potential products, markets, and stakeholders to address the commercialization of novel upcycled products. Product ideation and addressing market feasibility required thorough work due to the uniqueness of the process. The main challenge faced during the market feasibility was to assess the compatibility and the merchantable ability of the novel upcycled products that were physically unavailable until the actual project implementation phase.

10.5.2.7 Needs Assessment

Needs assessment aims to systematically determine the gaps between current conditions and desired conditions (Altschuld and Kumar 2010). Measuring the discrepancy between the current state (need) and the "wanted" state assists in evaluating the need appropriately. Assessing the needs of the local communities helped to construct the foundation of the project objectives. Simultaneously, it also served as an input to the project charter. Needs, wants, and most of the contextual attributes change over time while affecting the direction of a project (Camilleri 2018). Therefore, to address those discrepancies between the needs assessment and the feasibility study, the project team often reevaluated the decisions made during the pre-project period and ensured the adaptability and the validity of the project outcomes.

10.5.3 Setting Up Design Centers

Centralized hubs that facilitate university–community linkage are vital for disseminating community education. These hubs draw far-flung parts of the world closer together, enabling the movement of ideas and people (Owen-Smith 2018). The project established two design centers in Sri Lanka as central locations that could serve as a clearinghouse and an aggregator for knowledge, training, problems, and skills related to upcycling. The design centers act as an interface to facilitate university–industry–community partnerships to promote upcycling technology via a structured

FIGURE 10.19 Design center at the University of Moratuwa.

network. These two design centers are legally administrated by the University of Moratuwa in the Western Province (Figure 10.19) and the University of Jaffna in the Northern Province. Careful positioning of design centers on central locations increased geographical convenience and the general public's approachability. These design centers were also good sources with an absorptive capacity to allow warehouse information and combine it with existing knowledge to create new things. Design centers provided hands-on experience in upcycling to stakeholders in the academic, private, and informal waste sectors. These learning spaces simultaneously trained the "trainers" and the general public, which enabled the cocreation of knowledge between formal and informal sectors equitably with the active participation of leading research universities.

10.5.4 TECHNOLOGY AND MACHINERY

As described above, the waste upcycling technique introduced in this project adopts a simple technology to convert waste into quality consumer products without using any chemicals or adhesives, except for heat and pressure. The general processing technique is the same for most materials; however, the process slightly differs for different material compositions only in terms of the amount of heat and the pressure applied during the processing. Selecting the suitable material, identifying the ideal composition, and applying optimal temperature and pressure are the key strategies involved in this upcycling process. This upcycling process is unique, and it opens vast arrays of avenues to different kinds of materials, from simple consumer goods

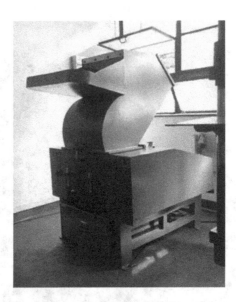

FIGURE 10.20 Shredder machine.

Note: The shredder (Figure 10.20) is a mechanical device used to shred waste materials into small particles. The size of the shredded materials can be adjusted as per the requirement. This machine has the capacity to shred different types of waste materials, from paper, plastic to even denser materials like wood and metal.

to advanced materials like nonstructural building materials. Major equipment and machinery involved in this process were locally built, and few were purchased from China.

The shredder is a mechanical device used to shred waste materials into small particles. The size of the shredded materials can be adjusted as per the requirement. This machine has the capacity to shred different types of waste materials, from paper, plastic to even denser materials like wood and metal (Figure 10.20).

The laminating press is the basic equipment needed in the product development process. The waste composite blend is placed between the beds of this press to apply heat and pressure. The temperature, pressure, and material processing time are adjustable. This machine is more suitable for developing lighter materials for consumer goods (Figure 10.21).

The hot press is the heavy-duty or the industrial version of the Laminating press. This equipment is used to process bulkier materials and to apply higher temperatures and pressure (Figure 10.22).

The cold press is used to cool down and consolidate the processed materials under controlled conditions. Alternative techniques can be used for the cooling process; however, it depends on the complexity of the composite material (Figure 10.23).

FIGURE 10.21   Laminating Press.

Note: The laminating press (Figure 10.21) is the basic equipment needed in the product development process. The waste composite blend is placed between the beds of this press to apply heat and pressure. The temperature, pressure, and material processing time are adjustable. This machine is more suitable for developing lighter materials for consumer goods.

FIGURE 10.22   Hot Press.

Note: The hot press (Figure 10.22) is the heavy-duty or the industrial version of the Laminating Press. This equipment is used to process bulkier materials and to apply higher temperatures and pressure.

FIGURE 10.23   Cold Press.

Note: The cold press (Figure 10.23) is used to cool down and consolidate the processed materials under controlled conditions. Alternative techniques can be used for the cooling process; however, it depends on the complexity of the composite material.

10.5.5   Community Projects

The collaborative project between Sri Lanka and Australia successfully established three sustainable waste upcycling community ventures in Sri Lanka. These case studies are some of the best examples of adopting upcycling in developing nations. The following section briefly describes two of those upcycling brands currently available in the market.

10.5.5.1   Yaal Fibre

Yaal Fibre upcycles banana fiber extracted from discarded banana trunks at a wholesale banana trading facility in Jaffna, Sri Lanka. This community venture is a women's group that started weaving eco-friendly products from banana fiber with startup support from a German not-for-profit organization. Initially, they had a basic product line comprised of woven hats, baskets, bags, pouches, and coasters. Weaving these products were extremely time consuming and required a certain level of skill to master the quality. Unfortunately, the value they received for these products was relatively low, primarily due to the unavailability of reliable market opportunities in the Northern Province. Aiming to add value to their business and to address prevailing market barriers through product diversification, the project team introduced a simple upcycling technique that can be used to develop composite sheets using banana fiber and waste polyethylene. The process involved using a machine that incorporates heat-press technology to create lightweight composite sheets that can be further

FIGURE 10.24   Products by Yaal Fibre.

developed into products like notebook jackets, placemats, coasters, and clipboards. They sourced waste plastic materials such as low density polythene (LDPE), high density polyethene (HDPE), and polypropylene (PP) from a local bakery and from a vehicle spare parts storage facility, which discarded these clean plastics, which initially served as packaging materials of their products. While diversifying the product line, the project team also diversified the markets, including islandwide retail and service-industry clients.

Yaal Fibre products (Figure 10.24) received higher demand from the service industry clients like hotel and restaurants, travel agencies, gift shops, and even corporate organizations for their corporate events. Yaal Fibre had to increase their capacities to meet the rising demand, which eventually caused strains in sourcing quality waste plastics due to the operational termination of the spare parts facility in Jaffna. The incorporation of waste plastic into these composites had its ups and downs. Plastic composites gave added advantages to the fibrous materials, making them durable and water-resistant. Yet, this new process is not without its critics. Fiber-plastic composites entrap biodegradable materials inside the plastic, a material that is considered non-biodegradable. Some argue that upcycling makes the end product complex by mixing different materials that could otherwise be recycled separately. However, prior to this upcycling project, those waste plastics were either burned or disposed of improperly. Yaal Fibre enabled direct benefits to the local community through employment, post-war revitalization, livelihoods, and, more importantly, intangible benefits like women empowerment. Since 2016, Yaal Fibre has upcycled nearly 240 kg of plastic waste into value-added products and provided financial aids for nearly 15 women who were previously unemployed.

10.5.5.2 Katana Upcycle

Katana Upcycle was a small-scale recycling business used to collect and segregate plastic waste generated in the local industrial zones. This entrepreneurial venture

FIGURE 10.25   Products by Katana Upcycles.

completely turned its direction from waste segregation to manufacturing upcycling products that present both green and aesthetic values. Katana Upcycle has a wide range of waste materials as inputs to their product manufacturing process. They combine LDPE, HDPE waste plastics with the materials like old sarees, obsolete clothing, used curtains, gunny bags, and tetra pack food wrappers usually neglected by most recyclers. Their product range includes files, folders, notebooks, stationery, pouches, and even tiles from waste food wrappers with mixed materials (Figure 10.25). While being a small-scale domestic manufacturer, Katana Upcycle facilitated employment for nearly ten women from the neighborhood under flexible working schedules. Those women preferred the convenience of flexible work schedules, which enabled the management of their household responsibilities. Since 2016, Katana Upcycle has transformed more than 700 kg of waste materials into refined products while running successful pop-up stalls in eco-friendly marketplaces like Good Market.

These types of waste-to-wealth businesses that source their raw materials locally have greater potential for the circular economy as they create more sustainable employment opportunities, often for marginalized communities. Directly and indirectly, these types of ventures provide many tangible and intangible benefits. Stimulating local economies by generating both income and job opportunities, empowering women, increasing the self-esteem of the informal waste sector, and addressing environmental degradation through value addition to waste are a few of the many benefits associated with waste-to-wealth businesses.

10.5.6   MARKET ANALYSIS

Marketing, branding, and business development play a significant role in introducing novel products to the market. Business development and brand management strategies must be unique and creative to attract the right customer from the right target audience. Entrepreneurs should carefully decide what kind of upcycling business/ product that they should choose. How they define their target audience will have a significant difference in marketing and branding tactics. Before officially launching Yaal Fibre and Katana Upcycle, the project team carefully scanned the existing market gaps and opportunities to identify the best positioning strategies. Business development and brand marketing plans were often cross-referenced to ensure the commercial viability of these community enterprises. Both Yaal Fibre and Katana Upcycle followed an eight-step brand marketing framework to position their products in existing markets: (1) identification of the target audience, (2) understanding the positioning, (3) situational analysis for the macroenvironment, (4) SWOT (strengths, weaknesses, opportunities, threats) analysis for microenvironment, (5) identification of potential competitors, (6) development of brand elements, (7) creation of awareness channels, and (8) evaluation of strategic goal and objectives. The brand positioning of these two enterprises slightly differed due to the distinctive arrays of product ranges they had. Existing markets that promoted eco-friendly products and services created a friendly atmosphere for these two entities. Less rigidness of administration and established reputation of these eco-friendly markets were significant enablers for entry-level upcycling entrepreneurs. The competitive advantage of these upcycling community ventures was their exciting storyline of the product and its life cycle.

10.5.7 CHALLENGES

Establishing novel resource management concepts is quite challenging, especially in developing countries due to context-specific aspects. The project team identified and faced many challenges during the course of the project, some of which were addressed and some yet to be solved.

10.5.7.1 Socioeconomic Inhibitors

Most of the developing countries, including Sri Lanka, have income disparity issues. Most of the time, people are more sensitive to their daily socioeconomic challenges than environmental issues, though long-term ecological problems are dire. It was challenging to establish the idea of waste being a resource among the general public. The project team had to execute an extensive series of education and communication programs to improve consumer perceptions and beliefs about upcycling and upcycled products. Using social media platforms and mass media channels attracted like-minded consumers, but influencing the purchasing intentions and buying behaviors were challenging due to the price-conscious economic context in Sri Lanka. Consumers' preference for the less expensive mass-produced products inhibited getting a proper market value for upcycled products. Although waste upcycling ventures positively impacted job creation and waste reduction, it still remains a niche market.

10.5.7.2 Technical Inhibitors

Lack of skillset and absence of affordable machinery were major drawbacks that hindered the effective execution of upcycling ventures. Technology played a significant role in the stages of product designing, prototyping, and manufacturing.

Purchasing new equipment and machinery was expensive due to their unavailability in the local market, requiring imports from China or India. The project team identified a local manufacturer who can fabricate and maintain the necessary machines for upcycling under a nominal fee eliminating the need for expensive spare parts and expertise. However, it indicated the importance of nourishing the local mechanical industry to overcome the unavailability of affordable technology for startup entrepreneurs.

10.5.7.3   Environmental Inhibitors

Globally, there is a growing interest in finding new ways to upcycle resources in response to environmental concerns. Though this could be currently recognizable in Sri Lanka, the trend is still in its early stages. Some individuals and organizations preferred buying upcycled products instead of purchasing mass-produced products as an ethical choice. Yet, it tends to be a slow trend due to socioeconomic barriers mentioned in the previous section. Improper disposal of waste materials has hindered the possibilities of reusing waste as inputs due to hygienic concerns. Therefore, it is vital to promote correct waste separating and segregating practices at both domestic and industrial waste sources. As a whole, the lack of awareness in anthropogenic climate change and barriers in public environmental sustainability education needs to be addressed to suppress the environmental inhibitors.

10.5.7.4   Financial and Economic Aspects

Identifying suitable markets and positioning the products were challenging due to the distinctive nature of upcycled products and the price-conscious customer base in Sri Lanka. An average Sri Lankan customer rarely chooses an expensive alternative, even if that product is eco-friendly or responsibly manufactured compared to the existing products. Customers' attitudes directly affect the longevity of the retailers and the social enterprises. Intensive target marketing and one-to-one promotion aided in altering consumers' perceptions to a certain level. Not having access to reliable funding options was a critical challenge faced by most social entrepreneurs on their starting-up or scaling-up stages. Although there are loan schemes, grants, and funding programs available for micro and small-scale enterprises, the lengthy application processes often discourage many entrepreneurs from seeking support through these mechanisms. Local development banks could design better applications and evaluation systems for upcycling businesses. Enabling sustainable microfinancing options will be beneficial for encouraging upcycling ventures in developing countries.

10.5.7.5   Political and Legal Aspects

Political and legal aspects play an essential role in developing sustainable businesses in any country. Sri Lanka lacks strict policies and regulations on environmental ethics, including waste management. Inadequate regulations for health and safety practices in reusing, recycling, and upcycling have caused insecurities among the informal waste sector. Social enterprises are not given any tax concessions or value-added tax (VAT) exemptions to advocate sustainable business. The Sri Lankan government also needs to update the current waste management frameworks to regulate responsible waste disposal and promote mindful consumption habits through policy until it is practiced as an established norm.

10.6   MOVING BEYOND CIRCULARITY

The emerging concept of waste upcycling has proved its benefits to the environment, economy, and society. Upcycling gained attention during the last decade due to growing concerns of escalated waste generation, resource scarcity (Herman, Sbarcea, and Panagopoulos 2018), and the tendency toward the Maker Movement (Unterfrauner et al. 2019). Also, upcycling extend the lifetimes of materials and products in the system, eventually reducing greenhouse gas emissions and energy consumptions (Sung et al, 2019). Fundamentally, upcycling slows down the material flow within the system, enabling material efficiency and reduces the need for raw material extraction (Sung et al, 2018). Thus, upcycling minimizes the stress caused due to resource scarcity and promotes optimal use of existing resources.

Upcycling is also a form of creative and inventive transformation of waste, which creates ecological, social, and economic benefits. Upcycled products are aesthetic while keeping the essence of their previous life. The relation between the previous life as waste material and the new status as an upcycled product creates the main object of interest. The best part of an upcycled product is not its appearance but the story it tells. Upcycled products often tell a story of how a material is collected, crafted into creative items by men and women in a home-based industry or a social enterprise and how the process is generating employment opportunities.

Successful upcycling ventures in Sri Lanka demonstrate that upcycling businesses can be economically sustainable from financial, employment, and poverty perspectives. The use of waste as a resource in upcycling delivers financial savings that could have been used to purchase new materials as inputs while generating more job opportunities for underprivileged people (Sung et al, 2018). In addition, upcycling also creates mindful business networks in industrial symbiosis, enabling resource sharing and waste trading within the local context. Furthermore, upcycling cultivates psychological and sociocultural benefits such as a sense of community, empowerment, identity, renewed self-esteem, self-confidence, and also educational avenues that share knowledge and skills (Bridgens et al. 2018). Innovation and value retention based upcycling business models are more sustainable because they create effective networks among entrepreneurs, consumers, stakeholders, and products, enabling a smoother transition to a more circular economy (Kozlowski, Searcy, and Bardecki 2018).

The role of the consumer and their buying behavior has a significant impact on the successful implementation of circular economy business models. Scholars argue that preference for upcycled products could impact consumer buying behavior, creating new identities for waste via stimulating ethical and responsible buying decisions (Santulli and Langella 2013; J. Singh et al. 2019). This can be a challenging approach in developing countries due to the limitations of purchasing power and income disparity. Case studies from Sri Lanka affirm these barriers caused by price-conscious purchasing decisions, which positions upcycled products as a niche category in developing nations. These barriers could be addressed through community education via promoting home-based/domestic upcycling to encourage sustainable waste management at the source (Sung et al, 2014). Scaling up upcycling ventures is the way forward for developing nations to embed circular economy into their economic models by valorizing "waste", which dismisses the prevailing challenges of social and economic resource accessibility. Therefore, embedding upcycling into economic models in developing nations could alleviate many hurdles toward a circular economy by simultaneously addressing sustainable waste management, economic diversification, employment, poverty reduction, and environmental protection.

10.7 CONCLUDING REMARKS

This chapter explored upcycling as a holistic approach to manage both waste and our thinking process by bringing together critical thinking, creativity, and action. It proposed that we need to find new ways to build a resource-sharing culture and promote conscious consumerism. We should take a step back and realize that these creative efforts are helping us move toward an awareness of the waste in our own lives, in all forms. As we redesign and recreate, we need to ask, "How can we close the loop? What will happen to the product in the end?"

Upcycling is a step toward a regenerative design culture. It is about finding new opportunities to connect discarded materials with a creative process to add more value. It is a creative approach where instead of throwing away a used product, it is converted into something unique with a higher aesthetic value. Using materials that are often overlooked, disregarded, and obsolete and giving them a new purpose can create awareness of waste management and sustainable consumption and production. In a broader sense, upcycling requires a collective effort from everyone—waste generators, designers, entrepreneurs, consumers, and the community to contribute to a circular waste economy. The scarcity of materials and products is felt greater today due to the COVID-19 global pandemic. The skills and knowledge to make our own products are important than ever. We hope this book and this chapter on upcycling will guide the reader to move in that direction.

REFERENCES

Adjei, Kofi, Kwame Opoku-Bonsu, and Emmanuel O. Asiamah. 2016. "Utilization of Cullets for the Production of Glass Tiles through Kiln Casting." *Journal of Science and Technology* 36 (3): 124–33. DOI: 10.4314/just.v36i3.12

Ali, Nawwar Shukriah, Nuur Farhana Khairuddin, and Shahriman Zainal Abidin. 2013. "Upcycling: Re-Use and Recreate Functional Interior Space Using Waste Materials." http://dornob.com/transforming-dumpster-home-for-camouflaged-urban-living/

Altschuld, James, and David Kumar. 2010. *Needs Assessment: An Overview.* Thousand Oaks, CA: SAGE Publications. https://doi.org/10.4135/9781452256795

Ayres, Robert U., and Allen V. Kneese. 1969. "Production, Consumption, and Externalities." *American Economic Review* 59 (3): 282–97. www.jstor.org/stable/1808958

Bause, Katharina, Aline Radimersky, Marinette Iwanicki, and Albert Albers. 2014. "Feasibility Studies in the Product Development Process." *Procedia CIRP* 21: 473–78. https://doi.org/https://doi.org/10.1016/j.procir.2014.03.128

Bridgens, Ben, Mark Powell, Graham Farmer, Claire Walsh, Eleanor Reed, Mohammad Royapoor, Peter Gosling, Jean Hall, and Oliver Heidrich. 2018. "Creative Upcycling: Reconnecting People, Materials and Place through Making." *Journal of Cleaner Production* 189: 145–54. https://doi.org/10.1016/j.jclepro.2018.03.317

Caldera, Savindi, Randika Jayasinghe, Cheryl Desha, Les Dawes, and Selena Ferguson. 2020. "Evaluating Barriers, Enablers and Opportunities for Closing the Loop through 'Waste Upcycling': A Systematic Literature Review." *Journal of Sustainable Development of Energy, Water and Environment Systems* 10 (1): 1–20. https://doi.org/http://dx.doi.org/10.13044/j.sdewes.d8.0367

Camacho-Otero, Juana, Casper Boks, and Ida N. Pettersen. 2018. "Consumption in the Circular Economy: A Literature Review." *Sustainability* 10 (8): 1–25. https://doi.org/10.3390/su10082758

Camara, Nuno Da. 2006. "The Relationship between Internal and External Stakeholders and Organisational Alignment." *Henley Manager Update* 18 (1): 41–52. https://doi.org/10.1177/174578660601800104

Camilleri, Mark Anthony. 2018. "Understanding Customer Needs and Wants." In *Travel Marketing, Tourism Economics and the Airline Product*, edited by Mark Anthony Camilleri, 29–50. Cham: Springer International Publishing. https://doi.org/10.1007/978-3-319-49849-2_2

Celik, Gokhan, Robert M. Kennedy, Ryan A. Hackler, Magali Ferrandon, Akalanka Tennakoon, Smita Patnaik, Anne M. LaPointe, Salai C. Ammal, Andreas Heyden, Frédéric A. Perras, and Marek Pruski. 2019. "Upcycling Single-Use Polyethylene into High-Quality Liquid Products." *ACS Central Science* 5 (11): 1795–803.

Cerminara, Giulia, and Raffaello Cossu. 2018. "Waste Input to Landfills." In *Solid Waste Landfilling*, edited by Raffaello Cossu and Rainer Stegmann, 15–39. Amsterdam: Elsevier. https://doi.org/10.1016/b978-0-12-407721-8.00002-4

Conlon, Katie, Randika Jayasinghe, and Ranahansa Dasanayake. 2019. "Circular Economy: Waste-to-Wealth, Jobs Creation, and Innovation in the Global South." *World Review of Science, Technology and Sustainable Development* 15 (2): 145–59. https://doi.org/10.1504/WRSTSD.2019.099377

Cuc, Sunhilde and Simona Tripa. 2018. "Re-Design and Upcycling-a Solution for the Competitiveness of Small and Medium-Sized Enterprises in the Clothing Industry." *Industria Textila* 69 (1): 31–6. doi: 10.35530/IT.069.01.1417

Cumming, Deb. 2016. "A Case Study Engaging Design for Textile Upcycling." *Journal of Textile Design Research and Practice* 4 (2): 113–28. https://doi.org/10.1080/20511787.2016.1272797

Dissanayake, Geetha, and Pammi Sinha. 2015. "An Examination of the Product Development Process for Fashion Remanufacturing." *Resources, Conservation and Recycling* 104: 94–102. https://doi.org/10.1016/j.resconrec.2015.09.008

Dodoo, Ambrose, Leif Gustavsson, and Roger Sathre. 2014. "Handbook of Recycling: Recycling of Lumber." In *Handbook of Recycling*, edited by Ernst Worrell and Markus A. Reuter, 151–63. Boston: Elsevier. https://doi.org/https://doi.org/10.1016/B978-0-12-396459-5.00011-8

Dyer, Thomas D. 2014. "Handbook of Recycling: Glass Recycling." In *Handbook of Recycling*, edited by Ernst Worrell and Markus A. Reuter, 191–209. Boston: Elsevier. https://doi.org/https://doi.org/10.1016/B978-0-12-396459-5.00014-3

Ellen MacArthur Foundation. 2013. *Towards the Circular Economy Vol. 1: An Economic and Business Rationale for an Accelerated Transition*. UK: Ellen MacArthur Foundation Publishing.

Ellen MacArthur Foundation. 2017. *A New Textiles Economy: Redesigning Fashion's Future.* UK: Ellen MacArthur Foundation Publishing. www.ellenmacarthurfoundation.org/publications

Goodman, Louis J. 1988. "Feasibility Analysis and Appraisal of Projects BT." In *Project Planning and Management: An Integrated System for Improving Productivity*, edited by Louis J. Goodman, 26–55. Boston: Springer US. https://doi.org/10.1007/978-1-4684-6587-7_3

Gregory, Amanda J., Jonathan P. Atkins, Gerald Midgley, and Anthony M. Hodgson. 2020. "Stakeholder Identification and Engagement in Problem Structuring Interventions." *European Journal of Operational Research* 283 (1): 321–40. https://doi.org/https://doi.org/10.1016/j.ejor.2019.10.044

Grossmann, Harald, Toni Handke, and Tobias Brenner. 2014. "Handbook of Recycling: Paper Recycling." In *Handbook of Recycling*, edited by Ernst Worrell and Markus A. Reuter, 165–78. Boston: Elsevier. https://doi.org/https://doi.org/10.1016/B978-0-12-396459-5.00012-X

Guerrero, Lilliana Abarca, Ger Maas, and William Hogland. 2013. "Solid Waste Management Challenges for Cities in Developing Countries." *Waste Management* 33 (1): 220–32. https://doi.org/https://doi.org/10.1016/j.wasman.2012.09.008

Hawken, Paul. 2018. *Drawdown: The Most Comprehensive Plan Ever Proposed to Reverse Global Warming.* UK: Penguin Books. https://books.google.lk/books?id=QhlADwAAQBAJ

Herman, Krzysztof, Madalina Sbarcea, and Thomas Panagopoulos. 2018. "Creating Green Space Sustainability through Low-Budget and Upcycling Strategies." *Sustainability* 10 (6): 1–15. https://doi.org/10.3390/su10061857

Hoque, Mohammed Ziaul. 2017. "Mental Budgeting and the Financial Management of Small and Medium Entrepreneurs." *Cogent Economics & Finance* 5 (1): 1–19. https://doi.org/10.1080/23322039.2017.1291474

Jayasinghe, Randika. 2015. *A Critical Study of the Wastescape in the Western Province of Sri Lanka: Pathways towards Alternative Approaches.* Australia: University of Western Australia.

Jayasinghe, Randika, and Caroline Baillie. 2017. "Engineering with People: A Participatory Needs and Feasibility Study of a Waste-Based Composite Manufacturing Project in Sri Lanka." In *Green Composites: Waste and Nature-Based Materials for a Sustainable Future*, second edition, 149–80. Elsevier. https://doi.org/10.1016/B978-0-08-100783-9.00007-1

Karagiannidis, Avraam, Martin Wittmaier, S. Langer, B. Bilitewski, and Apostolos Malamakis. 2009. "Thermal Processing of Waste Organic Substrates: Developing and Applying an Integrated Framework for Feasibility Assessment in Developing Countries." *Renewable and Sustainable Energy Reviews* 13 (8): 2156–62. https://doi.org/https://doi.org/10.1016/j.rser.2008.09.035

Kopnina, Helen. 2018. "Circular Economy and Cradle to Cradle in Educational Practice." *Journal of Integrative Environmental Sciences* 15 (1): 119–34. https://doi.org/10.1080/1943815X.2018.1471724

Kozlowski, Anika, Cory Searcy, and Michal Bardecki. 2018. "The ReDesign Canvas: Fashion Design as a Tool for Sustainability." *Journal of Cleaner Production* 183: 194–207. https://doi.org/10.1016/j.jclepro.2018.02.014

Krausmann, Fridolin, Dominik Wiedenhofer, Christian Lauk, Willi Haas, Hiroki Tanikawa, Tomer Fishman, Alessio Miatto, Heinz Schandl, and Helmut Haberl. 2017. "Global Socioeconomic Material Stocks Rise 23-Fold over the 20th Century and Require Half of

Annual Resource Use." *Proceedings of the National Academy of Sciences of the United States of America* 114 (8): 1880–85. https://doi.org/10.1073/pnas.1613773114

Leal Filho, Walter, Dawn Ellams, Sara Han, David Tyler, Valérie Julie Boiten, Arminda Paco, Harri Moora, and Abdul Lateef Balogun. 2019. "A Review of the Socio-Economic Advantages of Textile Recycling." *Journal of Cleaner Production* 218: 10–20. https://doi.org/10.1016/j.jclepro.2019.01.210

Mahpour, Amirreza. 2018. "Prioritizing Barriers to Adopt Circular Economy in Construction and Demolition Waste Management." *Resources, Conservation and Recycling* 134: 216–27. https://doi.org/10.1016/j.resconrec.2018.01.026

Maio, Francesco Di, Peter Carlo Rem, Francesco Di Maio, and Peter Carlo Rem. 2015. "A Robust Indicator for Promoting Circular Economy through Recycling." *Journal of Environmental Protection* 6 (10): 1095–104. https://doi.org/10.4236/jep.2015.610096

McDonough, William, and Michael Braungart. 2009. *Cradle to Cradle: Remaking the Way We Make Things*. London: Vintage.

McDonough, William, Michael Braungart, and Bill Clinton. 2013. *The Upcycle: Beyond Sustainability—Designing for Abundance*. New York: North Point Press.

Meadows, Dennis H., Donella L. Meadows, and Jørgen Randers. 1972. *The Limit to Growth*. New York: Universe Books.

Meadows, Dennis H., Jørgen Randers, and Dennis L. Meadows. 2004. *Limits to Growth: The 30 Year Update*. Vermont: Chelsea Green Publishing.

Nilsen, Per, and Susanne Bernhardsson. 2019. "Context Matters in Implementation Science: A Scoping Review of Determinant Frameworks That Describe Contextual Determinants for Implementation Outcomes." *BMC Health Services Research* 19 (1): 1–21. https://doi.org/10.1186/s12913-019-4015-3

Owen-Smith, Jason. 2018. *Research Universities and the Public Good*. Redwood City, CA: Stanford University Press.

Reike, Denise, Walter J. V. Vermeulen, and Sjors Witjes. 2018. "The Circular Economy: New or Refurbished as CE 3.0? Exploring Controversies in the Conceptualization of the Circular Economy through a Focus on History and Resource Value Retention Options." *Resources, Conservation and Recycling* 135: 246–64. https://doi.org/https://doi.org/10.1016/j.resconrec.2017.08.027

Santulli, Carlo, and Carla Langella. 2013. "'+ Design—Waste': A Project for Upcycling Refuse Using Design Tools." *International Journal of Sustainable Design* 2 (2): 105. https://doi.org/10.1504/ijsdes.2013.057121

Shen, Li, and Ernst Worrell. 2014. "Handbook of Recycling: Plastic Recycling." In *Handbook of Recycling*, edited by Ernst Worrell and Markus A. Reuter, 179–90. Boston: Elsevier. https://doi.org/https://doi.org/10.1016/B978-0-12-396459-5.00013-1

Singh, Jagdeep, Kyungeun Sung, Tim Cooper, Katherine West, and Oksana Mont. 2019. "Challenges and Opportunities for Scaling up Upcycling Businesses—The Case of Textile and Wood Upcycling Businesses in the UK." *Resources, Conservation and Recycling* 150: 1–15. https://doi.org/10.1016/j.resconrec.2019.104439

Steinhilper, R., and M. Hieber. 2001. "Remanufacturing-the Key Solution for Transforming 'Downcycling' into 'Upcycling' of Electronics." *IEEE International Symposium on Electronics and the Environment*, 161–66. https://doi.org/10.1109/isee.2001.924520

Sung, Kyungeun, Tim Cooper, and Sarah Kettley. 2014. "Individual Upcycling Practice: Exploring the Possible Determinants of Upcycling Based on a Literature Review." In *Proceedings of the Sustainable Innovation 2014 Conference, 3–4 November*, 237–44. Copenhagen. http://irep.ntu.ac.uk/id/eprint/2559

Sung, Kyungeun, Tim Cooper, and Sarah Kettley. 2019. "Factors Influencing Upcycling for UK Makers." *Sustainability* 11 (3): 1–26. https://doi.org/10.3390/su11030870

Sung, Kyungeun, Tim Cooper, and Sarah Kettley. 2018. "Emerging Social Movements for Sustainability: Understanding and Scaling Up Upcycling in the UK." In *The Palgrave Handbook of Sustainability: Case Studies and Practical Solutions*, edited by Robert Brinkmann and Sandra J Garren, 299–312. Cham: Springer International Publishing. https://doi.org/10.1007/978-3-319-71389-2_15

Sung, Kyungeun, Tim Cooper, and Sarah Kettley. 2019. "Developing Interventions for Scaling up UK Upcycling." *Energies* 12 (14): 1–31. https://doi.org/10.3390/en12142778

Sung, Kyungeun, Tim Cooper, Usha Ramanathan, and Jagdeep Singh. 2017. "Challenges and Support for Scaling up Upcycling Businesses in the UK: Insights from Small-Business Entrepreneurs." In *Product Lifetimes and the Environment Conference Proceedings*. Delft University of Technology and IOS Press.

Thamae, Thimothy, and Caroline Baillie. 2009. *Natural Fibre Composites: Turning Waste into Useful Materials*. Saarbrücken, Germany: VDM Publishing.

Unterfrauner, Elisabeth, Jing Shao, Margit Hofer, and Claudia M. Fabian. 2019. "The Environmental Value and Impact of the Maker Movement—Insights from a Cross-Case Analysis of European Maker Initiatives." *Business Strategy and the Environment* 28 (8): 1518–33. https://doi.org/https://doi.org/10.1002/bse.2328

Walls, Margaret. 2013. "Deposit-Refund Systems in Practice and Theory." *Encyclopedia of Energy, Natural Resource, and Environmental Economics* 3: 133–37. https://doi.org/10.1016/B978-0-12-375067-9.00035-8

Wegener, Charloett, and Marie K. Aakjær. 2016. "Upcycling—a New Perspective on Waste in Social Innovation." *Journal of Comparative Social Work* 11 (2): 1–19.

Wiedenhofer, Dominik, Tomer Fishman, Christian Lauk, Willi Haas, and Fridolin Krausmann. 2019. "Integrating Material Stock Dynamics Into Economy-Wide Material Flow Accounting: Concepts, Modelling, and Global Application for 1900–2050." *Ecological Economics* 156: 121–33. https://doi.org/10.1016/j.ecolecon.2018.09.010

World Economic Forum. 2016. *The New Plastics Economy—Rethinking the Future of Plastics*. Geneva: Ellen MacArthur Foundation, McKinsey & Company. www.ellenmacarthurfoundation.org/publications

Index

Note: Page numbers in *italics* indicate figures, those in **bold** indicate tables.